Oxford Graduate Texts in Mathematics

OXFORD GRADUATE TEXTS IN MATHEMATICS

Books in the series

1. Keith Hannabuss: *An Introduction to Quantum Theory*
2. Reinhold Meise and Dietmar Vogt: *Introduction to Functional Analysis*
3. James G. Oxley: *Matroid Theory*
4. N.J. Hitchin, G.B. Segal, and R.S. Ward: *Integrable Systems: Twistors, Loop Groups, and Riemann Surfaces*
5. Wulf Rossmann: *Lie Groups: An Introduction through Linear Groups*
6. Qing Liu: *Algebraic Geometry and Arithmetic Curves*
7. Martin R. Bridson and Simon M. Salamon (eds): *Invitations to Geometry and Topology*
8. Shmuel Kantorovitz: *Introduction to Modern Analysis*
9. Terry Lawson: *Topology: A Geometric Approach*
10. Meinolf Geek: *An Introduction to Algebraic Geometry and Algebraic Groups*
11. Alastair Fletcher and Vladimir Markovic: *Quasiconformal Maps and Teichmüller Theory*
12. Dominic Joyce: *Riemannian Holonomy Groups and Calibrated Geometry*
13. Fernando Villegas: *Experimental Number Theory*
14. Péter Medvegyev: *Stochastic Integration Theory*
15. Martin A. Guest: *From Quantum Cohomology to Integrable Systems*
16. Alan D. Rendall: *Partial Differential Equations in General Relativity*
17. Yves Félix, John Oprea and Daniel Tanré: *Algebraic Models in Geometry*
18. Jie Xiong: *Introduction to Stochastic Filtering Theory*
19. Maciej Dunajski: *Solitons, Instantons, and Twistors*
20. Graham R. Allan: *Introduction to Banach Spaces and Algebras*
21. James Oxley: *Matroid Theory, Second Edition*
22. Simon Donaldson: *Riemann Surfaces*
23. Clifford Henry Taubes: *Differential Geometry: Bundles, Connections, Metrics and Curvature*

Differential Geometry

Bundles, Connections, Metrics and Curvature

Clifford Henry Taubes
Department of Mathematics
Harvard University

OXFORD
UNIVERSITY PRESS

OXFORD
UNIVERSITY PRESS

Great Clarendon Street, Oxford OX2 6DP

Oxford University Press is a department of the University of Oxford.
It furthers the University's objective of excellence in research, scholarship,
and education by publishing worldwide in

Oxford New York

Auckland Cape Town Dar es Salaam Hong Kong Karachi
Kuala Lumpur Madrid Melbourne Mexico City Nairobi
New Delhi Shanghai Taipei Toronto

With offices in

Argentina Austria Brazil Chile Czech Republic France Greece
Guatemala Hungary Italy Japan Poland Portugal Singapore
South Korea Switzerland Thailand Turkey Ukraine Vietnam

Oxford is a registered trade mark of Oxford University Press
in the UK and in certain other countries

Published in the United States
by Oxford University Press Inc., New York

First published 2011

British Library Cataloguing in Publication Data

Data available

Library of Congress Cataloging in Publication Data

Data available

Typeset by SPI Publisher Services, Pondicherry, India
Printed and bound by
CPI Group (UK) Ltd, Croydon, CR0 4YY

ISBN 978–0–19–960588–0 (hbk.)
978–0–19–960587–3 (pbk.)

1 3 5 7 9 10 8 6 4 2

Preface

This book is meant to be an introduction to the subject of vector bundles, principal bundles, metrics (Riemannian and otherwise), covariant derivatives, connections and curvature. I am imagining an audience of first-year graduate students or advanced undergraduate students who have some familiarity with the basics of linear algebra and with the notion of a smooth manifold. Even so, I start with a review of the latter subject. I have tried to make the presentation as much as possible self-contained with proofs of basic results presented in full. In particular, I have supplied proofs for almost all of the background material either in the text or in the chapter appendices. Even so, you will most likely have trouble if you are not accustomed to matrices with real and complex number entries, in particular the notions of an eigenvalue and eigenvector. You should also be comfortable working with multi-variable calculus. At the very end of each chapter is a very brief list of other books with parts that cover some of the chapter's subject matter.

I have worked out many examples in the text, because truth be told, the subject is not interesting to me in the abstract. I for one need to *feel* the geometry to understand what is going on. In particular, I present in detail many of the foundational examples.

I learned much of the material that I present here from a true master, Raoul Bott. In particular, I put into this book the topics that I recall Raoul covering in his first-semester graduate differential geometry class. Although the choice of topics are those I recall Raoul covering, the often idionsyncratic points of view and the exposition are my own.

Contents

1 Smooth manifolds

Said briefly, differential geometry studies various constructions that can be built on a smooth manifold. Granted this, the task for Chapter 1 is to summarize various facts about smooth manifolds that are used, either explicitly or implicitly, in much of what follows. These facts are offered sans proof since proofs can be had in most any textbook on differential topology; my favorite being *Differential Topology* by Victor Guillemin and Alan Pollack (1974).

1.1 Smooth manifolds

The Ur smooth manifold of dimension n is the n-dimensional Euclidean space, $\mathbb{R}^n$. I think of $\mathbb{R}^n$ as coming with a standard set of coordinates, $(x_1, \ldots, x_n)$. The Euclidean space is the local model for all other manifolds.

1.1.1 Topological manifolds

A manifold of dimension n is a paracompact, Hausdorff space such that each point has a neighborhood that is homeomorphic to $\mathbb{R}^n$ or, what amounts to the same thing, to the interior of a ball in $\mathbb{R}^n$. Such a neighborhood is called a *coordinate chart* and the pull-back of the coordinate functions from $\mathbb{R}^n$ are called *local coordinates*. A collection

$$\mathcal{U} = \{(U, \varphi) : U \subset M \textit{ is open and } \varphi\colon U \to \mathbb{R}^n \textit{ is an embedding}\}$$

is said to be a coordinate atlas for M if $M = \cup_{(U,\varphi)\in\mathcal{U}} U$. A coordinate atlas is said to be *locally finite* if it is the case that each point $p \in M$ is contained in a finite collection of its open sets.

Two topological manifolds, M and M′, are said to be homeomorphic if there is a homeomorphism between them. This is a continuous, 1–1, surjective map from M to M′ whose inverse is continuous.

1.1.2 Smooth manifolds

Let $\mathcal{U}$ denote a coordinate atlas for M. Suppose that (U, φ_U) and $(U', \varphi_{U'})$ are two elements from $\mathcal{U}$. The map

$$\varphi_{U'} \circ \varphi_U^{-1}: \varphi_U(U' \cap U) \to \varphi_{U'}(U' \cap U)$$

is a homeomorphism between two open subsets of Euclidean space. This map is said to be the *coordinate transition function* for the pair of charts (U, φ_U) and $(U', \varphi_{U'})$.

A smooth structure on M is defined by an equivalence class of coordinate atlases with the following property: All transition functions are diffeomorphisms. This is to say that they have partial derivatives to all orders, as do their inverses. Coordinate atlases $\mathcal{U}$ and $\mathcal{V}$ are deemed to be equivalent when the following condition holds: Given any pairs $(U, \varphi_U) \in \mathcal{U}$ and $(V, \varphi_V) \in \mathcal{V}$, the compositions

$$\varphi_U \circ \varphi_V^{-1}: \varphi_V(V \cap U) \to \varphi_U(V \cap U), \ \varphi_V \circ \varphi_U^{-1}: \varphi_U(V \cap U) \to \varphi_V(V \cap U) \qquad (*)$$

are diffeomorphisms between open subsets of Euclidean space.

In what follows, a map with partial derivatives to all orders is said to be *smooth*. A manifold M with a smooth structure is said to be a *smooth* manifold. The point here is that one can do calculus on smooth manifolds, for it makes sense to say that a function $f: M \to \mathbb{R}$ is differentiable, or infinitely differentiable. The collection of smooth functions on M is denoted by $C^\infty(M; \mathbb{R})$ and consists of those functions with the following property: Let $\mathcal{U}$ denote any given coordinate atlas from the equivalence class that defines the smooth structure. If $(U, \varphi) \in \mathcal{U}$, then $f \circ \varphi^{-1}$ is a smooth function on $\varphi(U)$. This requirement defines the same set of smooth functions no matter the choice of representative atlas by virtue of the condition $(*)$ that defines the equivalence relation.

1.1.3 Maps between smooth manifolds

If M and N are smooth manifolds, then a map $h: M \to N$ is said to be smooth if the following is true: Let $\mathcal{U}$ denote a locally finite coordinate atlas from the equivalence class that gives the smooth structure to M; and let $\mathcal{V}$ denote a corresponding atlas for N. Then each map in the collection

$$\{\psi \circ h \circ \varphi^{-1}: h(U) \cap V \neq \emptyset\}_{(U,\varphi) \in \mathcal{U}, (V,\psi) \in \mathcal{V}}$$

is smooth as a map from one Euclidean space to another. Note again that the equivalence relation described by $(*)$ above guarantees that the definition of a smooth map depends only on the smooth structures of M and N, but not on the chosen representative coordinate atlases.

Two smooth manifolds are said to be diffeomorphic when there exists a smooth homeomorphism h: M → N with smooth inverse. This is to say that M is homeomorphic to N and that the given equivalence classes of atlases that define the respective smooth structures are one and the same.

1.2 The inverse function theorem and implicit function theorem

Various constructions of smooth manifolds rely on two theorems from differential topology, these being the *inverse function theorem* and the *implicit function theorem*. What follows here is a digression to state these theorems. Their proofs are sketched in Appendix 1.1.

The statement of these theorems introduces the notion of the *differential* of a map between one Euclidean space and another. Here is a reminder: Let $U \subset \mathbb{R}^m$ denote a given open set and let $\psi: U \to \mathbb{R}^n$ denote a given map. The differential of ψ is denoted by ψ_*; it is a matrix valued function on U with m columns and n rows whose entries come from the partial derivatives of the components of ψ. This is to say that if ψ has components $(\psi_1, \ldots, \psi_n)$, then the entry in the i'th row and j'th column is $\partial_j \psi_i = \frac{\partial \psi_i}{\partial x_j}$. (Here and throughout, ∂_j is shorthand for the partial derivative with respect to the coordinate x_j.)

Theorem 1.1 (the inverse function theorem): *Let* $U \subset \mathbb{R}^m$ *denote a neighborhood of the origin, and let* $\psi: U \to \mathbb{R}^m$ *denote a smooth map. Suppose that* $p \in U$ *and that the matrix* ψ_* *at* p *is invertible. Then there is a neighborhood* $V \subset \mathbb{R}^m$ *of* $\psi(p)$ *and a smooth map* $\sigma: V \to U$ *such that* $\sigma(\psi(p)) = p$ *and*

- $\sigma \circ \psi$ *is the identity on some neighborhood* $U' \subset U$ of p,
- $\psi \circ \sigma$ *is the identity on* V.

Conversely, if ψ_* *is not invertible at* p, *then there is no such set* V *and map* σ *with these properties.*

The key conclusion here is that a given map has an inverse on a sufficiently small neighborhood of any given point if and only if its differential has an inverse at the point.

The statement of the implicit function theorem (Theorem 1.2) refers to a *regular* value of a map between Euclidean spaces. I remind you of the definition: Let $U \subset \mathbb{R}^m$ denote a given open subset and let $\psi: U \to \mathbb{R}^k$ denote a given smooth map. A point $a \in \mathbb{R}^k$ is said to be a *regular value* of ψ if the matrix ψ_* is surjective at all points in $\psi^{-1}(a)$.

Theorem 1.2 (the implicit function theorem) *Fix nonnegative integers* $m \geq n$. *Suppose that* $U \subset \mathbb{R}^m$ *is an open set,* $\psi\colon U \to \mathbb{R}^{m-n}$ *is a smooth map, and* $a \in \mathbb{R}^{m-n}$ *is a regular value of* ψ. *Then* $\psi^{-1}(a) \subset U$ *has the structure of a smooth,* n-*dimensional manifold whose smooth structure is defined by coordinate charts of the following sort: Fix a point* $p \in \psi^{-1}(a)$. *Then there is a ball* $B \subset \mathbb{R}^m$ *centered at* p *such that the orthogonal projection from* B *to the kernel of* $\psi_*|_p$ *restricts to* $\psi^{-1}(a) \cap B$ *as a coordinate chart. In addition, there is a diffeomorphism,* $\varphi\colon B \to \mathbb{R}^m$ *such that* $\varphi(B \cap \psi^{-1}(a))$ *is a neighborhood of the origin in the* n-*dimensional linear subspace of points* $(x_1, \ldots, x_m)$ *with* $x_{n+1} = \cdots = x_m = 0$.

This theorem is, of course, vacuous if $m < n$. By the way, this theorem is often used in conjunction with

Theorem 1.3 (Sard's theorem) *Suppose that* $U \subset \mathbb{R}^m$ *is an open set,* $\psi\colon U \to \mathbb{R}^n$ *is a smooth map. Then the set of regular values of* ψ *have full measure.*

This means the following: Fix any ball in $\mathbb{R}^n$. Delete the all-nonregular values from this ball. The volume of the resulting set is the same as that of the ball. Thus, a randomly chosen point from the ball has probability 1 of being a regular value. (A special case of Sard's theorem (Theorem 14.14) is proved in Appendix 14.3.)

1.3 Submanifolds of $\mathbb{R}^m$

Let $f\colon \mathbb{R}^m \to \mathbb{R}$ denote a smooth function and let $a \in \mathbb{R}$ denote a regular value of f. Then $f^{-1}(a)$ is a smooth, (m−1)-dimensional manifold; this is a consequence of the implicit function theorem. This is an example of an (m−1)-dimensional submanifold of $\mathbb{R}^m$. For a concrete example, take the function $x \to f(x) = |x|^2$ where $|x|$ denotes the Euclidean norm of x. Any nonzero number is a regular value, but of course only positive regular values are relevant. In this case $f^{-1}(r^2)$ for $r > 0$ is the sphere in $\mathbb{R}^{n+1}$ of radius r. Spheres of different radii are diffeomorphic.

Definition 1.4 *A submanifold in* $\mathbb{R}^m$ *of dimension* $n < m$ *is a subset,* Σ, *with the following property: Let* p *denote any given point in* Σ. *There is a ball* $U_p \subset \mathbb{R}^m$ *around* p *and a map* $\psi_p\colon U_p \to \mathbb{R}^{m-n}$ *with* 0 *as a regular value and such that* $\Sigma \cap U_p = \psi_p^{-1}(0)$.

The implicit function theorem says that any such Σ is a smooth, n-dimensional manifold.

The following lemma can be used to obtain examples of submanifolds.

Lemma 1.5 *Suppose that* $n \leq m$ *and that* $B \subset \mathbb{R}^n$ *is an open ball centered on the origin. Let* $\varphi\colon B \to \mathbb{R}^m$ *denote a smooth, 1–1 map whose differential is everywhere injective. Let* $W \subset B$ *denote any given open set with compact closure. Then* $\varphi(W)$ *is a submanifold of* $\mathbb{R}^m$ *such that* $\varphi|_W\colon W \to \varphi(W)$ *is a diffeomorphism.*

Proof of Lemma 1.5 This can be proved using the inverse function theorem (Theorem 1.1). To say more, fix any given point $p \in \varphi(W)$. Needed is a smooth map, ψ_p from a ball centered at p to $\mathbb{R}^{m-n}$ that maps p to the origin, whose differential at p is surjective, and is such that $\varphi(W)$ near p is $\psi_p^{-1}(0)$. To obtain this data, let $z \in W$ denote the point mapped by φ to p. Let $K \subset \mathbb{R}^m$ denote the kernel of the adjoint of $\varphi_*|_p$. This is an $(m - n)$-dimensional subspace because the differential of φ at z is injective. Keeping this in mind, define $\lambda\colon W \times K \to \mathbb{R}^m$ by the rule $\lambda(x, v) = \varphi(x) + v$. Note that $\lambda(z, 0) = p$. As the differential of λ at $(z, 0)$ is an isomorphism, the inverse function theorem finds a ball, U_p, of p and a smooth map $\eta\colon U_p \to W \times K$ such that the following is true: First, $\eta(U_p)$ is an open neighborhood of $(z, 0)$. Second, $\varphi \circ \eta$ is the identity. Third, $\eta \circ \varphi$ is the identity on $\eta(U_p)$. Granted this, let $\mathfrak{p}\colon W \times K \to K$ denote the orthogonal projection. Identify K with $\mathbb{R}^{n-m}$ and define the map $\psi_p\colon U_p \to K = \mathbb{R}^{n-m}$ by the rule $\psi_p(x) = \mathfrak{p}\eta(x)$. By construction, $\psi_p^{-1}(0) = \varphi(\eta(U_p))$ and $\psi_{p*}|_p$ is surjective.

Here is an example of a dimension 2 submanifold in $\mathbb{R}^3$. Fix $\rho \in (0, 1)$. The submanifold is the set of points $(x_1, x_2, x_3) \in \mathbb{R}^3$ where

$$f(x_1, x_2, x_3) = ((x_1^2 + x_2^2)^{1/2} - 1)^2 + x_3^2 = \rho^2.$$

This is a torus of revolution; it is diffeomorphic to $S^1 \times S^1$ as can be seen by introducing an angle $\phi \in \mathbb{R}/2\pi\mathbb{Z}$ for the left-most S^1 factor, a second angle φ for the right-most factor, and then using these to parametrize the locus by the map that sends

$$(\phi, \varphi) \to \left(x_1 = (1 + \rho\cos\phi)\cos\varphi,\ x_2 = (1 + \rho\cos\phi)\sin\varphi,\ x_3 = \rho\sin\phi\right).$$

A torus sitting in $\mathbb{R}^3$ looks like the surface of a tire inner tube of the sort that you would inflate for a bicycle. Figure 1.1 is a depiction viewed from above.

Figure 1.1 *A torus sitting in* $\mathbb{R}^3$

A surface in $\mathbb{R}^3$ with some number holes when viewed from above can also appear as a submanifold of $\mathbb{R}^3$. The notion of what constitutes a hole has rigorous definition; see for example Chapter 12 in the book *Topology* by James Munkres (2000). Chapter 14 in the present volume gives two equivalent definitions using differential geometric notions. Suffice it to say here that the torus depicted above has genus 1, and a standard pretzel has genus 2. What follows depicts a surface in $\mathbb{R}^3$ with a given genus g. This construction was explained to the author by Curt McMullen. To start, fix $r \in (0, \frac{1}{100g})$ and a function $h: \mathbb{R}^2 \to \mathbb{R}$ with the following properties:

- $h(x, y) \leq 0$ *where* $|x|^2 + |y|^2 \geq 1$.
- *For each* $k = \{1, \ldots, g\}$, *require* $h(x, y) \leq 0$ *where* $|x - \frac{k}{2g}|^2 + |y|^2 \leq r^2$.
- $h(x, y) > 0$ *otherwise.*
- $|\frac{\partial h}{\partial x}| + |\frac{\partial h}{\partial y}| \neq 0$ *at any point where* $h = 0$, *thus on the circle of radius 1 about the origin, and on the* g *circles of radius* r *about the respective points in the set* $\left\{\left(x = \frac{k}{2g}, y = 0\right)\right\}_{k=1,\ldots,g}$.

Figure 1.2 is a schematic of how h might look like in the case $g = 2$. The black area is where $h > 0$; the white is where $h \leq 0$.

The corresponding surface in $\mathbb{R}^3$ is the set of points where the function $f = z^2 - h(x, y)$ is zero. The condition on h in the fourth bullet guarantees that 0 is a regular value of f and so $f^{-1}(0)$ is a smooth submanifold in $\mathbb{R}^3$. It is compact because $h(x, y) \leq 0$ when $x^2 + y^2$ is greater than 1.

Any n-manifold can be realized as a submanifold of $\mathbb{R}^m$ for some $m \gg n$. What follows is an argument for the case when the manifold in question is compact manifold. To start, fix a finite coordinate atlas for M. Label the elements in this atlas as $\{(U_1, \varphi_1), \ldots, (U_N, \varphi_N)\}$. Fix a subordinate partition of unity, $\{\chi_\alpha\}_{1\leq\alpha\leq N}$. This is a set of smooth functions on M such that any given χ_α has support in the corresponding set U_α and such that $\Sigma_{1\leq\alpha\leq N}\, \chi_\alpha = 1$ at each point. Appendix 1.2 gives a quick tutorial in such thing. Introduce the map

Figure 1.2 *h in the case g=2*

$$\psi: M \to \mathbb{R}^{Nm+N} = \times_{1 \le a \le N}(\mathbb{R} \times \mathbb{R}^m)$$

that is defined by the following rule: Send any given point $x \in M$ to the point in the α'th factor of $\mathbb{R}^{N(m+1)}$ given by $(1 - \chi_\alpha(x), \chi_\alpha(x)\phi_\alpha(x))$. The differential of this map is injective because there exists near any given point at least one index α such that $\phi_\alpha \neq 0$. If this is the case for α, then the map to $\mathbb{R}^{m+1}$ from a small neighbohood of this point given by $x \to (1 - \chi_\alpha(x), \chi_\alpha(x)\varphi_\alpha(x))$ is injective. This understood, the implicit function theorem can be used to infer that the image of ψ is a submanifold of $\mathbb{R}^{N(m+1)}$ and that ψ is a diffeomorphism onto its image.

1.3.1 Coordinates near a submanifold

The implicit function theorems can be used to construct "nice" coordinates for $\mathbb{R}^n$ near any given point of a submanifold. Here is what can be done: Let $Y \subset \mathbb{R}^n$ denote a submanifold of some dimension d, and let p denote any given point in Y. Then there is a neighborhood $U \subset \mathbb{R}^n$ of p and a diffeomorphism $\psi: U \to \mathbb{R}^n$ such that $\psi(Y)$ near p is the locus of points $(x_1, \ldots, x_n)$ where $x_{d+1} = x_{d+2} = \cdots = x_n = 0$.

1.4 Submanifolds of manifolds

Suppose that M is a smooth manifold of dimension m. A subset $Y \subset M$ is a submanifold of some given dimension $n \le m$ if the following is true: Fix any point $p \in Y$ and there is an open set $U \subset M$ that contains p, and a coordinate chart map $\phi: U \to \mathbb{R}^m$ such that $\phi(Y \cap U)$ is an n-dimensional submanifold of $\mathbb{R}^m$.

Note that if $\psi: M \to N$ is smooth, and if $Y \subset M$ is a smooth submanifold, then the restriction $\psi|_Y: Y \to N$ is smooth.

1.4.1 Immersions and submersions

The notion of a submanifold is closely related to the notions of immersion and submersion. To say more, let M denote a manifold of dimension m and let n denote one of dimension Y. Suppose first that $n \le m$. A smooth map $\psi: Y \to M$ is said to be an *immersion* when it has the following property: Fix any point $p \in Y$ and an open set $V \subset Y$ that contains p with a coordinate map $\phi_V: V \to \mathbb{R}^n$. Let $U \subset M$ denote an open set containing $\psi(p)$ with a coordinate map $\phi_U: U \to \mathbb{R}^m$. Then the differential of $\phi_U \circ \psi \circ \phi_V^{-1}$ at $\phi_V(p)$ is an injective linear map from $\mathbb{R}^n$ to $\mathbb{R}^m$. This implies that $\psi(V)$ is a n-dimensional submanifold of M on some small neighborhood of $\psi(p)$.

Suppose next that $n \geq m$. The map ψ is said to be a *submersion* when the following is true: Fix p, V, and U as above. Then the differential of $\phi_U \circ \psi \circ \phi^{-1}_V$ at p is a surjective linear map from $\mathbb{R}^n$ to $\mathbb{R}^m$. An application of the inverse function theorem leads to the conclusion that $\psi^{-1}(\psi(p))$ is a (n−m)-dimensional submanifold of Y near p.

1.5 More constructions of manifolds

What follows are some simple ways to obtain new manifolds from old ones.

1.5.1 Products of smooth manifolds

The product of two circles has the following generalization: Suppose that M, N are smooth manifolds. Then $M \times N$ has a canonical smooth structure. A coordinate atlas for the latter can be taken as follows: Fix coordinate atlases, $\mathcal{U}$ for M and $\mathcal{V}$ for N, that define the respective smooth structures. Then a coordinate atlas for $M \times N$ is the collection $\{(U \times V, \varphi \times \psi)\}_{(U,\varphi)\in\mathcal{U},\ (V,\psi)\in V}$. The dimension of $M \times N$ is the sum of the dimensions of M and N. For example, the circle is S^1; the n-fold product of the circle is called the n-dimensional torus.

1.5.2 Open subsets of manifolds

Let M denote a smooth manifold and $W \subset M$ any open subset. Then W inherits the structure smooth manifold. Indeed, fix a coordinate atlas for M that defines the smooth structure. Then a coordinate atlas for the smooth structure on W can be taken to be the collection $\{(U \cap W, \varphi)\}_{(U,\ \varphi)\in\mathcal{U}}$.

1.5.3 Quotients that are manifolds

Let M denote a smooth manifold. An equivalence relation on M is a collection of disjoint subsets in M whose union is the whole of M. I will use $M/\sim$ to denote this collection of sets. An element in the set $M/\sim$ is said to be an *equivalence class*; and when x and x′ are points in M, I write $x \sim x'$ when they lie in the same equivalence class. A map π: $M \to M/\sim$ is defined by sending any given point to its equivalence class. The set $M/\sim$ is given the quotient topology whereby a set $U \subset M/\sim$ is declared to be an open set if and only if $\pi^{-1}(U)$ is open in M.

It is often the case that $M/\sim$ inherits from M the structure of a smooth manifold. In particular, this occurs when the following is true: Let $x \in M$.

Then there is an open neighborhood $V \subset M$ of x such that $\pi|_V$ is a 1–1 and a homeomorphism from V to $\pi(V)$. If this is the case, then $M/\sim$ has a canonical smooth structure, this defined as follows: Let $\mathcal{U}$ denote a given coordinate atlas for M that defines the smooth structure. Fix an open cover, $\mathcal{V}$, of M such that π is 1–1 on any set from $\mathcal{V}$. A coordinate atlas for M in $\mathcal{U}$'s equivalence class is defined by the collection $\{(U \cap V, \varphi|_{U \cap V})\}_{(U,\varphi)\in\mathcal{U},\ V\in\mathcal{V}}$. With this understood, a coordinate atlas that defines the smooth structure for $M/\sim$ is given by the collection $\{(\pi(U \cap V), \varphi \circ \pi^{-1}\colon \pi(U \cap V) \to \mathbb{R}^n)\}_{(U,\varphi)\in\mathcal{U},\ V\in\mathcal{V}}$. Note in particular that the map π from M to $M/\sim$ is a smooth map.

Here is a first example: The circle S^1 can be viewed as $\mathbb{R}/\sim$ where the equivalence relation identifies points x and x′ if and only if $x = x' + 2\pi k$ for some $k \in \mathbb{Z}$. With S^1 viewed as the unit circle in $\mathbb{R}^2$, the map π becomes the map sending $t \in \mathbb{R}$ to the point $(\cos(t), \sin(t))$. When I think of S^1 in terms of this equivalence relation, I will write it as $\mathbb{R}/2\pi\mathbb{Z}$.

By the same token, the n-dimensional torus $\times_n S^1$ can be viewed as $\mathbb{R}^n/\sim$ where the equivalence relation identifies points x and x′ if each entry of x differs from the corresponding entry of x′ by 2π times an integer.

Here is a second example: The real projective space $\mathbb{RP}^n = S^n/\sim$ where the equivalence relation identifies unit vectors x and x′ in $\mathbb{R}^{n+1}$ if and only if $x = \pm x'$. As a parenthetical remark, note that $\mathbb{RP}^n$ can be viewed as the set of 1-dimensional linear subspaces in $\mathbb{R}^{n+1}$. The correspondence is as follows: A given 1-dimensional subspace has precisely two vectors of unit length, one is -1 times the other. These two vectors define a point in $\mathbb{RP}^n$. Conversely, a point in $\mathbb{RP}^n$ gives the linear subspace that is spanned by either of its corresponding points in S^n.

For the third example, fix two integers $n \geq 2$ and $m \geq 1$. Define an equivalence relation $\sim$ on S^{2n-1} as follows: The point $x = (x_1, x_2, \ldots, x_{2n})$ is equivalent to $x' = (x_1', x_2', \ldots, x_{2n}')$ if and only if for each $k \in \{1, \ldots, n\}$, the following is true:

$$x_{2k+1}' = \cos(2\pi j/m)x_{2k+1} - \sin(2\pi j/m)x_{2k},$$
$$x_{2k}' = \cos(2\pi j/m)x_{2k} + \sin(2\pi j/m)x_{2k+1}$$

with j some integer. Note that $\mathbb{RP}^{2n+1}$ is obtained when $m = 2$. Manifolds of this sort are examples of *lens spaces*.

1.6 More smooth manifolds: The Grassmannians

As noted above, the real projective space $\mathbb{RP}^n$ can be viewed as the set of 1-dimensional vector subspaces in $\mathbb{R}^{n+1}$. In general, if $m > n$, then the set of n-dimensional vector subspaces in $\mathbb{R}^m$ also has the structure of a smooth manifold. The latter manifold is denoted by Gr(m; n) and is called the *Grassmannian of n-planes in* $\mathbb{R}^m$. The manifold structure for Gr(m; n) can be obtained by exhibiting a coordinate atlas whose transition functions are diffeomorphisms. The first task, however, is to give Gr(m; n) a suitable topology. The latter is the metric topology defined as follows: Let $V \subset \mathbb{R}^m$ denote a given n-dimensional subspace. Let $\Pi_V: \mathbb{R}^m \to V$ denote the orthogonal projection. An open neighborhood of V is the set, $\mathcal{O}_V$, of subspaces V' such that $\Pi_V: V' \to V$ is an isomorphism. A basis of open neighborhoods of V is indexed by the positive numbers; and the $\varepsilon > 0$ member, $\mathcal{O}_{V,\varepsilon}$, consists of the n-dimensional subspaces $V' \subset \mathbb{R}^m$ such that $|\Pi_V(v) - v| < \varepsilon|v|$ for all $v \neq 0$ in V'.

As a manifold, Gr(m; n) has dimension n(m−n). The neighborhood $\mathcal{O}_V$ will serve as a coordinate chart centered around V; thus a map from $\mathcal{O}_V$ to $\mathbb{R}^{n(n-m)}$ that gives the local coordinates is needed. In order to give the coordinate map, first introduce $\mathbb{M}(m, n)$ to denote the vector space of $(m - n) \times m$ matrices. The entries of any given matrix $\mathfrak{p} \in \mathbb{M}(m, n)$ serve as coordinates and identify this space with $\mathbb{R}^{n(m-n)}$. Now, let $V \subset \mathbb{R}^m$ denote a given n-dimensional subspace and let $V^\perp \subset \mathbb{R}^m$ denote its orthogonal complement. Use $\Pi_{V^\perp}: \mathbb{R}^m \to V^\perp$ in what follows to denote the orthogonal projection to $V^\perp$. Fix a basis for V to identify it with $\mathbb{R}^m$ and fix a basis for $V^\perp$ to identify it with $\mathbb{R}^{m-n}$. Use these bases to identify the space of linear maps from V to $V^\perp$ (denoted by $\mathrm{Hom}(V; V^\perp)$) with $\mathbb{R}^{n(m-n)}$. Now, write $\mathbb{R}^m = V \times V^\perp$. Suppose that $V' \in \mathcal{O}_V$. Since $\Pi_V: V' \to V$ is an isomorphism, there is a linear inverse $L_{V,V}: V \to V'$. This understood, the composition $\Pi_{V^\perp} \circ L_{V,V'}$ is a linear map from V to $V^\perp$, thus an element in $\mathrm{Hom}(V; V') = \mathbb{R}^{n(m-n)}$. This point is defined to be $\varphi_V(V')$. It is an exercise to check that the map φ_V is continuous. To see that it is a homeomorphism, it is enough to exhibit a continuous inverse: The inverse sends a given $n \times (m-n)$ matrix $\mathfrak{p} \in \mathrm{Hom}(V; V^\perp) = \mathbb{R}^{n(m-n)}$ to the m-dimensional linear subspace $\varphi_V^{-1}(\mathfrak{p}) = \{(v, \mathfrak{p}v) : v \in V\}$.

It remains yet to verify that the transition function between any two intersecting charts defines a smooth map between domains in $\mathbb{R}^{n(m-n)}$. This task is deferred to Appendix 2.1.

It is left as a linear algebra exercise to verify that the transition functions between two charts $\mathcal{O}_V$ and $\mathcal{O}_{V'}$ are diffeomorphisms.

I also leave it as an exercise to verify that the smooth structure just defined for Gr(n+1; 1) is the same as that given previously for $\mathbb{RP}^n$.

Appendix 1.1 How to prove the inverse function and implicit function theorems

What follows are detailed outlines of proofs first of the inverse function theorem and then of the implicit function theorem. See, for example, *Differential Topology* by Victor Guillemin and Alan Pollack (1974) for unabridged proofs.

A1.1.1 Proof of the inverse function theorem (Theorem 1.1)

The first point to make is that no generality is lost by taking p and $\psi(p)$ to be the origin. Let $\mathfrak{m}$ denote the $n \times n$ matrix $\psi_*|_0$. I look for an inverse, σ, that has the form

$$\sigma(x) = \mathfrak{m}^{-1}x + \mathfrak{f}(x)$$

where $\mathfrak{f}$ is such that $|\mathfrak{f}(x)| \leq c_0\,|x|^2$. Here I have introduced a convention that is used in this chapter and all subsequent chapters: What is denoted by c_0 is always a constant, greater than 1, whose value increases on subsequent appearances. The use of this generic notation is meant to avoid the proliferation of different symbols or subscripts to denote positive numbers which are either greater than 1 or less than 1, but are such that the precise value is of no real consequence to the discussion at hand. The convention here takes c_0 to be greater than 1, so when it is important that a positive number have value less than 1, I use c_0^{-1}.

To find $\mathfrak{f}$, I use Taylor's theorem with remainder to write $\psi(x) = \mathfrak{m}x + \mathfrak{r}(x)$ where $\mathfrak{r}$ obeys $|\mathfrak{r}(x)| \leq c_0|x|^2$. Note also that any given partial derivative of $\mathfrak{r}$ obeys $|\partial_j\mathfrak{r}(x)| \leq c_0|x|$. In any event, the map σ is an inverse to ψ if an only if $\psi(\sigma(x)) = x$ for x near 0 in $\mathbb{R}^n$. In terms of $\mathfrak{f}$, this demands that

$$\mathfrak{m}\,\mathfrak{f}(x) + \mathfrak{r}(\mathfrak{m}^{-1}x + \mathfrak{f}(x)) = 0.$$

This is to say that

$$\mathfrak{f}(x) = -\mathfrak{m}^{-1}\mathfrak{r}(\mathfrak{m}^{-1}x + \mathfrak{f}(x)).$$

This sort of equation can be solved for any given x using the contraction mapping theorem (Theorem 1.6). The point being that the desired solution, $\mathfrak{f}(x)$, to the preceding equation is a fixed point of the map from $\mathbb{R}^n$ to $\mathbb{R}^n$ that sends y to $\mathcal{T}_x(y) = -\mathfrak{m}^{-1}\mathfrak{r}(\mathfrak{m}^{-1}x + y)$. The contraction mapping theorem asserts the following:

Theorem 1.6 (the contraction mapping theorem) *Let* $B \subset \mathbb{R}^n$ *denote a closed ball and* $\mathcal{T}$ *a map from* B *to itself. The map* $\mathcal{T}$ *has unique fixed point in* B *if* $|\mathcal{T}(y) - \mathcal{T}(y')| \leq (1 - \delta)\, |y - y'|$ *for some positive* δ *and for all* $y, y' \in B$.

Proof of Theorem 1.6 Start with $y_0 \in B$ and then construct a sequence $\{y_k = \mathcal{T}(y_{k-1})\}_{k=1,2,\ldots}$. This is a Cauchy sequence because $|y_k - y_{k-1}| = |\mathcal{T}(y_{k-1}) - \mathcal{T}(y_{k-2})| \leq (1 - \delta)|y_{k-1} - y_{k-2}|$. Let y_f denote the limit point. This point is in B and $\mathcal{T}(y_f) = y_f$. There can't be any other fixed point in B; were y' such a point, then $|y_f - y'| = |\mathcal{T}(y_f) - \mathcal{T}(y')| \leq (1 - \delta)|y_f - y'|$ which can be true only if $y_f = y'$.

In the case at hand, $|\mathcal{T}_x(y)| \leq c_0(|x|^2 + |y|^2)$ because $\mathfrak{r}$ comes from the remainder term in the Taylor's expansion for ψ. This implies that $\mathcal{T}_x$ maps any ball of radius $r < \frac{1}{2}c_0^{-1}$ to itself if $|x| \leq c_0^{-1}r$. For the same reason, $|\mathcal{T}_x(y) - \mathcal{T}_x(y')| \leq c_0(|x| + |y|)\, |y - y'|$, and so $\mathcal{T}_x$ is a contraction mapping on such a ball if $|x| \leq c_0^{-1}r$. The unique fixed point is $\mathfrak{f}(x)$ that obeys $|\mathfrak{f}(x)| \leq c_0 r^2$.

It remains yet to prove that the map $x \to \mathfrak{f}(x)$ is smooth. A proof that such is the case can also be had by successively using the contraction mapping theorem to analyze first the difference $|\mathfrak{f}(x) - \mathfrak{f}(x')|$, and then successively higher difference quotients.

For the first difference quotient, use the fact that $\mathfrak{f}(x)$ and $\mathfrak{f}(x')$ are respective fixed points of $\mathcal{T}_x$ and $\mathcal{T}_{x'}$ to write their difference as

$$\mathfrak{f}(x) - \mathfrak{f}(x') = \mathfrak{m}^{-1}(\mathfrak{r}(\mathfrak{m}^{-1}x + \mathfrak{f}(x))) - \mathfrak{r}(\mathfrak{m}^{-1}x' + \mathfrak{f}(x'))$$

Given that $\mathfrak{r}$ is the remainder term in Taylor's theorem, and given that x and x' lie in the ball of radius r, and that $\mathfrak{f}(x)$ and $\mathfrak{f}(x')$ lie in the ball of radius $c_0 r^2$, this last equation implies that $|\mathfrak{f}(x) - \mathfrak{f}(x')| \leq c_0 r\, (|x - x'| + |\mathfrak{f}(x) - \mathfrak{f}(x')|)$. In particular, if $r < \frac{1}{2}c_0^{-1}$, this tells us that $|\mathfrak{f}(x) - \mathfrak{f}(x')| \leq c_0^{-1}\, |x - x'|$.

A1.1.2 Proof of the implicit function theorem (Theorem 1.2)

What follows is a sketch of the proof of the implicit function theorem. The first step constructs a local coordinate chart centered on any given point in $\psi^{-1}(a)$. Let p denote such a point. Introduce $K \subset \mathbb{R}^m$ to denote the kernel of $\psi_*|_p$, and let $\mathfrak{p}$: $\mathbb{R}^m \to K$ denote the orthogonal projection. Now define a map, φ: $\mathbb{R}^m \to \mathbb{R}^m = K \times \mathbb{R}^{m-n}$ by sending any given point $x \in \mathbb{R}^m$ to $(\mathfrak{p}(x - p), \psi(x) - a)$. The differential of this map at p is an isomorphism, and so the inverse function theorem finds balls $B_K \subset K$ and $B' \subset \mathbb{R}^{m-n}$ about the respective origins with a map σ: $B_K \times B' \to \mathbb{R}^m$ that obeys $\sigma(0) = p$ and is such that $\varphi \circ \sigma =$ identity. In particular, this has the following implication: Let $y \in B_K$. Then $\psi(\sigma(y, 0)) = a$. As a consequence, the map from B_K to $\mathbb{R}^m$ that sends y to $\sigma(y, 0)$ is 1–1 and invertible. The inverse is the map $x \to \mathfrak{p}(x - p)$ from $\mathbb{R}^m$ to $K = \mathbb{R}^n$, and so the latter map gives local coordinates for $\psi^{-1}(a)$ near p.

Granted the preceding, it remains yet to prove that the transition functions for these coordinate charts are smooth and have smooth inverses. There is nothing terribly complicated about this, and so it is left to the reader.

Appendix 1.2 Partitions of unity

This appendix says something about how to construct a locally finite partition of unity for a coordinate atlas. The story starts on $[0, \infty)$. What follows directly describes a smooth function that is zero at points $t \in [0, \infty)$ with $t \geq 1$ and is nonzero at all points $t \in [0, \infty)$ with $t < 1$. The function, χ, is defined by the rule whereby

- $\chi(t) = e^{-1/(1-t)}$ *for* $t < 1$.
- $\chi(t) = 0$ *for* $t \geq 1$.

Granted this, suppose that $\mathcal{U}$ is a locally finite coordinate atlas for a given smooth, n-dimensional manifold with $0 \in \mathbb{R}^n$ in all chart images. Since $\mathcal{U}$ is locally finite, one can assign to each $(U, \varphi_U) \in \mathcal{U}$ a positive real number, r_U, such that the following is true: Let U_r denote the inverse image via φ_U of the ball of radius r_U about the origin in $\mathbb{R}^n$. Then the collection $\{U_r\}_{(U,\varphi)\in\mathcal{U}}$ is an open cover of M. This understood, associate to any given $(U, \varphi_U) \in \mathcal{U}$ the function σ_U on M that is defined as follows: If p is not in U, then $\sigma_U(p) = 0$. If p is in U, then $\sigma_U(p)$ is equal to $\chi(|\varphi_U(p)|/r_U)$. Note that $\sigma_U(p)$ is nonzero on U_r but zero on the complement of U.

The collection $\{\sigma_U\}_{(U,\varphi)\in\mathcal{U}}$ is almost, but not quite, the desired subordinate partition of unity function. It is certainly the case that $\sum_{(U,\varphi)\in\mathcal{U}}\sigma_U$ is nowhere zero, but this sum is not necessarily equal to 1. The desired subordinate partition of unity consists of the set $\{\chi_U\}_{(U,\varphi)\in\mathcal{U}}$ where $\chi_U = (\sum_{(U',\varphi')\in\mathcal{U}} \sigma_{U'})^{-1}\sigma_U$.

Additional reading

- *Differential Topology*, Victor Guillemin and Alan Pollack, Prentice-Hall, 1974.
- *An Introduction to Manifolds*, Loring Tu, Springer, 2008.
- *Elementary Differential Topology*, James R. Munkres, Princeton University Press, 1966.
- *An Introduction to Differentiable Manifolds and Riemannian Geometry*, William M. Boothby, Academic Press, 1975.

2 Matrices and Lie groups

Matrices and especially invertible matrices play a central role in most all of differential geometry. In hindsight, I don't find this to be surprising. The role stems ultimately from Taylor's theorem: A smooth map between two Euclidean spaces can be well approximated near any given point by a linear map, this the differential at the point. The inverse and implicit function theorems (Theorems 1.1 and 1.2) depend entirely on this observation. In any event, matrices and notions from linear algebra appear in all of the subsequent chapters. Constructions involving matrices also give some interesting and explicit examples of smooth manifolds.

2.1 The general linear group

I use $\mathbb{M}(n; \mathbb{R})$ to denote the vector space of $n \times n$ matrices with real entries. This is a copy of the Euclidean space $\mathbb{R}^{n^2}$ with the entries of a matrix giving the coordinate functions. Matrices m and m' can be multiplied, and matrix multiplication gives a smooth map, $(m, m') \to mm'$, from $\mathbb{M}(n; \mathbb{R}) \times \mathbb{M}(n; \mathbb{R})$ to $\mathbb{M}(n; \mathbb{R})$. This is smooth by virtue of the fact that the entries of mm' are linear functions of the coordinates of m, and also linear in the coordinates of m'. The space $\mathbb{M}(n; \mathbb{R})$ has two special functions. The first function, $\det: \mathbb{M}(n; \mathbb{R}) \to \mathbb{R}$ sends any given matrix to its determinant. This is a smooth function since it is a polynomial of degree n in the entries. The determinant of a matix, m, is denoted by $\det(m)$.The other function of interest is the trace. The trace is the sum of the diagonal entries and so a linear function of the entries. The trace of a matrix m is denoted by $\operatorname{tr}(m)$.

Let $Gl(n; \mathbb{R})$ denote the subspace in $\mathbb{M}(n; \mathbb{R})$ of invertible matrices. This is to say that $m \in Gl(n; \mathbb{R})$ if and only if the $\det(m) \neq 0$. This is an open subset of $\mathbb{M}(n; \mathbb{R})$, and so de facto $Gl(n; \mathbb{R})$ is a smooth manifold of dimension n^2. In particular, the n^2 entries of the matrix restrict to $Gl(n; \mathbb{R})$ near any given point to give local coordinates. The multiplication map restricts to $Gl(n; \mathbb{R})$ as a smooth map.

The manifold $Gl(n; \mathbb{R})$ has a canonical diffeomorphism, this the map to itself that sends any given matrix $\mathfrak{m}$ to the inverse matrix $\mathfrak{m}^{-1}$. This map is smooth because any given entry of $\mathfrak{m}^{-1}$ is the quotient by $\det(\mathfrak{m})$ of a polynomial in the entries of $\mathfrak{m}$. For example, in the 2×2 case, write

$$\mathfrak{m} = \begin{pmatrix} a & b \\ c & d \end{pmatrix},$$

and then its determinant is $ad - bc$ and its inverse is

$$\mathfrak{m}^{-1} = \frac{1}{ad - bc} \begin{pmatrix} d & -b \\ -c & a \end{pmatrix}.$$

The form just asserted for the entries of the inverse in the $n > 2$ case (a polynomial divided by the determinant) can be seen by using the reduced row echelon form algorithm to construct the inverse.

What follows gives a very useful formula for the directional derivatives of the map $\mathfrak{m} \to \mathfrak{m}^{-1}$. Let $a \in \mathbb{M}(n; \mathbb{R})$ denote any given matrix. Then the directional derivatives of the coordinates of map $\mathfrak{m} \to \mathfrak{m}^{-1}$ in the direction a are the entries of the matrix $-\mathfrak{m}^{-1}a\mathfrak{m}^{-1}$. Consider, for example, the coordinate given by the (i,j) entry, $(\mathfrak{m}^{-1})_{ij}$: The directional derivative in the direction a of this function on $Gl(n; \mathbb{R})$ is $-(\mathfrak{m}^{-1}a\mathfrak{m}^{-1})_{ij}$. In particular, the partial derivative of the function $\mathfrak{m} \to (\mathfrak{m}^{-1})_{ij}$ with respect to the coordinate $\mathfrak{m}_{rs}$ is $-(\mathfrak{m}^{-1})_{ir}(\mathfrak{m}^{-1})_{sj}$.

2.2 Lie groups

A group is a set with a fiducial point and two special maps. To say more let G denote the set. The fiducial point is denoted in what follows by ι; it is called the *identity.* The first of the maps is the *multiplication* map, this a map from $G \times G$ to G. It is customary to denote this map by $(\mathfrak{g}, \mathfrak{g}') \to \mathfrak{g}\mathfrak{g}'$. This map is constrained so that $\mathfrak{g}\iota = \mathfrak{g}$ and also $\iota\mathfrak{g} = \mathfrak{g}$. The other map sends G to G; it is the *inverse*. It is customary to write the inverse as $\mathfrak{g} \to \mathfrak{g}^{-1}$. The inverse map is constrained so that $\mathfrak{g}\mathfrak{g}^{-1} = \iota = \mathfrak{g}^{-1}\mathfrak{g}$.

A Lie group is a group with the structure of a smooth manifold such that both the multiplication map and the inverse map are smooth. The manifold $Gl(n; \mathbb{R})$ as just described is the Ur example of a Lie group. Note that $\mathbb{R}^n$ with the origin playing the role of ι, with addition playing the role of multiplication, and with the map $x \to -x$ playing the role of inverse is a group. The latter is an example of an *Abelian group*, this a group with the property that $\mathfrak{g}\mathfrak{g}' = \mathfrak{g}'\mathfrak{g}$ for all pairs $\mathfrak{g}$ and $\mathfrak{g}'$.

It is often the case that interesting groups occur as subgroups of larger groups. By way of reminder, a subgroup of a group G is a subset, H, that contains ι, is mapped to itself by the inverse map, and is such that multiplication maps $H \times H$ to H.

Lemma 2.1 *A subgroup of a Lie group that is also a submanifold is a Lie group with respect to the induced smooth structure.*

Proof of Lemma 2.1 This follows by virtue of the fact that the restriction to a submanifold of any smooth map to any given manifold defines a smooth map from the submanifold to the given manifold.

2.3 Examples of Lie groups

The group $Gl(n; \mathbb{R})$ contains sub-Lie groups that arise most often in differential geometric constructions.

2.3.1 The group $Sl(n; \mathbb{R})$

Let $Sl(n; \mathbb{R}) \subset Gl(n; \mathbb{R})$ denote the subset of matrices whose determinant is 1. This is a subgroup by virtue of the fact that $\det(mm') = \det(m)\det(m')$. It is a smooth submanifold of dimension $n^2 - 1$. This follows from the fact that the differential of this function can be written as

$$d(\det)|_m = \det(m) \sum_{i,j} (m^{-1})_{ij} dm_{ij} = \det(m)\ \mathrm{tr}(m^{-1} dm).$$

Here, dm denotes the $n \times n$ matrix whose (i, j)'th entry is the coordinate differential dm_{ij}. As can be seen from the preceding formula, the differential of det is not zero where det is nonzero.

As it turns out, the group $Sl(2; \mathbb{R})$ is diffeomorphic to $S^1 \times \mathbb{R}^2$. This can be seen by using a linear change of coordinates on $\mathbb{M}(2; \mathbb{R})$ that writes the entries in terms of coordinates (x, y, u, v) as follows:

$$m_{11} = x - u, \quad m_{22} = x + u, \quad m_{12} = v - y \ \textit{ and } \ m_{21} = v + y.$$

The condition that $\det(m) = 1$ now says that $x^2 + y^2 = 1 + u^2 + v^2$. This understood, the diffeomorphism from $S^1 \times \mathbb{R}^2$ to $Sl(2; \mathbb{R})$ sends a triple (θ, a, b) to the matrix determined by $x = (a^2 + b^2)^{1/2} \cos(\theta)$, $y = (a^2 + b^2)^{1/2} \sin\theta$, $u = a$, $v = b$. Here, $\theta \in [0, 2\pi]$ is the angular coordinate for S^1.

2.3.2 The orthogonal groups O(n) and SO(n)

The orthogonal group O(n) is the set of matrices $\mathfrak{m} \in \mathrm{Gl}(n; \mathbb{R})$ such that $\mathfrak{m}^T\mathfrak{m} = 1$. Here, $(\mathfrak{m}^T)_{ik} = \mathfrak{m}_{ki}$. Since $(\mathfrak{a}\mathfrak{b})^T = \mathfrak{b}^T\mathfrak{a}^T$, the set of orthogonal matrices forms a subgroup. The group SO(n) sits in O(n) as the subgroup of orthogonal matrices with determinant 1. To see that SO(n) and O(n) are manifolds, introduce $\mathrm{Sym}(n; \mathbb{R})$ to denote the vector subspace of $n \times n$ symmetric matrices. This is to say that a matrix $\mathfrak{h}$ is in $\mathrm{Sym}(n; \mathbb{R})$ when $\mathfrak{h}^T = \mathfrak{h}$; thus when $\mathfrak{h}_{ij} = \mathfrak{h}_{ji}$. This vector space has dimension $n(n+1)/2$. The vector space $\mathrm{Sym}(n; \mathbb{R})$ is a linear subspace of the Euclidean space $\mathbb{M}(n; \mathbb{R})$, and so a version of $\mathbb{R}^{n(n+1)/2}$. Let ψ: $\mathbb{M}(n; \mathbb{R}) \to \mathrm{Sym}(n; \mathbb{R})$ denote the map $\mathfrak{m} \to \mathfrak{m}^T\,\mathfrak{m}$. This map is quadratic in the entries of the matrix, and so a smooth map between Euclidean spaces. As is explained in the next paragraph, the identity matrix $\iota \in \mathrm{Sym}(n; \mathbb{R})$ is a regular value of ψ. This being the case, it follows from the implicit function theorem that $\psi^{-1}(\iota)$ is a submanifold of $\mathrm{Gl}(n; \mathbb{R})$ and so a Lie group.

To see that ι is a regular value, one must verify that the differential of ι at any given matrix $\mathfrak{m} \in O(n)$ is surjective. To see this, note that the differential of ψ at a given matrix $\mathfrak{m}$ can be written as $\psi^*|_{\mathfrak{m}} = d\mathfrak{m}^T\mathfrak{m} + \mathfrak{m}^Td\mathfrak{m}$, where $d\mathfrak{m}$ is the matrix of coordinate differentials with (i, j) entry equal to $d\mathfrak{m}_{ij}$. To explain this notation, I should tell you how to interpret $d\mathfrak{m}^T\mathfrak{m} + \mathfrak{m}^Td\mathfrak{m}$ as a linear map from the Euclidean space $\mathbb{M}(n; \mathbb{R})$ to the Euclidean space $\mathrm{Sym}(n; \mathbb{R})$. This linear map acts as follows: Let $\mathfrak{a} \in \mathbb{M}(n; \mathbb{R})$ denote any given matrix. Then $\psi_*|_{\mathfrak{m}}(\mathfrak{a})$ is the symmetric matrix $\mathfrak{a}^T\mathfrak{m} + \mathfrak{m}^T\mathfrak{a}$. The rule here is to replace $d\mathfrak{m}$ with $\mathfrak{a}$. Note that this notation is used throught this book, so it is best to come to terms with it. (It was used above to write $d(\det)|_{\mathfrak{m}}$ as $\det(\mathfrak{m})\,\mathrm{tr}(\mathfrak{m}^{-1}d\mathfrak{m})$.) In any event, granted this identification of $\psi_*|_{\mathfrak{m}}$ it is necessary to prove the following: Given a symmetric matrix $\mathfrak{h}$, there exists a matrix $\mathfrak{a}$ such that $\mathfrak{a}^T\mathfrak{m} + \mathfrak{m}^T\mathfrak{a} = \mathfrak{h}$. Given that $\mathfrak{m}\mathfrak{m}^T = \iota$, this equation is solved by taking $a = \frac{1}{2}\,\mathfrak{m}\,\mathfrak{h}$.

Note that SO(3) is diffeomorphic to the quotient space $\mathbb{RP}^3 = S^3/{\sim}$ where $\sim$ equates any given x with $\pm$ x. Meanwhile, SO(4) is diffeomorphic to $(S^3 \times S^3)/{\sim}$ where the equivalence relation equates a pair $(x, y) \in SO(3)$ only with itself and with $(-x, -y)$.

2.4 Some complex Lie groups

The group $\mathrm{Gl}(n; \mathbb{R})$ has as its analog the group $\mathrm{Gl}(n; \mathbb{C})$ of $n \times n$ invertible matrices with complex number entries. There are two equivalent ways to view $\mathrm{Gl}(n; \mathbb{C})$ as a Lie group. The first deals with complex numbers right off the bat by introducing the vector space (over $\mathbb{C}$) of $n \times n$ complex matrices. The latter is denoted by $\mathbb{M}(n; \mathbb{C})$. This vector space can be viewed as a version of

$\mathbb{R}^{2n^2}$ by using the real and imaginary parts of the n^2 entries as coordinates. Viewed in this light, multiplication of matrices defines a smooth map from $\mathbb{M}(n; \mathbb{C}) \times \mathbb{M}(n; \mathbb{C})$ to $\mathbb{M}(n; \mathbb{C})$ by virtue of the fact that the entries of a product $\mathfrak{m}\mathfrak{m}'$ is linear in the entries of $\mathfrak{m}$ and linear in those of $\mathfrak{m}'$. The determinant function maps $\mathbb{M}(n; \mathbb{C})$ to $\mathbb{C}$, and this is a smooth map since it is a polynomial of degree n in the entries. The latter map is also denoted by $\det(\cdot)$, with it understood that this version has values in $\mathbb{C}$ while the version defined on the space of matrices with real entries has values in $\mathbb{R}$.

The group $Gl(n; \mathbb{C})$ appears as the open set in the Euclidean space $\mathbb{M}(n; \mathbb{C})$ where the determinant is nonzero. The matrix multiplication for $\mathbb{M}(n; \mathbb{C})$ restricts to a smooth map from $\times_2 Gl(n; \mathbb{C})$ to $Gl(n; \mathbb{C})$. Meanwhile, the inverse $\mathfrak{m} \to \mathfrak{m}^{-1}$ is smooth because any given entry of $\mathfrak{m}^{-1}$ is the quotient of an n'th order polynomial in the entries of $\mathfrak{m}$ by the nonzero, smooth, $\mathbb{C}$-valued function given by the determinant function $\det(\cdot)$.

The second and equivalent way to view $Gl(n; \mathbb{C})$ as a manifold does not introduce complex numbers until late in the game. This view is ultimately useful because it introduces some constructions that play a role in later chapters. This approach views $Gl(n; \mathbb{C})$ as a subgroup in $Gl(2n; \mathbb{R})$. To see $Gl(n; \mathbb{C})$ in this light, I first introduce the notion of an *almost complex structure*, this an element $j \in \mathbb{M}(2n; \mathbb{R})$ with $j^2 = -\iota$ with ι again denoting the identity matrix. The canonical example is j_0 whose $n = 1$ and $n = 2$ versions are

$$\begin{pmatrix} 0 & -1 \\ 1 & 0 \end{pmatrix}, \begin{pmatrix} 0 & -1 & 0 & 0 \\ 1 & 0 & 0 & 0 \\ 0 & 0 & 0 & -1 \\ 0 & 0 & 1 & 0 \end{pmatrix}.$$

In the $n > 2$ case, j_0 is given as follows: Let e_k denote the basis vector for $\mathbb{R}^{2n}$ that has 1 in the entry k and zero in all other entries. Then j_0 is defined by the rules $j_0 e_{2n-1} = e_{2n}$ and $j_0 e_{2n} = -e_{2n-1}$. Use $\mathbb{M}_j$ for the moment to denote the vector subspace of $\mathbb{M}(2n; \mathbb{R})$ of matrices $\mathfrak{m}$ such that $\mathfrak{m} j_0 - j_0 \mathfrak{m} = 0$. This is to say that the entries of $\mathfrak{m}$ are such that

$$\mathfrak{m}_{2k,2i} = \mathfrak{m}_{2k-1,2j-1} \textit{ and } \mathfrak{m}_{2k,2j-1} = \mathfrak{m}_{2k-1,2j}$$

for each pair $k, j \in \{1, \ldots, n\}$.

As is explained momentarily, the vector space $\mathbb{M}_j$ is $\mathbb{M}(n; \mathbb{C})$. In any event, $\mathbb{M}_j$ is a vector space over $\mathbb{R}$, and so a Euclidean space. Introduce for the moment $\mathbb{G}_j \subset \mathbb{M}_j$ to denote the subset of invertible elements. Note that $\mathbb{G}_j$ is not empty because the identity matrix ι and also j_0 are in $\mathbb{G}_j$. As an open subset of a Euclidean space, $\mathbb{G}_j$ is a smooth manifold. It is also a group by virtue of the fact that $\mathfrak{m}\mathfrak{m}' j_0 = \mathfrak{m} j_0 \mathfrak{m}' = j_0 \mathfrak{m}\mathfrak{m}'$ when $\mathfrak{m}$ and $\mathfrak{m}'$ are both in $\mathbb{G}_j$. Thus it is a Lie group. I shall explain in a moment how to identify $\mathbb{G}_j$ with $Gl(n; \mathbb{C})$.

To see that $\mathbb{M}_j$ is $\mathbb{M}(n;\mathbb{C})$, note first that the eigenvalues of j_0 are $\pm i$, and any matrix that commutes with j_0 must preserve the eigenspaces. Indeed, let $\{e_1,\ldots,e_{2n}\}$ denote the standard basis for $\mathbb{R}^{2n}$, and then a basis of eigenvectors for the eigenvalue i are $\{v_1 = e_1 - ie_2,\ldots, v_n = e_{2n-1} - ie_{2n}\}$. The corresponding basis of eigenvectors for the eigenvalue $-i$ is given by the complex conjugates of $\{v_1,\ldots, v_n\}$. If m commutes with j_0, then it must act as $m(v_k) = a_{kj}v_j$ (note the summation convention for repeated indices) where $a_{kj} \in \mathbb{C}$. Likewise, $m(\bar{v}_k) = \bar{a}_{kj}\,\bar{v}_j$. Here and in what follows, the overscore indicates complex conjugation. Note that the number $a_{kj} \in \mathbb{C}$ can be written in terms of the entries of m as $a_{kj} = m_{2k-1,2j-1} + im_{2k,2j-1}$.

The preceding implies that m is determined by n^2 complex numbers and so the dimension of $\mathbb{M}_j$ is $2n^2$. The identification between $\mathbb{M}_j$ and $\mathbb{M}(n;\mathbb{C})$ comes about by associating to any given $m \in \mathbb{M}_j$ the $n \times n$ matrix with complex entries given by the numbers $\{a_{kj}\}$. The identification between $\mathbb{M}_j$ and $\mathbb{M}(n;\mathbb{C})$ identifies the subset $\mathbb{G}_j \subset \mathbb{M}_j$ with the group $Gl(n;\mathbb{C})$. To elaborate, suppose that a is a given element in $\mathbb{M}_j$ and let $m_{\mathbb{C}}$ denote the corresponding element in $\mathbb{M}(n;\mathbb{C})$. As an element in $\mathbb{M}(2n;\mathbb{R})$, the matrix m has a real valued determinant; and as an element in $\mathbb{M}(n;\mathbb{C})$, the matrix $m_{\mathbb{C}}$ has a $\mathbb{C}$-valued determinant. These two notions of a determinant are such that

$$\det(m) = |\det(m_{\mathbb{C}})|^2.$$

Thus, the real $2n \times 2n$ matrix a is an invertible element in $\mathbb{M}_j$ if and only if the corresponding complex, $n \times n$ matrix $m_{\mathbb{C}}$ is an invertible element in $\mathbb{M}(n;\mathbb{C})$. This correspondence between $\mathbb{G}_j$ and $Gl(n;\mathbb{C})$ has the following property: If m' and $m'_{\mathbb{C}}$ are paired, then the correspondence pairs mm' with $m_{\mathbb{C}}m'_{\mathbb{C}}$. As a consequence, if the correspondence pairs $m \in \mathbb{G}_j$ with $m_{\mathbb{C}} \in Gl(n;\mathbb{C})$, it then pairs m^{-1} with $m_{\mathbb{C}}^{-1}$.

2.5 The groups Sl(n; ℂ), U(n) and SU(n)

The three most commonly seen subgroups of $Gl(n;\mathbb{C})$ are the subject of this part of the chapter.

2.5.1 The group Sl(n; ℂ)

This is the subset of matrices in $\mathbb{M}(n;\mathbb{C})$ with determinant 1. To be precise, the determinant here is that of an $n \times n$ matrix with complex entries. For example, if the matrix is an $n \times n$ diagonal matrix, the determinant is the product of the diagonal entries. In particular, the determinant function on $\mathbb{M}(n;\mathbb{C})$ is to be viewed as a map, $\det\colon \mathbb{M}(n;\mathbb{C}) \to \mathbb{C}$.

As a parenthetical remark, when $\mathbb{M}(n; \mathbb{C})$ is viewed as the subvector space $\mathbb{M}_j \subset \mathbb{M}(2n; \mathbb{R})$, it inherits a determinant function from $\mathbb{M}(2n; \mathbb{R})$, this mapping to $\mathbb{R}$. Let me call the latter $\det_{\mathbb{R}}$ for the moment, and call the aforementioned determinant function $\det_{\mathbb{C}}$. These two functions on $\mathbb{M}_j$ are related by the rule $\det_{\mathbb{R}} = |\det_{\mathbb{C}}|^2$. This can be seen as follows: A matrix $\mathfrak{m} \in \mathbb{M}(n; \mathbb{C})$ has n complex eigenvalues $\{\lambda_1, \ldots, \lambda_n\}$. When viewed as an element in $\mathbb{M}_j$ and thus in $\mathbb{M}(2n; \mathbb{R})$, this matrix has 2n eigenvalues. The first n are $\{\lambda_1, \ldots, \lambda_n\}$, and the remaining n eigenvalues are the latter's complex conjugates.

In any event, to show that $Sl(n; \mathbb{C})$ is a submanifold, I should prove that the complex determinant $\det: \mathbb{M}(n; \mathbb{C}) \to \mathbb{C}$ has 1 as a regular value. Here is how to do this: The directional derivative of det at an invertible matrix $\mathfrak{m} \in \mathbb{M}(n; \mathbb{C})$ in the direction of a given matrix $\mathfrak{a} \in \mathbb{M}(n; \mathbb{C})$ is $\det(\mathfrak{m})\,\mathrm{tr}(\mathfrak{m}^{-1}\mathfrak{a})$. To see that any given value in $\mathbb{C}$ can be had by choosing an appropriate matrix $\mathfrak{a}$, let $c \in \mathbb{C}$ denote the desired value, and then take $\mathfrak{a} = c \det(\mathfrak{m})^{-1}\, \mathfrak{m}$.

2.5.2 The group U(n)

This is the subset of matrices $\mathfrak{m} \in \mathbb{M}(n; \mathbb{C})$ that obey $\mathfrak{m}^\dagger \mathfrak{m} = \iota$. This is to say that $U(n)$ is the subgroup of matrices $\mathfrak{m}$ with $\mathfrak{m}^{-1} = \mathfrak{m}^\dagger$. Here, $\mathfrak{m}^\dagger$ is the matrix whose (i, j) entry is the complex conjugate of the (j, i) entry of $\mathfrak{m}$.

To prove that $U(n)$ is a Lie group, I need only verify that it is a submanifold on $\mathbb{M}(n; \mathbb{C})$.

To start, introduce the vector space $\mathrm{Herm}(n)$ of $n \times n$ matrices $\mathfrak{h}$ that obey $\mathfrak{h}^\dagger = \mathfrak{h}$. I remind you that such a matrix is said to be Hermitian. This is a vector space over $\mathbb{C}$ of dimension $\frac{1}{2}n(n+1)$; its complex coordinates can be taken to be the entries on and above the diagonal. Define a map $\psi: \mathbb{M}(n; \mathbb{C}) \to \mathrm{Herm}(n)$ by the rule $\psi(\mathfrak{m}) = \mathfrak{m}^\dagger \mathfrak{m}$. Since $U(n) = \psi^{-1}(\iota)$, this subset is a submanifold if the identity matrix is a regular value of ψ. To see that such is the case, I note that the differential ψ_* at a matrix $\mathfrak{m}$ sends a given matrix a to $\psi_*|_{\mathfrak{m}}(\mathfrak{a}) = \mathfrak{m}^\dagger \mathfrak{a} + \mathfrak{a}^\dagger \mathfrak{m}$. The differential $\psi_*|_{\mathfrak{m}}$ is surjective if and only if any given $\mathfrak{h} \in \mathrm{Herm}(n)$ can be written as $\mathfrak{m}^\dagger \mathfrak{a} + \mathfrak{a}^\dagger \mathfrak{m}$ for some $\mathfrak{a} \in \mathbb{M}(n; \mathbb{C})$. Given that $\mathfrak{m}$ is invertible, then so is $\mathfrak{m}^\dagger$; its inverse is $(\mathfrak{m}^{-1})^\dagger$. This understood, take $\mathfrak{a} = \frac{1}{2}(\mathfrak{m}^\dagger)^{-1}\mathfrak{h}$.

2.5.3 The group SU(n)

This is the subgroup in $U(n)$ of matrices $\mathfrak{m}$ with $\det(\mathfrak{m}) = 1$. Note that if $\mathfrak{m} \in U(n)$, then $|\det(\mathfrak{m})| = 1$, since the equation $\mathfrak{m}^\dagger \mathfrak{m} = \iota$ implies that $\det(\mathfrak{m}^\dagger \mathfrak{m}) = \det(\iota) = 1$, and so $\det(\mathfrak{m}^\dagger)\det(\mathfrak{m}) = 1$. Given that the determinant of $\mathfrak{m}^\dagger$ is the complex conjugate of the determinant of $\mathfrak{m}$, this implies that $|\det(\mathfrak{m})|^2 = 1$. To see that $SU(n)$ is a Lie group, it is sufficient to prove that it is a submanifold of $\mathbb{M}(n; \mathbb{C})$.

For this purpose, define a map ϕ: $\mathbb{M}(n; \mathbb{C}) \to \text{Herm}(n) \times \mathbb{R}$ by the rule $\phi(m) = (m^\dagger m, -\frac{i}{2}(\det(m) - \det(m^\dagger))$. Noting that $\det(m^\dagger)$ is the complex conjugate of $\det(m)$, it follows that SU(n) is the component of $\phi^{-1}((\iota, 0))$ that consists of matrices with determinant equal to 1. The other component consists of matrices with determinant equal to -1.

To see that ϕ is surjective, I must verify that its differential at any given $m \in SU(n)$ is surjective. The latter at a matrix m sends any given matrix $a \in \mathbb{M}(n; \mathbb{C})$ to the pair in $\text{Herm}(n) \times \mathbb{R}$ whose first component is $m^\dagger a + a^\dagger m$ and whose second component is the imaginary part of $\det(m)\,\text{tr}(m^{-1}a)$. Let $(\mathfrak{h}, t)$ denote a given element in $\text{Herm}(n) \times \mathbb{R}$. Given that $m \in SU(2)$, then $a = \frac{1}{2} m\mathfrak{h} - i\, t m$ is mapped by $\phi_*|_m$ to $(\mathfrak{h}, t)$.

2.6 Notation with regards to matrices and differentials

Suppose that f is a function on $\mathbb{M}(n; \mathbb{R})$. Since the entries of a matrix serve as the coordinates, the differential of f can be written as

$$df = \sum_{1 \le i,j \le n} \left(\frac{\partial}{\partial m_{ij}} f\right) dm_{ij}.$$

I remind you that this differential form notation says the following: The directional derivative of f in the direction of a matrix a is

$$\sum_{1 \le i,j \le n} \left(\frac{\partial}{\partial m_{ij}} f\right) a_{ij}.$$

The latter can be written compactly as $\text{tr}\,(\frac{\partial}{\partial m} f a^T)$ if it is agreed to introduce $\frac{\partial}{\partial m} f$ to denote the matrix whose (i, j) component is $\frac{\partial}{\partial m_{ij}} f$. Here, $(\cdot)^T$ denotes the transpose of the indicated matrix.

The preceding observation suggests viewing the collection of coordinate differentials $\{dm_{ij}\}$ as the entries of a *matrix of differentials*, this denoted by dm. Introducing this notation allows us to write df above as

$$\text{tr}\left(\left(\frac{\partial}{\partial m} f\right) dm^T\right).$$

Here, dm^T is the matrix of differentials whose (i, j) entry is dm_{ji}. This notation is commonly used, and in particular, it will be used in subsequent parts of this book.

An analogous short-hand is used when f is a function on $\mathbb{M}(n; \mathbb{C})$. In this case, the entries of a matrix serve as the complex coordinates for $\mathbb{M}(n; \mathbb{C})$. This is to say that their real and imaginary parts give $\mathbb{M}(n; \mathbb{C})$ its Euclidean coordinates as a manifold. The differential of the function f can be written concisely as

$$\mathrm{tr}\Big(\Big(\frac{\partial}{\partial \mathfrak{m}} f\Big) d\mathfrak{m}^T\Big) + \mathrm{tr}\Big(\Big(\frac{\partial}{\partial \bar{\mathfrak{m}}} f\Big) d\mathfrak{m}^\dagger\Big).$$

Note that this notation makes sense even when f is complex valued. For that matter, it makes sense when f is replaced by a map, ψ, from either $\mathbb{M}(n; \mathbb{R})$ or $\mathbb{M}(n; \mathbb{C})$ to a given vector space, V. In the case of $\mathbb{M}(n; \mathbb{R})$, the differential of ψ can be written succinctly as

$$\Psi_* = \mathrm{tr}\Big(\Big(\frac{\partial}{\partial \mathfrak{m}} \psi\Big) d\mathfrak{m}^T\Big).$$

To disentangle all of this, suppose that V has dimension q, and one is given a basis with respect to which ψ has coordinates $(\psi^1, \ldots, \psi^q)$. Then what is written above says the following: The differential of ψ at a given matrix $\mathfrak{m}$ is the linear map from $\mathbb{M}(n; \mathbb{R})$ to V that sends a matrix $\mathfrak{a}$ to the vector with coordinates

$$\Big(\sum\nolimits_{1\le i,j\le n} \frac{\partial}{\partial \mathfrak{m}_{ij}} \psi^1 \mathfrak{a}_{ij}, \ldots, \sum\nolimits_{1\le i,j\le n} \frac{\partial}{\partial \mathfrak{m}_{ij}} \psi^q \mathfrak{a}_{ij}\Big).$$

There is, of course, the analogous notation for the case when ψ maps $\mathbb{M}(n; \mathbb{C})$ to a vector space.

Appendix 2.1 The transition functions for the Grassmannians

This appendix ties up a loose end from Chapter 1.6; it supplies a proof that the transition functions between any two coordinate charts in a given Grassmannian is a smooth map. To set the stage, fix integers $m > 1$ and $n \in \{1, \ldots, m\}$ so to define the Grassmannian Gr(m; n) as described in Chapter 1.6. Let V and V′ denote two elements in Gr(m; n) and let $\mathcal{O}_V$ and $\mathcal{O}_{V'}$ denote the respective coordinate charts from Chapter 1.6. Coordinate maps $\varphi_V{}^{-1}: \mathbb{R}^{n(m-n)} \to \mathcal{O}_V$ and $\varphi_{V'}{}^{-1}: \mathbb{R}^{n(m-n)} \to \mathcal{O}^{V'}$ are described. What follows explains why $\varphi_{V'} \circ \varphi_V{}^{-1}$ is smooth on its domain of definition.

To start this task, introduce $\{e_i\}_{1\le i\le n}$ and $\{u_a\}_{1\le a\le m-n}$ to denote the respective orthonormal bases chosen for V and $V^\perp$. Use $\{e'_i\}_{1\le i\le n}$ and $\{u'_a\}_{1\le a\le m-n}$ to denote their respective V′ counterparts. Now let $W \subset \mathcal{O}_V \cap \mathcal{O}_{V'}$ and let $\mathfrak{p} \in \mathbb{R}^{n(m-n)}$ and $\mathfrak{p}' \in \mathbb{R}^{n(m-n)}$ denote respective images under φ_V and $\varphi_{V'}$. Identify $\mathbb{R}^{n(m-n)}$ with the vector space of matrices with n rows and m columns so as to write the components of $\mathfrak{p}$ as $\{\mathfrak{p}_{ia}\}_{1\le i\le n, 1\le a\le m-n}$ and those of $\mathfrak{p}'$ as $\{\mathfrak{p}'_{ia}\}_{1\le i\le n, 1\le a\le m-n}$. By definition, the coordinate $\mathfrak{p} = \varphi_V(W)$ is such that any given $w \in W$ can be parametrized by $\mathbb{R}^n$ as

$$w = \sum\nolimits_{1\le i\le n} v_i\Big(e_i + \sum\nolimits_{1\le a\le m-n} \mathfrak{p}_{ia}\, u_a\Big)$$

with $v \in \mathbb{R}^n$. Likewise, the coordinate $\mathfrak{p}' = \varphi_{V'}(W)$ is defined so that

$$w = \sum\nolimits_{1\le i\le n} v'_i \Big(e'_i + \sum\nolimits_{1\le a\le m-n} \mathfrak{p}'_{ia}\, u'_a\Big)$$

with $v' \in \mathbb{R}^n$. Note in particular that any given $w \in W$ is parametrized in two ways by points in $\mathbb{R}^n$. This being the case, the two vectors v and v' must be functionally related. To be explicit, the i'th component of v' can be written in terms of v by taking the inner product of e'_i with w and using the preceding equations to first identify the latter with v'_i, and then using the other equation above to write this inner product in terms of v. The result of equating these two expressions for $e'_i{\cdot}w$ is

$$v'_i = \sum\nolimits_{1\le k\le n} v_k \Big(e'_i{\cdot}e_k + \sum\nolimits_{1\le a\le m-n} \mathfrak{p}_{ka} e'_j{\cdot}u_a\Big)$$

where $x{\cdot}x'$ denotes the Euclidean inner product on $\mathbb{R}^m$. This last equation asserts that $v'_i = \sum_{1\le k\le n} G_{ik} v_k$ where the numbers $\{G_{ik}\}_{1\le i,k\le n}$ are

$$G_{ik} = e'_i{\cdot}e_k + \sum\nolimits_{1\le a\le n} \mathfrak{p}_{ka}\; e'_i{\cdot}e_k.$$

These numbers can be viewed as the entries of an $n \times n$ matrix G. Note that this matrix G is invertible when the respective projections Π_V and $\Pi_{V'}$ map W isomorphically to V and V'. This is to say that $G \in Gl(n; \mathbb{R})$ when $W \in \mathcal{O}_V \cap \mathcal{O}_{V'}$.

The assignment of $\mathfrak{p} \in \mathbb{R}^{n(m-n)}$ to G defines a smooth map from $\varphi_V(\mathcal{O}_V \cap \mathcal{O}_{V'})$ in $\mathbb{R}^{n(m-n)}$ into $Gl(n; \mathbb{R})$ since the expression above for G's entries define an affine function of the coordinate functions on $\mathbb{R}^{m(n-m)}$. Save the matrix G for the moment.

Use the expression that parametrizes $w \in W$ in terms of $v' \in \mathbb{R}^n$ to write $u'_a{\cdot}w$ in terms of v' as

$$\sum\nolimits_{1\le i\le n} v'_i \mathfrak{p}'_{ia}.$$

Meanwhile, $u'_a{\cdot}w$ can also be written in terms of v using the expression that writes w in terms of v:

$$\sum\nolimits_{1\le i\le n} v_k \Big(u'_a{\cdot}e_k + \sum\nolimits_{1\le b\le m-n} \mathfrak{p}_{kb}\; u'_a{\cdot}u_b\Big).$$

Keeping in mind that these two expressions are equal, take v' to be i'th basis vector in $\mathbb{R}^n$ to see that

$$\mathfrak{p}'_{ia} = \sum\nolimits_{1\le k\le n} (G^{-1})_{ik} \Big(u'_a{\cdot}e_k + \sum\nolimits_{1\le b\le m-n} \mathfrak{p}_{kb} u'_a{\cdot}u_b\Big).$$

Here, G^{-1} is the inverse of the matrix G. With the entries of G viewed as a function of $\mathfrak{p} \in \varphi_V(\mathcal{O}_V \cap \mathcal{O}_{V'})$, the latter expression writes $\mathfrak{p}'$ as a function of $\mathfrak{p}$. This function is the transition function $\varphi_{V'} \circ \varphi_V^{-1}$. This expression for $\mathfrak{p}'$ as a function on $\varphi_V(\mathcal{O}_V \cap \mathcal{O}_{V'})$ defines a smooth function on $\varphi_V(\mathcal{O}_V \cap \mathcal{O}_{V'})$ because the inverse map from $Gl(n; \mathbb{R})$ to itself is a smooth map.

Additional reading

- *Matrix Groups, An Introduction to Lie Group Theory*, Andrew Baker, Springer, 2002.
- *Lie Groups, An Introduction through Linear Groups*, Wulf Rossmann, Oxford University Press, 2002.
- *Matrix Groups for Undergraduates*, Kristopher Tapp, American Mathematical Society, 2006.
- *Lie Groups, Beyond an Introduction*, Anthony W. Knapp, Birkhauser Boston, 1966.
- *Differential Geometry, Lie Groups and Symmetric Spaces*, Sigurdur Helgason, American Mathematical Society, 1978.
- *Lie Groups, Lie Algebras and Representations, An Elementary Introduction*, Brian C. Hall, Springer, 2003.

3 Introduction to vector bundles

Euclidean space is a vector space, but the typical manifold is not. Vector bundles bring vector spaces back into the story. To a first approximation, a vector bundle (over a given manifold) is a family of vector spaces that is parametrized by the points in the manifold. Some sort of vector bundle appears in most every aspect of differential geometry. Chapter 3 introduces these objects and gives the first few examples. By way of a warning, I give my favorite definition first; the standard definition is given subsequently in Chapter 3.2. (These definitions describe the same objects.)

3.1 The definition

Let M denote a smooth manifold of dimension m. A *real vector bundle* over M of fiber dimension n is a smooth manifold, E, of dimension n + m with the following additional structure:

- *There is a smooth map* $\pi: E \to M$.
- *There is a smooth map* $\hat{o}: M \to E$ *such that* $\pi \circ \hat{o}$ *is the identity.*
- *There a smooth map* $\mu: \mathbb{R} \times E \to E$ *such that*
  a) $\pi(\mu(r, v)) = \pi(v)$
  b) $\mu(r, \mu(r', v)) = \mu(rr', v)$
  c) $\mu(1, v) = v$
  d) $\mu(r, v) = v$ *for* $r \neq 1$ *if and only if* $v \in \mathrm{im}(\hat{o})$.
- *Let* $p \in M$. *There is a neighborhood,* $U \subset M$, *of* p *and a map* $\lambda_U: \pi^{-1}(U) \to \mathbb{R}^n$ *such that* $\lambda_U: \pi^{-1}(x) \to \mathbb{R}^n$ *for each* $x \in U$ *is a diffeomorphism obeying* $\lambda_U(\mu(r, v)) = r\lambda_U(v)$.

The first four remarks that follow concern notation. The remaining two remarks are more substantive.

Remark 3.1 The map from E to E given by $v \to \mu(r, v)$ for any given $r \in \mathbb{R}$ is called *multiplication by r*; this is because the map μ defines an action on E of the multiplicative group of real numbers. This $\mathbb{R}$-action is reflected in the notation used henceforth that writes $\mu(r, v)$ as rv.

Remark 3.2 If $W \subset M$ is any given set, then $\pi^{-1}(W) \subset E$ is denoted by $E|_W$. The inverse images via π of the points in M are called the *fibers of E*; and if $p \in M$ is any given point, then the fiber $\pi^{-1}(p)$ is said to be the *fiber over p*. The map π itself is called the *bundle projection map*, or just the *projection map*.

Remark 3.3 The map $\hat{o}$ is called the *zero section*. The image of $\hat{o}$ is also called the zero section; but it is almost always the case that these two meanings are distinguished by the context. This terminology reflects the fact that item (d) of the third bullet and the fourth bullet require $\lambda_U \circ \hat{o} = 0 \in \mathbb{R}^n$.

Remark 3.4 The map λ_U defines a diffeomorphism,

$$\varphi_U\colon E|_U \to U \times \mathbb{R}^n$$

given by $\varphi_U(v) = (\pi(v), \lambda_U(v))$. It is customary to talk about φ_U rather than λ_U. The map φ_U is said to be a *local trivialization* for E.

Remark 3.5 Let $U \subset M$ denote an open set where there is a map $\lambda_U\colon E|_U \to \mathbb{R}^n$ as above. The map λ_U can be used to give a local coordinate chart map for E in the following way: Let $(W, \varphi\colon W \to \mathbb{R}^{\dim(M)})$ denote a coordinate chart with $W \subset U$. Then the pair

$$(E|_W, (\varphi, \lambda_U)\colon E|_W \to \mathbb{R}^{\dim(M)} \times \mathbb{R}^n = \mathbb{R}^{\dim(M)+n})$$

is a coordinate chart for E.

Remark 3.6 The definition given above endows each fiber of E with a canonical vector space structure. To see how this comes about, note first that each fiber is diffeomorphic to $\mathbb{R}^n$; this being an obvious necessary condition. Second, the $\mathbb{R}$-action on E gives an $\mathbb{R}$-action on each fiber, and this gives $\mathbb{R}$-action for the vector space structure. Third, the zero section $\hat{o}$ gives each fiber an origin for the vector space structure.

Granted these points, I need only describe the notion of vector addition on each fiber. This is done using a local trivialization. To elaborate, let $p \in M$ and let $U \subset M$ containing p denote an open set with a local trivializing map $\varphi_U\colon E|_U \to U \times \mathbb{R}^n$. Write the latter map as $v \to (\pi(v), \lambda_U(v))$. If v and v′ are in $E|_p$, define $v + v'$ to be $\lambda_{U,p}^{-1}(\lambda_{U,p}(v) + \lambda_{U,p}(v'))$ where $\lambda_{U,p}^{-1}$ is the inverse of the diffeomorphism from $E|_p$ to $\mathbb{R}^n$ that is defined by λ_U.

To see that this makes sense, I must demonstrate that the addition so defined does not depend on the chosen set U. The point here is that a given point p may lie in two sets, U and U′, that come with corresponding maps λ_U and $\lambda_{U'}$. I need to prove that the same point $v + v'$ is obtained when using $\lambda_{U'}$. Now, this follows if and only if

$$\lambda_{U',p}(\lambda_{U,p}^{-1}(e + e')) = \lambda_{U',p}(\lambda_{U,p}^{-1}(e)) + \lambda_{U',p}(\lambda_{U,p}^{-1}(e'))$$

for all pairs e and $e' \in \mathbb{R}^n$. This is to say that the map

$$\psi_{U',U} = \lambda_{U'} \circ (\lambda_U^{-1}) \colon (U \cap U') \times \mathbb{R}^n \to \mathbb{R}^n$$

must be linear in the $\mathbb{R}^n$ factor; it must have the form $(p, v) \to g_{U',U}(p)v$ where $g_{U',U}$ is a smooth map from $U \cap U'$ to $Gl(n; \mathbb{R})$. That such is the case is a consequence of the next lemma.

Lemma 3.7 *A map* $\psi \colon \mathbb{R}^n \to \mathbb{R}^n$ *with the property that* $\psi(rv) = r\psi(v)$ *for all* $r \in \mathbb{R}$ *is linear.*

Proof of Lemma 3.7 Use the Chain rule to write $\frac{\partial}{\partial t}\psi(tv) = \psi_*|_{tv}(v)$. This is also equal to $\psi(v)$ since $\psi(tv) = t\psi(v)$. Thus, $\psi_*|_{tv}(v) = \psi(v)$. Since the right-hand side of this last equality is independent of t, so is the left-hand side. Thus, the map $t \to \psi_*|_{tv}(v)$ is independent of t, and so equal to $\psi_*|_0(v)$. As a consequence, $\psi(v) = \psi_*|_0(v)$.

The map $g_{U',U} \colon U \cap U' \to Gl(n; \mathbb{R})$ is said to be a *bundle transition function* for the vector bundle.

3.2 The standard definition

The observation that the map $\psi_{U',U}$ depicted above is linear in the $\mathbb{R}^n$ factor suggests an alternate definition of a vector bundle. This one, given momentarily, is often the more useful of the two; and it is the one you will find in most other books.

In this definition, a real vector bundle of fiber dimension n is given by the following data: First, a locally finite open cover, $\mathfrak{U}$, of M. Second, an assignment of a map, $g_{U',U} \colon U \cap U' \to Gl(n; \mathbb{R})$ to each pair $U, U' \in \mathfrak{U}$ subject to two constraints. First, $g_{U,U'} = g_{U',U}^{-1}$. Second, if U, U′ and U″ are any three sets from $\mathfrak{U}$ with $U \cap U' \cap U'' \neq \emptyset$, then $g_{U',U}\, g_{U,U''}\, g_{U'',U'} = \iota$. By way of notation, the map $g_{U',U}$ is deemed the *bundle transition function*. Meanwhile, the constraint $g_{U',U}\, g_{U,U''}\, g_{U'',U} = \iota$ is said to be a *cocycle constraint*.

Granted this data, the bundle E is defined as the quotient of the disjoint union $\bigcup_{U \in \mathfrak{U}} (U \times \mathbb{R}^n)$ by the equivalence relation that puts $(p', v') \in U' \times \mathbb{R}^n$ equivalent to $(p, v) \in U \times \mathbb{R}^n$ if and only if $p = p'$ and $v' = g_{U',U}(p)v$. To connect this definition with the previous one, define the map π to send the equivalence class of any given (p, v) to p. The multiplication map is defined by the rule that has $r \in \mathbb{R}$ sending the equivalence class of the point (p, v) to (p, rv). Finally, any given $U \in \mathfrak{U}$ has the corresponding map λ_U that is defined as follows: If $p \in U$, then any given equivalence class in $\pi^{-1}(p)$ has a unique representative $(p, v) \in U \times \mathbb{R}^n$. This equivalence class is sent by λ_U to $v \in \mathbb{R}^n$.

The smooth structure on a vector bundle E defined this way is that associated to the following sort of coordinate atlas: If necessary, subdivide each set $U \in \mathfrak{U}$ so that the resulting open cover consists of a set $\mathfrak{U}'$ where each set $U \in \mathfrak{U}'$ comes with a smooth coordinate map $\varphi_U: U \to \mathbb{R}^{\dim(M)}$ that embeds U as the interior of a ball. This is to say that the collection $\{(U, \varphi_U : U \to \mathbb{R}^{\dim(M)})\}_{U \in \mathfrak{U}'}$ consists of a coordinate atlas for M. This understood, then the collection $\{(E|_U, (\varphi_U, \lambda_U): E|_U \to \mathbb{R}^{\dim(M)} \times \mathbb{R}^n)\}$ defines a coordinate atlas for E. Note that the coordinate transition function for the intersection of two such charts, thus for $E|_{U' \cap U}$, sends a pair $(x, v) \in \varphi_U(U \cap U') \times \mathbb{R}^n$ to the point with $\mathbb{R}^{\dim(M)}$ coordinate $(\varphi_{U'} \circ \varphi_U^{-1})(x) \in \varphi_{U'}(U \cap U')$, and $\mathbb{R}^n$ coordinate $g_{U',U}(x)v$. Thus, the bundle transition functions $\{g_{U',U}\}_{U,U' \in \mathfrak{U}'}$ help to determine the smooth structure on E.

The data consisting of the cover $\mathfrak{U}$ with the bundle transition functions $\{g_{U',U}\}_{U,U' \in \mathfrak{U}}$ is called *cocycle data* for the bundle E.

3.3 The first examples of vector bundles

What follows are some first examples of vector bundles.

3.3.1 The trivial bundle

The vector bundle $M \times \mathbb{R}^n$ over M with fiber $\mathbb{R}^n$ is called the *trivial* $\mathbb{R}^n$ bundle for what I hope is an obvious reason. It is also often called the *product* bundle.

3.3.2 The Mobius bundle

Introduce the standard angular coordinate $\theta \in \mathbb{R}/(2\pi\mathbb{Z})$ for S^1. With S^1 viewed as the unit radius circle about the origin in $\mathbb{R}^2$, a given angle θ maps to the point $(\cos(\theta), \sin(\theta))$. Let $E \subset S^1 \times \mathbb{R}^2$ denote the subset of points $(\theta, (v_1, v_2))$ such that $\cos(\theta)v_1 + \sin(\theta)v_2 = v_1$ and $\sin(\theta)v_1 - \cos(\theta)v_2 = v_2$. This defines a bundle over S^1 with fiber dimension 1.

3.3.3 The tautological bundle over $\mathbb{RP}^n$

Define $\mathbb{RP}^n$ as $S^n/\{\pm 1\}$. A bundle $E \to \mathbb{RP}^n$ with fiber $\mathbb{R}$ is defined to be the quotient $(S^n \times \mathbb{R})/\{\pm 1\}$. Alternately, E is the set (x, γ) such that $x \in \mathbb{RP}^n$ and γ is the line through the origin with tangent vector proportional to $\pm x$.

3.3.4 Bundles over the Grassmannians

Fix positive integer $m > n$ and introduce from Chapter 1.6 the manifold $Gr(m; n)$ whose points are the n-dimensional subspaces of $\mathbb{R}^m$. Over $Gr(m; n)$ sits an essentially tautological $\mathbb{R}^n$-bundle that is defined to be the set of pairs $\{(V, v) : V \subset \mathbb{R}^m$ *is an* n-*dimensional subspace and* $v \in V\}$. Let E denote this bundle. The projection map $\pi: E \to Gr(m; n)$ sends a given pair (V, v) to the n-dimensional subspace $V \in Gr(m; n)$. The $\mathbb{R}$-action on E has $r \in \mathbb{R}$ sending the pair (V, v) to the pair (V, rv). The map $\hat{o}$ sends $V \in Gr(m; n)$ to the pair $(V, 0)$ where $0 \in \mathbb{R}^m$ is the origin. Let $\mathcal{O}_V \subset Gr(m; n)$ denote the subspace of n-dimensional subspaces V' which are mapped isomorphically via the orthogonal projection $\Pi_V: \mathbb{R}^m \to V$. The set $U = \mathcal{O}_V$ has the required map $\lambda_U: \pi^{-1}(U) \to \mathbb{R}^n$, this defined as follows: Fix a basis for V so as to identify V with $\mathbb{R}^n$. Granted this identification, define the map $\lambda_{\mathcal{O}_V}$ so as to send any given pair $(V', v) \in E|_{\mathcal{O}_V}$ to $\Pi_V(v') \in V = \mathbb{R}^n$. I leave it as an exercise to verify that E as just defined has a suitable smooth structure. This bundle E is called the *tautological* bundle over $Gr(m; n)$.

3.3.5 Bundles and maps from spheres

The next example uses cocycle data to define a vector bundle. To start, let M denote the manifold in question and let m denote its dimension. Fix a finite set $\Lambda \subset M$. Assign to each $p \in \Lambda$ the following data: First, a smooth map $g_p: S^{m-1} \to Gl(n; \mathbb{R})$. Second, a coordinate chart $(U, \varphi_p: U \to \mathbb{R}^m)$ with $p \in U$ such that $\varphi_p(p) = 0 \in \mathbb{R}^m$. Fix $r > 0$ and small enough so that the open ball, B, of radius r about the origin in $\mathbb{R}^m$ is in the image of each $p \in \Lambda$ version of φ_p; and so that the collection $\{U_p = \varphi_p^{-1}(B)\}_{p \in \Lambda}$ consists of pairwise disjoint sets.

To construct a vector bundle from this data, use the cover given by the open sets $U_0 = M - \Lambda$ and $\{U_p\}_{p \in \Lambda}$. The only nonempty intersections for this cover are those of the form $U_0 \cap U_p = \varphi_p^{-1}(B - \{0\})$. Define the bundle transition function $g_{0,p}: U_0 \cap U_p \to Gl(n; \mathbb{R})$ on such a set to be the map $x \to g_p(|\varphi_p(x)|^{-1}\varphi_p(x))$. As there are no nonempty intersections between three distinct sets in the cover, there are no cocycle conditions to satisfy. As a consequence, this data is sufficient to define a vector bundle over M having fiber dimension n.

3.4 The tangent bundle

Every smooth manifold comes a priori with certain canonical vector bundles, one of which is the *tangent bundle*, this a bundle whose fiber dimension is that of the manifold in question. The tangent bundle of a manifold M is denoted by TM. What follows is a cocycle definition of TM.

Let n denote the dimension of M. Fix a locally finite coordinate atlas, $\mathcal{U}$, for M. The collection of open sets from the pairs in $\mathcal{U}$ is used to define the tangent bundle by specifying the bundle transition functions $\{g_{U',U}: U' \cap U \to Gl(n; \mathbb{R})\}_{(U,\varphi),(U',\varphi') \in \mathcal{U}}$.

Fix a pair (U, φ_U) and $(U', \varphi_{U'})$ from the atlas $\mathcal{U}$. To define $g_{U',U}$, introduce the coordinate transition function $\psi_{U',U} = \varphi_{U'} \circ \varphi_U^{-1}$, a diffeomorphism from $\varphi_U(U' \cap U) \subset \mathbb{R}^n$ to $\varphi_{U'}(U' \cap U) \subset \mathbb{R}^n$. The differential of this map, $\psi_{U',U*}$, is a map from $\psi_U(U' \cap U)$ to $Gl(n; \mathbb{R})$. This understood, set $g_{U',U}(p) = \psi_{U',U*}|_{\varphi_U(p)}$ for $p \in U' \cap U$. The cocycle condition $g_{U',U} g_{U,U''} g_{U'',U'} = \iota$ is guaranteed by the Chain rule by virtue of the fact that $\psi_{U',U} \circ \psi_{U,U''} \circ \psi_{U'',U'}$ is the identity map from $\psi_{U'}(U' \cap U)$ to itself.

This definition appears to depend on the choice of the atlas $\mathcal{U}$ from the equivalence class that defines the smooth structure. As explained next, the same bundle TM appears for any choice of atlas from this equivalence class. To see why this is, agree for the moment to use $T\mathcal{U}$ to denote the version of TM defined as above using the atlas $\mathcal{U}$. Suppose that $\mathcal{V}$ is a coordinate atlas that is equivalent to $\mathcal{U}$. Let $T\mathcal{V}$ denote the version of M's tangent bundle that is defined as above from $\mathcal{V}$. I now define a coordinate atlas $\mathcal{W}$ as follows: The collection $\mathcal{W}$ is the union of two coordinate atlases. The first, $\mathcal{W}_{\mathcal{U}}$ consists of all pairs of the form $(U \cap V, \varphi_U)$ such that (U, φ_U) is from $\mathcal{U}$ and (V, φ_V) is from $\mathcal{V}$. The second atlas, $\mathcal{W}_{\mathcal{V}}$, consists of all pairs of the form $(U \cap V, \varphi_V)$ with U, V and φ_V as before. Thus, if $(U, \varphi_U) \in \mathcal{U}$ and $(V, \varphi_V) \in \mathcal{V}$, then the open set $U \cap V$ appears in two pairs from $\mathcal{W}$, one with the map φ_U and the other with the map φ_V. Use $T\mathcal{W}_{\mathcal{U}}$, $T\mathcal{W}_{\mathcal{V}}$ and $T\mathcal{W}$ to denote the versions of the tangent bundle that are defined respectively by $\mathcal{W}_{\mathcal{U}}$, $\mathcal{W}_{\mathcal{V}}$ and $\mathcal{W}$. It follows directly from the definition of the equivalence relation that $T\mathcal{W} = T\mathcal{W}_{\mathcal{U}} = T\mathcal{W}_{\mathcal{V}}$; indeed, such is the case because any given point in M is contained in some version of $U \cap V$. It also follows directly from the equivalence relation that $T\mathcal{W}_{\mathcal{U}} = T\mathcal{U}$ and $T\mathcal{W}_{\mathcal{V}} = T\mathcal{V}$.

What follows is an important point to make with regards to the definition of TM: If $U \subset M$ is any given coordinate chart $\varphi_U: U \to \mathbb{R}^n$, then there is a canonically associated bundle trivialization map from $TM|_U$ to $U \times \mathbb{R}^n$. The latter is denoted by φ_{U*}.

3.4.1 A geometric interpretation of TM

Suppose that $m \geq n$ and that $\sigma: M \to \mathbb{R}^m$ is an immersion. Recall that this means the following: Let $(U, \varphi: U \to \mathbb{R}^n)$ denote any given coordinate chart on M. Then $\sigma \circ (\varphi^{-1}): \mathbb{R}^n \to \mathbb{R}^m$ has injective differential. The map σ gives the tangent bundle to M a geometric interpretation whose description follows. To start, fix a locally finite coordinate atlas $\mathcal{U}$ for M. Given $p \in M$, fix a chart

$(U, \varphi) \in \mathcal{U}$ with $p \in U$; and let $\Pi_p \subset \mathbb{R}^m$ denote the n-dimensional linear subspace given by the differential of $\sigma \circ (\varphi^{-1})$. The subspace Π_p does not depend on the chosen chart containing p, nor on the atlas $\mathcal{U}$ representing the smooth structure. This understood, it follows that Π_p is canonically isomorphic as a vector space to $TM|_p$. As a consequence, TM can be viewed as sitting in $M \times \mathbb{R}^N$ as the set of pairs $\{(p, v) \in M \times \mathbb{R}^N : v \in \Pi_p\}$.

This view of TM identifies the projection π: TM $\to$ M with a restriction of the tautological projection from $M \times \mathbb{R}^m$ to M. Meanwhile, the $\mathbb{R}$-action on TM and the vector space structure on each fiber is that inherited from the $\mathbb{R}$-action on $\mathbb{R}^m$ and the vector space structure of $\mathbb{R}^m$.

Suppose that $m \geq n$ and that $\psi: \mathbb{R}^m \to \mathbb{R}^{m-n}$ is a smooth map. Suppose, in addition, that $a \in \mathbb{R}^{m-n}$ is a regular value of ψ. Recall from Chapter 1 that $M = \psi^{-1}(a)$ is a smooth, n-dimensional submanifold of $\mathbb{R}^m$. If $p \in M$, then the linear subspace Π_p is the kernel of the $\psi_*|_p$. Thus, the tangent bundle TM is the subset $\{(p, v) : p \in M \textit{ and } \psi_*|_p(v) = 0\}$.

As a parenthetical remark, note that the map from $\mathbb{R}^m \times \mathbb{R}^m$ to $\mathbb{R}^{m-n} \times \mathbb{R}^{m-n}$ that sends any given pair (p, v) to $\Psi(p, v) = (\psi(p), \psi_*|_p v)$ has $(a, 0) \in \mathbb{R}^{m-n} \times \mathbb{R}^{m-n}$ as a regular value. As noted in the previous paragraph, $\Psi^{-1}(a, 0)$ is TM. Thus, TM sits in $\mathbb{R}^m \times \mathbb{R}^m$ as a submanifold. It is left as an exercise to check that its smooth structure as M's tangent bundle is the same as that coming from the identification just given as the inverse image of a regular value of Ψ.

3.5 Tangent bundle examples

What follows are some tangent bundle examples.

3.5.1 The manifold $\mathbb{R}^n$

The manifold $\mathbb{R}^n$ has a canonical coordinate atlas, this the atlas with one chart, and with the associated map to $\mathbb{R}^n$ being the identity. Using this coordinate atlas $\mathcal{U} = \{(\mathbb{R}^n, \textit{the identity map})\}$ as a fiducial atlas identifies $T\mathbb{R}^n$ with the product bundle $\mathbb{R}^n \times \mathbb{R}^n$ with the bundle projection π given by projection to the left $\mathbb{R}^n$ factor.

3.5.2 An open subset of $\mathbb{R}^n$

An open set $M \subset \mathbb{R}^n$ has the fiducial coordinate atlas $\mathcal{U} = \{(M \cap \mathbb{R}^n, \textit{the identity map } M \to \mathbb{R}^n)\}$. Using this chart identifies M's tangent bundle with $M \times \mathbb{R}^n$

also. For example, the group $Gl(n; \mathbb{R})$ sits as an open set in the Euclidean space $\mathbb{M}(n; \mathbb{R})$. Thus, $TGl(n; \mathbb{R})$ can be canonically identified with $Gl(n; \mathbb{R}) \times \mathbb{M}(n; \mathbb{R})$. Likewise, $TGl(n; \mathbb{C})$ can be canonically identified with $Gl(n; \mathbb{C}) \times \mathbb{M}(n; \mathbb{C})$.

3.5.3 The sphere

The sphere $S^n \subset \mathbb{R}^{n+1}$ is the set of vectors $x \in \mathbb{R}^{n+1}$ with $|x| = 1$. This is $\psi^{-1}(1)$ where $\psi: \mathbb{R}^{n+1} \to \mathbb{R}$ is the map sending x to $|x|^2$. The differential of this map at a given point $x \in \mathbb{R}^{n+1}$ sends $v \in \mathbb{R}^{n+1}$ to $x \cdot v$ where $\cdot$ here denotes the Euclidean inner product. This being the case, TS^n sits in $\mathbb{R}^{n+1} \times \mathbb{R}^{n+1}$ as the set of pairs $\{(x, v) : |x| = 1$ *and* v *is orthogonal to* x$\}$.

3.5.4 A surface in $\mathbb{R}^3$

Take $g \geq 0$ and construct a surface of genus g embedded in $\mathbb{R}^3$ as described in Chapter 1.3 using a suitable function $h: \mathbb{R}^2 \to \mathbb{R}$. Recall that this surface is the level set with value 0 for the function $(x, y, z) \to z^2 - h(x, y)$. The tangent bundle to this surface sits in $\mathbb{R}^3 \times \mathbb{R}^3$ as the set of pairs $(p = (x, y, z), v = (v_x, v_y, v_z))$ with

$$z^2 - h(x,y) = 0 \quad \textit{and also} \quad 2zv_z - \left.\frac{\partial h}{\partial x}\right|_{(x,y)} v_x - \left.\frac{\partial h}{\partial y}\right|_{(x,y)} v_y = 0.$$

3.5.5 The group $Sl(n; \mathbb{R})$

This manifold sits inside $\mathbb{M}(n; \mathbb{R})$ as $\psi^{-1}(1)$ where ψ is the map ψ from $\mathbb{M}(n; \mathbb{R})$ to $\mathbb{R}$ given by $\psi(m) = \det(m)$. The differential of this map at an invertible matrix m sends $a \in \mathbb{M}(n; \mathbb{R})$ to $\det(m)\,\mathrm{tr}(m^{-1}a)$. The kernel of this is the vector subspace in $\mathbb{M}(n; \mathbb{R})$ consisting of matrices $\{a = mc : \mathrm{tr}(c) = 0\}$. Thus, the tangent bundle to $Sl(n; \mathbb{R})$ sits in $\mathbb{M}(n; \mathbb{R}) \times \mathbb{M}(n; \mathbb{R})$ as the submanifold consisting of the pairs $\{(m, mc) : \det(m) = 1$ *and* $\mathrm{tr}(c) = 0\}$.

3.5.6 The group O(n)

The group $O(n)$ sits in $\mathbb{M}(n; \mathbb{R})$ as $\psi^{-1}(\iota)$ where ψ is the map from $\mathbb{M}(n; \mathbb{R})$ to the Euclidean space $Sym(n; \mathbb{R})$ of $n \times n$ symmetric matrices that sends a matrix m to $m^T m$. The differential of ψ at m sends $a \in \mathbb{M}(n; \mathbb{R})$ to $m^T a + a^T m$. The kernel of this map at $m \in SO(n)$ is the vector subspace $\{a = mc : c^T = -c\}$. Thus, $TO(n)$ sits in $\mathbb{M}(n; \mathbb{R}) \times \mathbb{M}(n; \mathbb{R})$ as the submanifold of pairs $\{(m, mc) : m^T m = 1$ *and* $c^T = -c\}$.

3.5.7 The groups U(n) and SU(n)

The group U(n) sits in $\mathbb{M}(n; \mathbb{C})$ as the inverse image of ι via the map to the space Herm(n) of $n \times n$ Hermitian matrices that sends a matrix m to $m^\dagger m$. The kernel of the differential of this map at $m \in U(n)$ consists of the matrices $a \in \mathbb{M}(n; \mathbb{C})$ that can be written as $a = mc$ where $c^\dagger = -c$. This understood, TU(n) sits in $\mathbb{M}(n; \mathbb{C}) \times \mathbb{M}(n; \mathbb{C})$ as the submanifold of pairs $\{(m, mc): m^\dagger m = \iota \textit{ and } c^\dagger = -c\}$. Meanwhile, TSU(n) sits in $\mathbb{M}(n; \mathbb{C}) \times \mathbb{M}(n; \mathbb{C})$ as the submanifold of pairs $\{(m, mc) : \textit{both } m^\dagger m = \iota \textit{ and } \det(m) = 1, \textit{ and both } c^\dagger = -c \textit{ and } \mathrm{tr}(c) = 0\}$.

3.6 The cotangent bundle

The cotangent bundle is the second of the two canonical bundles that are associated to M. This one is denoted T^*M; and it has fiber dimension $n = \dim(M)$. What follows is the cocycle definition of T^*M.

Fix a coordinate atlas $\mathcal{U}$ for M. As with the tangent bundle, the collection of open sets that appear in the pairs from $\mathcal{U}$ supply the open cover for the cocycle definition. To define the bundle transition functions, suppose that (U, φ_U) and $(U', \varphi_{U'})$ are two charts from $\mathcal{U}$. Reintroduce the coordinate transition function $\psi_{U',U} = \varphi_{U'} \circ \varphi_U^{-1}$. As noted a moment ago, this is a diffeomorphism from one open set in $\mathbb{R}^n$ to another. In particular its differential $\psi_{U',U*}$ maps $\psi_U(U' \cap U)$ to $Gl(n; \mathbb{R})$. This understood, the bundle transition function $g_{U',U}$ for the cotangent bundle is defined so that its value at any given point $p \in U' \cap U$ is the transpose of the inverse of the matrix $\psi_{U',U*}|_{\varphi U(p)}$. Said differently, any given bundle transition function for T^*M is obtained from the corresponding bundle transition function for TM by taking the transpose of the inverse of the latter.

The same argument that proves TM to be independent of the representative coordinate atlas proves that T^*M is likewise independent of the representative atlas.

Here is a final remark: Let $U \subset M$ denote any given coordinate chart, and let φ_U denote the corresponding map from U to $\mathbb{R}^n$. It follows from the definition of T^*M that the latter has a canonically associated bundle trivialization map from $T^*M|_U$ to $U \times \mathbb{R}^n$. The inverse of this trivialization map, a map from $U \times \mathbb{R}^n$ to $T^*M|_U$, is denoted by ${\varphi_U}^*$.

The only example of the cotangent bundle I'll give now is that of $\mathbb{R}^n$, in which case the canonical coordinate atlas $\mathcal{U} = \{(\mathbb{R}^n, \textit{the identity map})\}$ identifes $T\mathbb{R}^n$ with the product bundle $\mathbb{R}^n \times \mathbb{R}^n$. More cotangent bundles examples are given in later chapters.

3.7 Bundle homomorphisms

Let $\pi: E \to M$ and $\pi': E' \to M$ denote a pair of vector bundles. A homomorphism $\mathfrak{h}: E \to E'$ is a smooth map with the property that if $p \in M$ is any given point, then $\mathfrak{h}$ restricts to $E|_p$ as a linear map to $E'|_p$. An *endomorphism* of a given bundle E is a homomorphism from E to itself.

An *isomorphism* between given bundles E and E′ is a bundle homomorphism that maps each $p \in M$ version of $E|_p$ isomorphically to the vector space $E'|_p$. For example, the definition of a vector bundle given in Chapter 1.1 implies the following: If E is a vector bundle with fiber dimension n and point $p \in M$ a given point, then p is contained in an open set U such that there is a bundle isomorphism $\varphi_U: E|_U \to U \times \mathbb{R}^n$.

These local trivializations can be used to give a somewhat different perspective on the notion of a bundle homomorphism. To set the background, suppose that E and E′ are two vector bundles over M of fiber dimensions n and n′, respectively. Let $p \in M$ a given point. There is then an open neighborhood $U \subset M$ of p with respective bundle isomorphisms $\varphi_U: E|_U \to U \times \mathbb{R}^n$ and $\varphi'_U: E'|_U \to U \times \mathbb{R}^{n'}$. A given bundle homomorphism $\mathfrak{h}: E \to E'$ is such that the composition $\varphi'_U \circ \mathfrak{h} \circ \varphi_U^{-1}$ is a map from $U \times \mathbb{R}^n$ to $U' \times \mathbb{R}^{n'}$ that can be written as $(x, v) \to (x, \mathfrak{h}_U(x)v)$ where $x \to \mathfrak{h}_U(x)$ is a smooth map from the set U to the vector space, $\mathrm{Hom}(\mathbb{R}^n; \mathbb{R}^{n'})$, of linear maps from $\mathbb{R}^n$ to $\mathbb{R}^{n'}$.

Note that if U′ is another open set in M with $U' \cap U \neq \emptyset$ and with corresponding bundle isomorphisms $\varphi_{U'}$ for $E|_{U'}$ and $\varphi'_{U'}$ for $E'|_{U'}$, then there is a corresponding $\mathfrak{h}_{U'}$. The latter is determined by $\mathfrak{h}_U$ on $U' \cap U$ as follows: Introduce the respective bundle transition functions, $\mathfrak{g}_{U',U}: U' \cap U \to \mathrm{Gl}(n; \mathbb{R})$ for E and $\mathfrak{g}'_{U',U}: U' \cap U \to \mathrm{Gl}(n'; \mathbb{R})$ for U′. Then $\mathfrak{h}_{U'} = \mathfrak{g}_{U',U}\, \mathfrak{h}_U (\mathfrak{g}'_{U',U})^{-1}$ at the points in $U' \cap U$.

Example 3.8 This second picture of a vector bundle isomorphism can be used to tie up a loose end that concerns Chapter 3.2's cocycle definition of a vector bundle. This loose end concerns the notion of the *refinement* of an open cover. To set the stage, let $\mathfrak{U}$ denote a given open cover of M. A second open cover, $\mathfrak{U}'$, is said to be a refinement of $\mathfrak{U}$ when each set from $\mathfrak{U}'$ is contained entirely in at least one set from $\mathfrak{U}$. This notion of refinement is often used when comparing constructions that are defined using distinct open covers. The point being that if $\mathfrak{U}$ and $\mathfrak{V}$ are any two open covers, then the collection $\{U \cap V\}_{U \in \mathfrak{U},\, V \in \mathfrak{V}}$ is a simultaneous refinement of both.

In any event, suppose that $n \geq 0$ is a given integer, that $\mathfrak{U}$ is a finite open cover, and that $\{\mathfrak{g}_{U,U'}: U \cap U' \to \mathrm{Gl}(n; \mathbb{R})\}_{U,U' \in \mathfrak{U}}$ supply the cocycle data needed to define a vector bundle over M with fiber $\mathbb{R}^n$. Let $E \to M$ denote the resulting vector bundle. Now suppose that $\mathfrak{V}$ is a second finite open cover that refines $\mathfrak{U}$.

Fix a map $\mathfrak{u}: \mathfrak{V} \to \mathfrak{U}$ such that $V \subset \mathfrak{u}(V)$ for all $V \in \mathfrak{V}$. This done, then the data $\{g_{U(V),U(V')}: V \cap V' \to Gl(n; \mathbb{R})\}_{V,V' \in \mathfrak{V}}$ supplies the cocycle data for a vector bundle, $E' \to M$. As it turns out, there is a canonical isomorphism from E' to E. To see this isomorphism, recall that the cocycle definition of E' defines the latter as the quotient of $\times_{V \in \mathfrak{V}} (V \times \mathbb{R}^n)$ by an equivalence relation. Likewise, E is the quotient of $\times_{U \in \mathfrak{U}} (U \times \mathbb{R}^n)$. Let $V \in \mathfrak{V}$. The isomorphism in question takes the equivalence class defined by $(p, v) \in V \times \mathbb{R}^n$ to that defined by the pair (p, v) with the latter viewed as a pair in $\mathfrak{u}(V) \times \mathbb{R}^n$. It is an exercise to unwind the definitions to see that this map is, as claimed, an isomorphism.

This canonical isomorphism is used implicitly to identify the bundle E with the bundle E' that is defined in this way using a refinement of the cover $\mathfrak{U}$.

Example 3.9 What follows describes a Lie group application of the notion of vector bundle isomorphism. To start, suppose that G is one of the Lie groups that were introduced in Chapter 2.3 or Chapter 2.5. In each case, the tangent bundle TG is isomorphic to the product bundle $G \times \mathbb{R}^{\dim(G)}$. In the case $G = Gl(n; \mathbb{R})$ or $Gl(n; \mathbb{C})$, this comes about because G is an open set in the Euclidean space $\mathbb{M}(n; \mathbb{R})$ or $\mathbb{M}(n; \mathbb{C})$ as the case may be.

The tangent bundle to $Sl(n; \mathbb{R})$ was identified in Chapter 3.5 with the submanifold in $Sl(n; \mathbb{R}) \times \mathbb{M}(n; \mathbb{R})$ of pairs of the form $(\mathfrak{m}, \mathfrak{m}\mathfrak{c})$ where $\mathfrak{c}$ has trace equal to 0. Let $\mathbb{M}_0(n; \mathbb{R})$ denote the vector space of such matrices. This is a Euclidean space, and the map from $Sl(n; \mathbb{R}) \times \mathbb{M}_0(n; \mathbb{R})$ to $TSl(n; \mathbb{R})$ that sends a pair $(\mathfrak{m}, \mathfrak{c})$ to $(\mathfrak{m}, \mathfrak{m}\mathfrak{c})$ is a bundle isomorphism. The tangent space to SO(n) was identified with the set of pairs $(\mathfrak{m}, \mathfrak{m}\mathfrak{c}) \in SO(n) \times \mathbb{M}(n; \mathbb{R})$ such that $\mathfrak{c}^T = -\mathfrak{c}$. Let $\mathbb{A}(n; \mathbb{R})$ denote the vector space of such $n \times n$, skew-symmetric matrices. The map that sends a pair $(\mathfrak{m}, \mathfrak{c})$ in the product vector bundle $SO(n) \times \mathbb{A}(n; \mathbb{R})$ to $(\mathfrak{m}, \mathfrak{m}\mathfrak{c}) \in TSO(n)$ is also a bundle isomorphism.

There are analogous isomorphisms for the groups U(n) and SU(n). To describe the former case, let $\mathbb{A}(n; \mathbb{C})$ denote the vector space of anti-Hermitian matrices. Thus, $\mathfrak{c} \in \mathbb{A}(n; \mathbb{C})$ if and only if $\mathfrak{c}^\dagger = -\mathfrak{c}$. The map that sends $(\mathfrak{m}, \mathfrak{c}) \in U(n) \times \mathbb{A}(n; \mathbb{C})$ to the pair $(\mathfrak{m}, \mathfrak{m}\mathfrak{c}) \in TU(n)$ is a bundle isomorphism. To consider the case of SU(n), introduce $\mathbb{A}_0(n; \mathbb{C})$ to denote the subvector space in $\mathbb{A}(n; \mathbb{C})$ of trace zero matrices. The same map restricts to the subspace $SU(n) \times \mathbb{A}_0(n; \mathbb{C})$ to give a bundle isomorphism from the latter space to TSU(n).

3.8 Sections of vector bundles

Let $\pi: E \to M$ denote a given vector bundle. A *section* of E is a smooth map $\mathfrak{s}: M \to E$ such that $\pi \circ \mathfrak{s}$ is the identity map on M. Thus a section assigns to each point $p \in M$ a point in $E|_p$. For example, the zero section $\hat{o}$ that appears in the definition given in Chapter 3.1 is an example. Note that the space of sections is a linear space. Moreover, it is linear over $C^\infty(M; \mathbb{R})$ in the following sense: If $\mathfrak{s}$ is any given section and $f\colon M \to \mathbb{R}$ any given function, then $f\mathfrak{s}$ is also a section; this is the section that assigns $f(p)\mathfrak{s}(p) \in E|_p$ to $p \in M$. The space of sections of E is denoted by $C^\infty(M; E)$ in what follows.

To see what a section looks like, suppose that $U \subset M$ is an open set with an isomorphism $\varphi_U\colon E|_U \to U \times \mathbb{R}^n$. Here, n is the fiber dimension of E. Let $\mathfrak{s}$ denote any given section of E. Then $\varphi_U \circ \mathfrak{s}\colon U \to U \times \mathbb{R}^n$ has the form $x \to (x, \mathfrak{s}_U(x))$ where $\mathfrak{s}_U$ is a smooth map from U to $\mathbb{R}^n$. Note that if $U' \subset M$ is an open set with a trivialization $\varphi_{U'}$ from $E|_{U'} \to U' \times \mathbb{R}^n$, then there is a corresponding map $\mathfrak{s}_{U'}$, with the latter determined by $\mathfrak{s}_U$ over $U' \cap U$ by the rule $\mathfrak{s}_{U'} = \mathfrak{g}_{U',U}\mathfrak{s}_U$ where $\mathfrak{g}_{U',U}\colon U' \cap U \to Gl(n; \mathbb{R})$ again denotes the bundle transition function.

This identification can be used to see why the vector space of sections of any given bundle E is infinite dimensional. Indeed, fix $U \subset M$ where there is a bundle isomorphism $\varphi_U\colon E|_U \to U \times \mathbb{R}^n$. Let $\mathfrak{s}_U\colon U \to \mathbb{R}^n$ denote any given map which is zero on the complement of a compact set in U. A section, $\mathfrak{s}$, of E can be defined now as follows: Set $\mathfrak{s} = \hat{o}$ on $E|_{M-U}$ and set $\mathfrak{s}$ on U to be the section given by $p \to \varphi_U^{-1}((p, \mathfrak{s}_U(p))$.

By the way, a comparison of the definition of a section with that of a bundle homomorphism finds that a section of E gives a bundle homomorphism from the trivial bundle $M \times \mathbb{R}$ to E, and vice versa. The correspondence identifies any given section $\mathfrak{s}$ with the homomorphism that send $(p, r) \in M\times\mathbb{R}$ to $r\mathfrak{s}(p)$. Meanwhile, a homomorphism $\mathfrak{h}\colon M \times \mathbb{R} \to E$ gives the section $p \to \mathfrak{h}|_p(1)$ where $1 \in \mathbb{R}$.

If $U \subset M$ is any open set, a set $\{\mathfrak{s}_k\}_{1\le k\le n}$ of sections of $E|_U$ are said to define a *basis* of sections of $E|_U$ if the set $\{\mathfrak{s}_k|_p\}_{1\le k\le n} \in E|_p$ is linearly independent for each $p \in U$ where E has fiber dimension n. Thus they define a basis of $E|_p$. A basis of sections gives an isomorphism from the product bundle $U\times\mathbb{R}^n$ to $E|_U$, this the homomorphism that sends a given $(p, (r_1, \ldots, r_n)) \in U\times\mathbb{R}^n$ to the point $\sum_{1\le k\le n} r_k\, \mathfrak{s}_k(p)$. Conversely, a basis of sections for $E|_U$ defines an isomorphism from $E|_U$ to $U \times \mathbb{R}^n$, this the homomorphism that sends any given point $v \in E|_p$ to $(p, (r_1(v), \ldots, r_n(v)) \in U\times\mathbb{R}^n$ where $v = \sum_{1\le k\le n} r_k\, \mathfrak{s}_k(p)$. A bundle $E \to M$ is isomorphic to the product bundle if and only if there is a basis of sections for the whole of M.

3.9 Sections of TM and T*M

A section of the tangent bundle of M is called a *vector field*. The space of sections of TM, thus $C^\infty(M; TM)$, can be viewed as the vector space of *derivations* on the algebra of functions on M. To elaborate, remark that the space $C^\infty(M; \mathbb{R})$ of smooth functions on M is an algebra with addition given by

$(f + g)(p) = f(p) + g(p)$ and multiplication given by $(fg)(p) = f(p)g(p)$. A derivation is by definition a map, $\mathfrak{L}$, from $C^\infty(M; \mathbb{R})$ to itself that obeys $\mathfrak{L}(f + g) = \mathfrak{L}f + \mathfrak{L}g$ and $\mathfrak{L}(fg) = \mathfrak{L}f\, g + f\, \mathfrak{L}g$. Let $\mathfrak{v}$ denote a given section of TM. The action on a function f of the derivation defined by $\mathfrak{v}$ is denoted by $\mathfrak{v}(f)$. To see what this looks like, go to a coordinate chart $U \subset M$ with coordinate map $\varphi_U: U \to \mathbb{R}^n$ with n here set equal to the dimension of M. Use $\varphi_{U*}: TM|_U \to U \times \mathbb{R}^n$ to denote the associated isomorphism between $TM|_U$ and the product bundle. Write $\varphi_U \circ \mathfrak{v}$ as the section of the product bundle given by $p \to (p, \mathfrak{v}_U = (v_1, \ldots, v_n): U \to \mathbb{R}^n)$. Meanwhile, write $f_U = f \circ \varphi_U{}^{-1}$. The analogous $(\mathfrak{v}f)_U$ is given by $(\mathfrak{v}f)_U = \sum_{1\le i\le n} v_i(\frac{\partial}{\partial x_i} f_U)$. The fact that this definition is consistent across coordinate charts follows using the Chain rule and the bundle transition function for TM.

The function $\mathfrak{v}f$ is often called the *Lie derivative* of f along the vector field $\mathfrak{v}$. It is also called the *directional* derivative of f along $\mathfrak{v}$ because the formula that writes $(\mathfrak{v}f)_U$ as $\sum_{1\le i\le n} v_i(\frac{\partial}{\partial x_i} f_U)$ identifies $(\mathfrak{v}f)_U$ with the directional derivative of the function f_U on $\mathbb{R}^n$ along the vector field with components $(v_1, \ldots, v_n)$.

The observation that vector fields give derivations motivates the notation whereby $\mathfrak{v}_U$ is written in terms of its components as $\sum_{1\le i\le n} v_i \frac{\partial}{\partial x_i}$; and it motivates the notation that uses $\left\{\frac{\partial}{\partial x_i}\right\}_{1\le i\le n}$ as a basis of sections for $T\mathbb{R}^n$. The map $(\varphi_U^{-1})_*$ takes this basis to a basis of sections of TM over U.

To see how a derivation of $C^\infty(M; \mathbb{R})$ gives a vector field, let $\mathfrak{L}$ denote a given derivation. Let $U \subset M$ denote an open set where there is a coordinate map φ_U to $\mathbb{R}^n$. For each $k \in \{1, \ldots, n\}$, let $f_k = x_k \circ \varphi_U$. Define c_k to be the function on U given by $\mathfrak{L}f_k$. It then follows using Taylor's theorem with remainder and the Chain rule that $\mathfrak{L}$ acts on any given function, $\mathfrak{L}f$, on U as $\mathfrak{L}f = \sum_{1\le k\le n} c_k \mathfrak{v}_k$, where $\mathfrak{v}_k$ is the vector field on U that corresponds via the trivialization $(\varphi_U)_*: TM|_U \to U \times \mathbb{R}^n$ to the basis $\left\{\frac{\partial}{\partial x_i}\right\}_{1\le i\le n}$.

A section of T*M is called a *1-form*. Any given function, f, on M defines a 1-form, this denoted by df. The latter is said to be the *exterior* derivative of f. There is an associated linear function on TM defined as follows: Let $p \in M$ and let $v \in TM|_p$. Let $\mathfrak{v}$ denote any given vector field on M with $\mathfrak{v}|_p = v$. The linear function df send v to $(\mathfrak{v}f)|_p$. To see what this looks like, let $U \subset M$ denote a coordinate chart and let $\varphi_U: U \to \mathbb{R}^n$ denote the corresponding map. Write $f \circ \varphi_U^{-1}$ as f_U again, this a function on a domain in $\mathbb{R}^n$. Meanwhile, φ_U defines a canonical trivialization of $T^*M|_U$. Then the image of the section df via this trivialization is the section $x \to \left(x, \left(\frac{\partial f_U}{\partial x_1}, \ldots, \frac{\partial f_U}{\partial x_n}\right)x\right)$ of the product bundle $\varphi(U) \times \mathbb{R}^n$.

This last observation motivates the writing of dh as $\sum_{1\le i\le n} \frac{\partial h}{\partial x_i} dx_i$ when h is a given function on $\mathbb{R}^n$. The notation indicates that the exterior derivatives of the coordinates functions $\{x_i\}_{1\le i\le n}$ define the basis $\{dx_i\}_{1\le i\le n}$ of sections for $T^*\mathbb{R}^n$. This basis corresponds via φ_U's trivialization of $T^*M|_U$ to a basis of sections of $T^*M|_U$.

Additional reading

- *Introduction to Smooth Manifolds*, John M. Lee, Springer, 2003.
- *Tensor Analysis on Manifolds*, Richard L. Bishop and Samuel Goldberg, Dover Publications, 1980.
- *Characteristic Classes*, John W. Milnor and James D. Stasheff, Princeton University Press, 1974.
- *Manifolds and Differential Geometry*, Jeffrey M. Lee, American Mathematical Society, 2009.
- *Vector Bundles, Volume 1: Foundations and Stiefel–Whitney classes*, Howard Osborn, Academic Press, 1983.
- *Differential and Riemannian Manifolds*, Serge Lange, Springer, 1995.

4 Algebra of vector bundles

Any linear operations that can be done to generate a new vector space from some given set of initial vector spaces can be done fiber-wise with an analogous set of vector bundles to generate a new vector bundle. What follows describes the most important examples.

4.1 Subbundles

Let π: $E \to M$ denote a vector bundle. A submanifold $E' \subset E$ is said to be a *subbundle* of E if, for each $p \in M$, the subspace $E'|_p = (E' \cap E|_p)$ is a vector subspace. An equivalent definition is as follows: The submanifold $E' \subset E$ is a subbundle if it is the image via a bundle homomorphism that is injective on each fiber.

To see that the structure of a subbundle $E' \subset E$ near any given point $p \in M$, fix an open set $U \subset M$ that contains p and that comes with an isomorphism $\varphi_U : E|_U \to U \times \mathbb{R}^n$. The image via φ_U of $E'|_U$ is an n′-dimensional subbundle of the product bundle $U \times \mathbb{R}^n$. Let $V' \subset \mathbb{R}^n$ denote the φ_U image of $E'|_p$, this an n′-dimensional vector subspace of $\mathbb{R}^n$. Fix a basis for V' so as to identify it with $\mathbb{R}^{n'}$. Let $\pi_{V'}$: $\mathbb{R}^n \to V' = \mathbb{R}^{n'}$ denote the orthogonal projection. Then there is a neighborhood $U' \subset U$ of p such that the composition $\pi_{V'} \circ \varphi_U$ maps $E'|_{U'}$ isomorphically to the trival bundle $U' \times V' = U' \times \mathbb{R}^{n'}$.

By way of an example, suppose that M sits in $\mathbb{R}^N$ as a submanifold. Then TM sits in the product bundle $M \times \mathbb{R}^N$ as a subbundle. To see a second example, fix positive integers $m > n$ and introduce from Chapter 1.6 the Grassmannian $Gr(m; n)$ whose points are the n-dimensional vector subspaces of $\mathbb{R}^m$. Introduce from Chapter 3.3 the tautological $\mathbb{R}^n$-bundle $E \to Gr(m; n)$, this the set of pairs $\{(V, v) \in Gr(m; n) \times \mathbb{R}^n : v \in V\}$. This bundle is defined to be a subbundle of the product bundle $Gr(m; n) \times \mathbb{R}^n$.

As it turns out, any given bundle π: $E \to M$ can be viewed as a subbundle of some product bundle $M \times \mathbb{R}^N$ for N very large. What follows explains why this is true in the case when M is compact. To start, fix a finite open cover $\mathfrak{U}$ of M such that each set $U \subset \mathfrak{U}$ has an associated bundle isomorphism $\varphi_U: E|_U \to U \times \mathbb{R}^n$. By enlarging the set $\mathfrak{U}$ if necessary, one can also assume that there is an associated, subordinate partition of unity—this is a set of functions $\{\chi_U: U \to [0, \infty)\}_{U \in \mathfrak{U}}$ such that any given χ_U is zero on M−U and such that $\sum_{U \in \mathfrak{N}} \chi_U = 1$ on M. Label the elements in $\mathfrak{U}$ as $\{U_\alpha\}_{1 \le \alpha \le K}$. Now define a bundle homomorphism $\psi: E \to M \times \mathbb{R}^{Kn}$ by the following rule: Write $\mathbb{R}^{Kn}$ as K factors of $\mathbb{R}^n$. A given point $v \in E$ is sent to the point in $M \times \mathbb{R}^{Kn}$ whose K'th $\mathbb{R}^n$ factor is $\chi_{U_k}(\pi(v))\,\varphi_{U_k}(v)$. By construction, the map so defined is fiber-wise injective, and so $\psi(E)$ sits in $M \times \mathbb{R}^{Kn}$ as a subbundle.

4.2 Quotient bundles

Suppose that V is a vector space and $V' \subset V$ a vector subspace. One can then define the quotient space V/V', this the set of equivalence classes that are defined by the rule that has $v \sim u$ when $v - u \in V'$. The quotient V/V' is a vector space in its own right, its dimension is $\dim(V) - \dim(V')$. Note that in the case when V is identified with $\mathbb{R}^n$, the orthogonal projection from $\mathbb{R}^n$ to the orthogonal complement of V' identifies the quotient bundle V/V' with this same orthogonal complement.

There is the analogous quotient construction for vector bundles over a given manifold. To elaborate, suppose that $\pi: E \to M$ is a vector bundle with fiber dimension n and $E' \subset E$ is a subbundle with fiber dimension $n' \le n$. The quotient bundle E/E' is defined so that its fiber over any given point $p \in M$ is $E|_p/E'|_p$. The projection from E/E' is induced by that of E to M, and the $\mathbb{R}$-action is induced from the corresponding action on E.

To see the structure of E/E' near a given point $p \in M$, suppose that $U \subset M$ is an open set that contains p and that comes with a bundle isomorphism $\varphi_U : E|_U \to U \times \mathbb{R}^n$. The latter restricts to $E'|_U$ so as to map $E'|_U$ isomorphically onto some n'-dimensional subbundle of the product bundle $U \times \mathbb{R}^n$. Set $\varphi_U(E'|_p) = V'$, this an n'-dimensional vector subspace of $\mathbb{R}^n$. Let $\pi_{V'}$ as before denote the orthogonal projection from $\mathbb{R}^n$ to V'. As noted in Chapter 4.1, there is a neighborhood $U' \subset U$ of p such that $\varphi'_{U'} = \pi_{V'} \circ \varphi_U$ maps $E'|_{U'}$ isomorphically to the trivial bundle $U \times V'$. Let $V'^{\perp} \subset \mathbb{R}^n$ denote the orthogonal complement of V', and let $\pi^{\perp}_{V'}$ denote the orthogonal projection from $\mathbb{R}^n$ to $V'^{\perp}$. Introduce $\mathfrak{p}: U \times V' \to V'^{\perp}$ to denote the composition $\pi^{\perp}_{V'} \circ (\varphi'_{U'})^{-1}$ and let $\varphi''_{U'} = \pi^{\perp}_{V'} \circ \varphi_U - \mathfrak{p} \circ \varphi'_{U'}$, this a map from $E|_{U'}$ to $U' \times V'^{\perp}$. By construction, this map sends vectors in $E'|_{U'}$ to zero. Meanwhile, it is surjective at p, and this implies that it is surjective on some neighborhood $U'' \subset U'$ of p. This understood, it induces an isomorphism $(E/E')|_{U'} \to U' \times V'^{\perp}$. Choose a basis for $V'^{\perp}$ to identify the latter with $\mathbb{R}^{n-n'}$ and the resulting map from $(E/E')|_{U'}$ to $U' \times \mathbb{R}^{n-n'}$ gives the requisite local isomorphism.

By way of an example, suppose that M is realized as a submanifold in $\mathbb{R}^m$ for some $m > n = \dim(M)$. As explained in Chapter 3.4, the tangent bundle TM appears as a subbundle of $M \times \mathbb{R}^n$. The quotient bundle $(M \times \mathbb{R}^n)/TM$ is said to be the *normal* bundle of M in $\mathbb{R}^n$. Consider the case when $M \subset \mathbb{R}^3$ is a surface of genus g as constructed in Chapter 1.3 using a function $h: \mathbb{R}^2 \to R$. Recall that M in this case is the locus of points where $z^2 - h(x, y) = 0$. The normal bundle in this case has fiber dimension 1, and orthogonal projection in $\mathbb{R}^3$ identifies it with the subbundle in $M \times \mathbb{R}^3$ given by

$$\left\{\left((x, y, z), v = r\left(z\frac{\partial}{\partial z} - \frac{\partial h}{\partial x}\bigg|_{(x,y)}\frac{\partial}{\partial x} - \frac{\partial h}{\partial y}\bigg|_{(x,y)}\frac{\partial}{\partial y}\right)\right) : r \in \mathbb{R}\right\}.$$

4.3 The dual bundle

Let V denote a vector space of dimension n. The dual vector space, V^*, is the vector space of linear maps from V to $\mathbb{R}$. There is a corresponding notion of the dual of a vector bundle. To say more, let $\pi: E \to M$ denote the vector bundle in question. Its dual is denoted by E^*; it is defined so that the fiber of E^* over any given point $p \in M$ is the dual of the vector space $E|_p$. To elaborate, E^* is, first of all, the set of pairs $\{(p, \mathfrak{l}) : p \in M$ *and* $\mathfrak{l}: E|_p \to \mathbb{R})$ *is a linear map*$\}$. The bundle projection $\pi: E^* \to M$ sends any given pair $(p, \mathfrak{l}) \in E^*$ to the point p. The $\mathbb{R}$ action is such that $r \in \mathbb{R}$ acts to send $(p, \mathfrak{l}) \in E^*$ to $(p, r\,\mathfrak{l})$.

The definition also requires a suitable map to $\mathbb{R}^n$ near any given point; a map that restricts to each fiber as an $\mathbb{R}$-equivariant diffeomorphism. For this purpose, let $U \subset M$ denote an open set where there is a smooth map $\lambda_U: E|_U \to \mathbb{R}^n$ that restricts to each fiber as an $\mathbb{R}$-equivarant diffeomorphism. The corresponding map λ_U^* for $E^*|_U$ is best viewed as a map to the vector space of linear functions on $\mathbb{R}^n$. (This, of course, is a copy of $\mathbb{R}^n$.) Granted this view, the map λ_U^* is defined by the following requirement: Let $p \in U$ and let $e \in E|_p$. Then λ_U^* maps $(p, \mathfrak{l})$ to the linear functional on $\mathbb{R}^n$ whose value on $\lambda_U(\mathfrak{e})$ is the same as that of $\mathfrak{l}$ on $\mathfrak{e}$.

Note that this definition of λ_U^* has the following consequence: Let $U' \subset M$ denote an open set with $U \cap U' \neq \emptyset$ and with an analogous map $\lambda_{U'}: E|_{U'} \to \mathbb{R}^n$. Introduce $g_{U',U}: U' \cap U \to Gl(n; \mathbb{R})$ to denote the corresponding bundle transition function. Given $\lambda_{U'}$, there is a corresponding $\lambda_{U'}^*: E^*|_{U'} \to \mathbb{R}^n$, and a corresponding bundle transition function for E^* that maps $U' \cap U$ to $Gl(n; \mathbb{R})$. The latter is the map $(g_{U',U}^{-1})^T$, the transpose of the inverse of $g_{U',U}$.

The description of E^* is not complete without saying something about its smooth structure. The latter comes about as follows: Let $\mathcal{U} = \{(U, \varphi: U \to \mathbb{R}^{\dim(M)})\}$ denote a coordinate atlas for M. No generality is lost by requiring that each set U from this atlas come with a map $\lambda_U: E|_U \to \mathbb{R}^n$ that restricts to each fiber as an $\mathbb{R}$-equivariant diffeomorphism. As noted earlier, the set $E|_U$ with the map (φ, λ_U) gives a coordinate chart for E. This understood, the smooth structure for E^* is that defined by the atlas given by $\{(E^*|_U, (\varphi, \lambda_U^*))\}_{(U,\varphi)\in\mathcal{U}}$. The coordinate transition functions for this atlas are smooth. Indeed, if $(U', \varphi') \in \mathcal{U}$ and $U' \cap U \neq \emptyset$, then the coordinate transition function for the intersection of the charts $E^*|_U \cap E^*|_{U'} = E^*|_{U\cap U'}$ is the map

$$(\varphi' \circ (\varphi)^{-1}, (g_{U',U}^{-1})^T(\cdot))$$

where $g_{U',U}: U' \cap U \to Gl(n; \mathbb{R})$ is the bundle transition function for E over $U \cap U'$. This is a smooth map, and so the map to $Gl(n; \mathbb{R})$ given by the transpose of its inverse is also a smooth map.

Note that if V is a vector space, then $(V^*)^*$ has a canonical identification with V. Indeed, a vector $v \in V$ gives the linear function $\mathfrak{l} \to \mathfrak{l}(v)$ on V^*. By the same token, if E is a vector bundle, then $(E^*)^* = E$.

Perhaps the canonical example is the case when E is the tangent bundle TM; in this case, E^* is the cotangent bundle, T^*M. This can be seen by comparing the bundle transition functions.

4.4 Bundles of homomorphisms

The notion of the dual of a vector space, $Hom(V; \mathbb{R})$, has a generalization as follows: Let V and V′ denote two vector spaces. Let $Hom(V; V')$ denote the set of linear maps from V to V′. This also has the structure of a vector space, one with dimension equal to the product of the dimensions of V and of V′. This can be seen by choosing a basis for V to identify it with $\mathbb{R}^{\dim(V)}$ and likewise for V′ to identify it with $\mathbb{R}^{\dim(V')}$. A linear map from V to V′ appears now as a matrix with dim(V) columns and dim(V′) rows.

The construction $Hom(\cdot; \cdot)$ can be applied fiber-wise to a pair of vector bundles, E and E′, over M, to give a new vector bundle, this denoted by $Hom(E; E')$. The fiber of $Hom(E; E')$ over any given point $p \in M$ is the vector space $Hom(E|_p; E'|_p)$. If E has fiber dimension n and E′ fiber dimension n′, then the fiber dimension of $Hom(E; E')$ is nn′. To continue with the definition of the vector bundle structure, note that $\mathbb{R}$-action is such that $r \in \mathbb{R}$ acts on $Hom(E; E')$

so as to send a given point $(p, \mathfrak{l}: E|_p \to E'|_p)$ to $(p, r\mathfrak{l})$. The required neighborhood maps to $\mathbb{R}^{nn'}$ are obtained as follows: Let $p \in M$ denote some given point and $U \subset M$ a neighborhood of p with R-equivariant maps $\lambda_U: E|_U \to \mathbb{R}^n$ and $\lambda'_U : E' \to \mathbb{R}^{n'}$. The requisite map $\lambda_{Hom,U} : Hom(E; E')|_U \to \mathbb{R}^{nn'}$ is defined so as to send any given point $(p', \mathfrak{l})$ to the point matrix $\lambda'_U \circ \mathfrak{l} \circ \lambda_U^{-1}$ where here λ_U^{-1} is viewed as a linear map from $\mathbb{R}^n$ to $E|_{p'}$.

As for the smooth structure, take U as above so that it comes with a coordinate map $\varphi: U \to \mathbb{R}^{\dim(M)}$. This understood, then $Hom(E; E')|_U$ is a coordinate chart for $Hom(E; E')$ with the map that sends any given pair $(x, \mathfrak{l})$ to $(\varphi(p), \lambda'_U \circ \mathfrak{l} \circ \lambda_U{}^{-1}) \in \mathbb{R}^{\dim(M)} \times \mathbb{R}^{nn'}$. To see that the coordinate transition functions are smooth, it proves useful to view any given point $\mathfrak{m} \in \mathbb{R}^{nn'}$ as a matrix $\mathfrak{m} \in Hom(\mathbb{R}^n; \mathbb{R}^{n'})$. This understood, suppose that $U' \subset M$ is another coordinate chart with map $\varphi': U' \to \mathbb{R}^{\dim(M)}$ and with corresponding bundle maps $\lambda_{U'}: E_{U'} \to \mathbb{R}^n$ and $\lambda'_{U'}: E'|_{U'} \to \mathbb{R}^{n'}$. Then the transition function sends a given point $(x, \mathfrak{m}) \in \varphi(U \cap U') \times Hom(\mathbb{R}^n; \mathbb{R}^{n'})$ to the point with coordinates $((\varphi' \circ \varphi^{-1})(x), g'_{U',U}(x)\, \mathfrak{m}\, g_{U',U}(x)^{-1}) \in \varphi'(U \cap U') \times Hom(\mathbb{R}^n; \mathbb{R}^{n'})$.

Note that a section of $Hom(E; E')$ over M is nothing more than a bundle homomorphism from E to E′.

4.5 Tensor product bundles

Fix vector spaces V and V′ with respective dimensions n and n′. The vector space $Hom(V^*; V')$ of linear maps from the dual space V^* to V′ is denoted by $V' \otimes V$. What follows gives a pedestrian way to view this vector space. Fix a basis $\{v_a\}_{1 \le a \le n}$ for V and a basis $\{v'_a\}_{1 \le a \le n'}$ for V′. A basis for the nn′-dimensional vector space $V' \otimes V$ can be written as $\{v'_a \otimes v_b\}_{1 \le a \le n', 1 \le b \le n}$ where the element $v'_a \otimes v_b$ represents the linear map from V^* to V′ that assigns to any given linear function $\mathfrak{l} \in V^*$ the element $\mathfrak{l}(v_b)\, v'_a \in V'$. Note that a *decomposable* element in $V' \otimes V$ is given by a pair $v' \in V'$ and $v \in V$; it assigns to any given $\mathfrak{l} \in V^*$ the element $\mathfrak{l}(v)v' \in V'$. A decomposable element of this sort is written as $v' \otimes v$. The vector space $V' \otimes V$ is spanned by its decomposable elements.

By analogy, if $E' \to M$ and $E \to M$ are any two vector bundles, the corresponding tensor product bundle is $E' \otimes E = Hom(E^*; E')$.

4.6 The direct sum

Let V and V′ denote a pair of vector spaces with respective dimensions n and n′. The direct sum $V \oplus V'$ is the vectors space of pairs $(v, v') \in V \times V'$ where addition is defined by $(v, v') + (u, u') = (u + v, u' + v')$ and the $\mathbb{R}$-action has $r \in \mathbb{R}$ sending (v, v') to (rv, rv').

To see the bundle analog, let $\pi: E \to M$ and $\pi': E' \to M$ denote a pair of vector bundles. Their direct sum, $E \oplus E'$, is also a bundle over M. It can be viewed as the subset in $E \times E'$ of pairs $\{(v, v') : \pi(v) = \pi'(v')\}$. As explained momentarily, this subset is a submanifold and thus $E \oplus E'$ is a smooth manifold. The bundle projection from $E \oplus E'$ to M sends (v, v') to $\pi(v)$, this by definition being the same point as $\pi'(v')$. The $\mathbb{R}$-action has $r \in \mathbb{R}$ acting on (v, v') as (rv, rv'). To complete the definition, I need to exhibit a local trivialization on some neighborhood of any given point. To do this, let $U \subset M$ denote a given open set with a pair of maps $\lambda_U: E|_U \to \mathbb{R}^n$ and $\lambda'_U: E'|_{U'} \to \mathbb{R}^{n'}$ that restrict to the respective fibers as an $\mathbb{R}$-equivariant diffeomorphism. The corresponding map for $(E \oplus E')|_U$ sends the latter to $\mathbb{R}^n \times \mathbb{R}^{n'}$ and is given by (λ_U, λ'_U). It is a consequence of these definitions that the fiber over any given point $p \in M$ of $E \oplus E'$ is the vector space direct sum $E|_p \oplus E'|_p$.

To tie up a loose end, I need to explain why $E \oplus E'$ sits in $E \times E'$ as a submanifold. For this purpose, take U as above, and let $W \subset U$ denote an open subset with a smooth coordinate embedding $\varphi: W \to \mathbb{R}^{\dim(M)}$. The pair (φ, λ_U) gives a coordinate embedding from $E|_W \to \mathbb{R}^{\dim(M)} \times \mathbb{R}^n$. Likewise, the pair (φ, λ'_W) gives a coordinate embedding from $E'|_W \to \mathbb{R}^{\dim(M)} \times \mathbb{R}^n$. This understood, then $(E|_W \times E|_{W'})$ has the coordinate embedding to $(\mathbb{R}^{\dim(M)} \times \mathbb{R}^n) \times (\mathbb{R}^{\dim(M)} \times \mathbb{R}^{n'})$ given by $((\varphi, \lambda_W), (\varphi, \lambda'_W))$. The composition of this map with the map to $\mathbb{R}^{\dim(n)}$ given by $((x, v), (x', v')) \to x - x'$ has 0 as a regular value. This, the inverse image of 0, is a smooth submanifold. The latter is, by definition $(E \oplus E')|_W$. As a parenthetical remark, it follows as a consequence that the map from $(E \oplus E')|_W$ to $\mathbb{R}^{\dim(M)} \times (\mathbb{R}^n \times \mathbb{R}^{n'})$ given by $(\varphi, (\lambda_U, \lambda'_U))$ is a coordinate embedding.

4.7 Tensor powers

Let V denote a given vector space. Recall that V^* is the vector space of *linear* maps from V to $\mathbb{R}$. Fix $k \geq 1$. As it turns out, the vector space $V^* \otimes V^*$ is canonically isomorphic to the vector space of bilinear maps from $V \times V$ to $\mathbb{R}$. This is to say the vector space of maps $(v_1, v_2) \to f(v_1, v_2) \in \mathbb{R}$ with the property that $f(\cdot, v_2)$ is linear for any fixed $v_2 \in V$ and $f(v_1, \cdot)$ is linear for any fixed $v_1 \in V$. The identification is such as to send any given decomposable element $\iota_1 \otimes \iota_2 \in V^* \otimes V^*$ to the function $(v_1, v_2) \to \iota_1(v_1)\iota_2(v_2)$.

By the same token, if $k \geq 1$ is any given integer, then $\bigotimes_k V^* = V^* \otimes \cdots \otimes V^*$ is canonically isomorphic as a vector space to the vector space of k-linear maps from $\bigotimes_k V$ to $\mathbb{R}$. The identification is such as to send a decomposable element $\iota_1 \otimes \ldots \otimes \iota_k$ to the map that sends any given k-tuple $(v_1, \ldots, v_k)$ to $\iota_1(v_1) \cdots \iota_k(v_k)$.

There is an analogous way to view the k-fold tensor power of the dual of a given bundle $\pi\colon E \to M$. Indeed, a section $\bigotimes_k E^*$ defines a k-linear, fiber preserving map from $E \oplus \cdots \oplus E$ to $M \times \mathbb{R}$. Conversely, a map $f\colon E \oplus \cdots \oplus E \to M \times \mathbb{R}$ that preserves the fibers and is linear in each entry defines a section of $\bigotimes_k E^*$.

4.7.1 Symmetric powers

A k-bilinear map, f, from $\times_k V$ to $\mathbb{R}$ is said to be *symmetric* if

$$f(v_1, \ldots, v_i, \ldots, v_j, \ldots, v_k) = f(v_1, \ldots, v_j, \ldots, v_i, \ldots, v_k)$$

for any pair of indices $i, j \in \{1, \ldots, k\}$ and vectors $(v_1, \ldots, v_k) \in \times_k V$. The vector space of symmetric, k-bilinear maps from $\times_k V$ to $\mathbb{R}$ is a vector subspace of $\bigotimes_k V^*$, this denoted by $\mathrm{Sym}^k(V^*)$. Note that this vector space is isomorphic to the vector space of homogeneous, k'th order polynomials on V. For example, an isomorphism sends a symmetric map f to the polynomial $v \to f(v, \ldots, v)$. The inverse of this isomorphism sends a homogeneous polynomial $\wp$ to the symmetric, k-bilinear map that sends a given $(v_1, \ldots, v_k) \in \times_k V$ to $(\frac{\partial}{\partial t_1} \cdots \frac{\partial}{\partial t_k} \wp(t_1 v_1 + \cdots + t_k v_k))|_{t_1 = \cdots = t_k = 0}$.

By the way, a smooth map $f\colon V \to \mathbb{R}$ such that $f(rv) = r^k f(v)$ for all $r \in \mathbb{R}$ and $v \in V$ is, de facto, a homogeneous polynomial of degree k. Indeed, any such map must obey $f(v) = \frac{1}{k!}(\frac{\partial^k}{\partial t^k} f(tv))|_{t=0}$, and so is given by the order k term in f's Taylor's expansion at the origin in V.

Let $\pi\colon E \to M$ again denote a vector bundle. The bundle $\mathrm{Sym}^k(E^*)$ is the subbundle of $\bigotimes_k E^*$ whose restriction to any given fiber defines a symmetric, k-linear function on the fiber. Said differently, the fiber of $\mathrm{Sym}^k(E^*)$ over any given $p \in M$ is the vector space of maps from $E|_p$ to $\mathbb{R}$ that obey $f(rv) = r^k f(v)$ for each $r \in \mathbb{R}$ and $v \in E|_p$. A standard convention sets $\mathrm{Sym}^0(E) = M \times \mathbb{R}$.

4.7.2 Antisymmetric powers

Fix $k \in \{1, \ldots n\}$. A k-linear map from $\times_k V$ to $\mathbb{R}$ is said to be *antisymmetric* when

$$f(v_1, \ldots, v_i, \ldots v_j, \ldots, v_k) = -f(v_1, \ldots, v_j, \ldots v_i, \ldots, v_k)$$

for any pair of indices $i, j \in \{1, \ldots, k\}$ and vectors $(v_1, \ldots, v_k) \in \times_k V$. The vector space of antisymmetric, k-linear maps from $\times_k V$ to $\mathbb{R}$ is also a vector subspace of $\otimes_k V^*$, this denoted by $\wedge^k V^*$. Note that $\wedge^k V^* = \{0\}$ if $k > n$, and that $\wedge^n V^* = \mathbb{R}$. If $k \in \{1, \ldots, n\}$, then $\wedge^k V^*$ has dimension $\frac{n!}{k!(n-k)!}$.

Suppose that $\{v_1, \ldots, v_n\}$ is a given basis for V. Introduce $\{\iota_1, \ldots, \iota_n\}$ to denote the dual basis for V^*. This set determines a basis of sections of $\wedge^k V^*$, where any given basis element is labeled by a k-tuple $(\alpha_1, \ldots, \alpha_k)$ of integers that obey $1 \leq \alpha_1 < \cdots < \alpha_k \leq n$. The corresponding basis element is written as

$$\iota_{\alpha_1} \wedge \iota_{\alpha_2} \wedge \cdots \wedge \iota_{\alpha_k};$$

this corresponds to the antisymmetric map from $\times_k V$ to $\mathbb{R}$ that assigns 1 to the element $(v_{\alpha_1}, \ldots, v_{\alpha_k})$ and zero to basis elements that have an entry whose label is not from $\{\alpha_1, \ldots, \alpha_k\}$. Note that there is a canonical homomorphism from $\wedge^k V^* \otimes \wedge^{k'} V^*$ to $\wedge^{k+k'} V^*$ given by antisymmetrization. The image of a pair (ω, μ) under this homomorphism is written as $\omega \wedge \mu$. This homomorphism is sometimes called the *wedge product* and other times the *exterior product*. The wedge/exterior product is such that $\omega \wedge \mu = (-1)^{kk'} \mu \wedge \omega$.

Let π: $E \to M$ denote a given vector bundle. If $k \in \{1, \ldots, n\}$, the bundle $\wedge^k E^*$ is the subbundle in $\otimes_k E^*$ whose fiber over any given point $p \in M$ consists of the elements that define antisymmetric, k-linear functions on $E|_p$. The bundle $\wedge^k E^*$ is the 0-dimensional bundle $M \times \{0\}$ if $k > n$. Meanwhile $\wedge^n E^*$ is a real line bundle. However, it need not be isomorphic to the trivial line bundle! To say more about the latter, suppose that $U \subset M$ is an open set and that φ_U: $E|_U \to U \times \mathbb{R}^n$ is a local trivialization of E over U. Let $U' \subset M$ denote a second open set, and suppose that $\varphi_{U'}$: $E|_{U'} \to U' \times \mathbb{R}^n$ is a local trivialization over U'. Assume that $U' \cap U \neq \emptyset$ so that there is a well-defined bundle transition function $g_{U',U}$: $U \cap U' \to Gl(n; \mathbb{R})$. The corresponding bundle transition function for $\wedge^n E^*$ is $\det(g_{U',U})$. The bundle $\wedge^n E$ is denoted by det(E).

The exterior/wedge product applied fiber-wise defines a bundle homomorphism from $\wedge^k E^* \otimes \wedge^{k'} E^*$ to $\wedge^{k+k'} E^*$ which is also called the exterior or wedge product.

Of special interest is the case of when $E = TM$ in which case a section of $\wedge^k T^*M$ is called a *k*-form.

Additional reading

- *Analysis and Algebra on Differentiable Manifolds*, P. M. Gadea and J. Munoz, Springer, 2001.
- *Introduction to Differential Topology*, T. Bröcker and K. Jänich, Cambridge University Press, 1972.

- *Tensor Analysis on Manifolds*, Richard L. Bishop and Samuel Goldberg, Dover Publications, 1980.
- *K-Theory, An Introduction*, Max Karoubi, Springer, 1978.
- *Introduction to Smooth Manifolds*, John M. Lee, Springer, 2003.
- *Fiber Bundles*, Dale Husemöller, Springer, 1993.

5 Maps and vector bundles

Suppose that M, N are smooth manifolds. As explained in what follows, any given map ψ: M $\to$ N can be used to associate a vector bundle on M to one on N.

5.1 The pull-back construction

Let π: E $\to$ N denote a vector bundle with fiber dimension n. There is a canonical bundle, denoted ψ^*E, over M, also with fiber dimension n, which comes with a canonical map $\hat{\psi}$: $\psi^*E \to E$ that restricts to the fiber over any given point $p \in M$ as a vector space isomorphism from $(\psi^*E)|_p$ to $E|_{\psi(p)}$. The bundle ψ^*E is the submanifold in $M \times E$ of pairs $\{(m, e) : \psi(m) = \pi(e)\}$. This is to say that the fiber of ψ^*E over $m \in M$ is the fiber of E over $\psi(m)$. The map $\hat{\psi}$ sends a pair $(m, e) \in \psi^*E$ (thus $m \in M$ and $e \in N$ with $\pi(e) = \psi(m)$) to the point $\hat{\psi}(m, e) = e$. To be somewhat more explicit, suppose that $U \subset N$ is an open set and that φ_U: $E|_U \to U \times \mathbb{R}^n$ is a bundle isomorphism. Then the ψ^*E is isomorphic over $\psi^{-1}(U) \subset M$ to $\psi^{-1}(U) \times \mathbb{R}^n$, the isomorphism taking a pair (m, e) to the pair $(m, \lambda_U(e))$ where λ_U: $E|_U \to \mathbb{R}^n$ is defined by writing $\varphi_U(e)$ as the pair $(\pi(e), \lambda_U(e))$. Let $U' \subset N$ denote a second open set with nonempty intersection with U and with a corresponding bundle trivialization $\varphi_{U'}$: $E|_{U'} \to U' \times \mathbb{R}^n$. The associated transition function $g_{U',U}$: $U' \cap U \to Gl(n; \mathbb{R})$ determines the transition function of ψ^*E for the sets $\psi^{-1}(U') \cap \psi^{-1}(U)$ as the latter is $g_{U',U} \circ \psi$.

Sections of bundles also pull-back. To elaborate, suppose that π: E $\to$ N is a smooth bundle and $\mathfrak{s}$: N $\to$ E is a section. Let ψ: M $\to$ N denote a smooth map. Then $\mathfrak{s} \circ \psi$ defines a section of ψ^*E.

Example 5.1 Let $\Sigma \subset \mathbb{R}^3$ denote a surface. Define a map from Σ to S^2 by associating to each point $p \in \Sigma$ the vector normal to Σ (pointing out from the region that is enclosed by Σ). This map is called the *Gauss* map. Denote it by n. Then $T\Sigma$ is isomorphic to n^*TS^2. The isomorphism is obtained as follows: View $TS^2 \subset \mathbb{R}^3 \times \mathbb{R}^3$ as the set of pairs of the form (v, w) with $|v|^2 = 1$ and $v{\cdot}w = 0$. Here, $v{\cdot}w$ denotes the dot product in $\mathbb{R}^3$. Granted this, then n^*TS^2 can be viewed as the set $\{(p, w) \in \Sigma \times \mathbb{R}^3 : n(p)\cdot w = 0\}$. As explained in Chapter 2.4, this is identical to $T\Sigma$.

A concrete example is had using a surface of genus g as constructed in Chapter 1.3 from a function, h, on $\mathbb{R}^2$. Recall that the surface is given by the set of points in $\mathbb{R}^3$ where $z^2 - h(x, y) = 0$. The corresponding Gauss map is given by

$$n|_{(x,y,z)} = \left(-\frac{\partial}{\partial x}h,\ \frac{\partial}{\partial y}h,\ 2z\right) \bigg/ (|dh|^2 + 4z^2)^{1/2}.$$

Example 5.2 Pick a point q in the region of $\mathbb{R}^3$ bounded by Σ and define a map, ψ, from Σ to S^2 by taking the direction from the given point on Σ to q. Is ψ^*TS^2 isomorphic to $T\Sigma$? What if the point q is outside Σ instead of inside?

Example 5.3 Let M denote a given m-dimensional manifold. What follows constructs a map from M to S^m that is relevant to one of the constructions in Chapter 3.3.5. To begin the story, fix a finite set $\Lambda \subset M$. Assign to each point $p \in \Lambda$ a coordinate chart $(U, \varphi_p : U \to \mathbb{R}^m)$ with $p \in U$ such that $\varphi_p(p) = 0 \in \mathbb{R}^m$. Fix $r > 0$ and small enough so that the open ball, B, of radius r about the origin in $\mathbb{R}^m$ is in the image of each $p \in \Lambda$ version of φ_p; and so that the collection $\{U_p = \varphi_p^{-1}(B)\}_{p\in\Lambda}$ consists of pair-wise disjoint sets. Use this data to define a map, ψ, from M to $S^m \subset \mathbb{R}^{m+1}$ as follows: The map ψ send all of the set $M(\bigcap_p U_p)$ to the point $(0, \ldots, 0, -1)$. To define ψ on a given version of U_p, write $\mathbb{R}^{m+1} = \mathbb{R}^m \times \mathbb{R}$. Introduce from Appendix 1.2 the function χ and set $\chi_p : U_p \to [0, 1]$ to equal $\chi(|\varphi_p|)$. This done, the map ψ sends a $p \in U_p$ to the point $(2\chi_p\varphi_p, \chi_p^2 - |\varphi_p|^2)/(\chi_p^2 + |\varphi_p|^2)^{-1}$.

To continue, fix a positive integer n and a map $g : S^{m-1} \to Gl(n; \mathbb{R})$. Define a rank n vector bundle $E \to S^m$ by using the coordinate patches $U_+ = S^m - (0, \ldots, -1)$ and $U_- = S^m - (0, \ldots, +1)$ with the transition function that $g_{U_-U_+}$ send a point $(x, x_{n+1}) \in S^n \subset \mathbb{R}^n \times \mathbb{R}$ to $g(x/|x|)$. The resulting bundle $\psi^*E \to M$ is isomorphic to the bundle constructed in Chapter 3.3.5 using the data consisting of Λ and the set $\{(U_p, \varphi_p, g_p = g)\}_{p\in\Lambda}$.

5.2 Pull-backs and Grassmannians

As it turns out, any given vector bundle is isomorphic to the pull-back of a tautological bundle over a Grassmannian, these as described in Chapter 3.3.4. The next proposition makes this precise.

Proposition 5.4 *Let* M *denote a smooth manifold, let* n *denote a positive integer, and let* π: E $\to$ M *denote a given rank* n *vector bundle. If* m *is sufficiently* *large, there exists a map* ψ_m: M → Gr(m; n) *and an isomorphism between* E *and the pull-back via* ψ_m *of the tautological bundle over* Gr(m; n).

Proof of Proposition 5.4 As noted in Chapter 4.1, the bundle E is isomorphic to a subbundle of some $N \gg 1$ version of the product bundle $M \times \mathbb{R}^N$. Use an isomorphism of this sort to view E as a subbundle in just such a product bundle. Let λ: $M \times \mathbb{R}^N \to \mathbb{R}^N$ denote the projection map. The assignment to $x \in M$ of the n-dimensional vector subspace $\lambda(E|_x) \subset \mathbb{R}^N$ defines a smooth map ψ: M → Gr(N; n). Let $E_N \to$ Gr(N; n) denote the tautological $\mathbb{R}^n$-bundle. This is to say that the fiber of E_N over a given subspace $\Pi \in$ Gr(N; m) is the vector space Π. Thus, the fiber of E_N over $\psi(x)$ is $E|_x$ and so E is the pull-back via ψ of the tautological bundle.

5.3 Pull-back of differential forms and push-forward of vector fields

There is a canonical, vector space homomorphism that maps $C^\infty(N; \wedge^k T^*N)$ to $C^\infty(M; \wedge^k T^*M)$ for any $k \in \{0, 1, \ldots, \}$, this is also denoted by ψ^*. It is defined so as to commute with wedge product, exterior multiplication and to factor with regards to compositions. This map is defined by the following rules

- $\psi^* f = f \circ \psi$ *and* $\psi^* df = d(\psi^* f)$ *when* $f \in C^\infty(N) = C^\infty(N; \wedge^0 T^*N)$.
- $\psi^*(f\omega) = (\psi^* f)\, \psi^*\omega$ *when* $f \in C^\infty(N)$ *and* $\omega \in C^\infty(N; \wedge^k T^*N)$ *for any* k.

Induction on k using these two rules defines ψ^* as an algebra homomorphism from $\bigoplus_{k\geq 0} C^\infty(N; \wedge^k T^*N)$ to $\bigoplus_{k\geq 0} C^\infty(M; \wedge^k T^*M)$ that commutes with d. Moreover, if Z is a third manifold and φ: N → Z is a smooth map, then $(\varphi \circ \psi)^* = \varphi^* \circ \psi^*$. The homomorphism ψ^* is called the *pull-back* homomorphism, and $\psi^*\omega$ is said to be the *pull-back* of ω.

The picture of the pull-back homomorphism in local coordinates shows that the latter is a fancy version of the Chain rule. To elaborate, fix positive integers m and n; and let ϕ denote a smooth map between $\mathbb{R}^m$ and $\mathbb{R}^n$. Introduce coordinates $(x_1, \ldots, x_m)$ for $\mathbb{R}^m$ and coordinates $(y_1, \ldots, y_n)$ for $\mathbb{R}^n$. Use $(\phi_1, \ldots, \phi_n)$ to denote the components of the map ϕ. Then

$$\phi^*(dy_k) = \Sigma_{1\leq i\leq m} \left(\frac{\partial \phi_k}{\partial x_i}\right)\bigg|_x dx_i.$$

Note that $d(\phi^* dy_k) = 0$ because $\frac{\partial^2 f}{\partial x_i \partial x_j} = \frac{\partial^2 f}{\partial x_j \partial x_i}$ for any given function f and indices i and j.

Use of local coordinate charts gives a picture of this map ψ^* which is essentially that given above for the cases $M = \mathbb{R}^m$ and $N = \mathbb{R}^n$. To elaborate, let M and N again be smooth manifolds and $\psi: M \to N$ a smooth map. Let $U \subset M$ denote a coordinate chart with coordinate map $\varphi_U: U \to \mathbb{R}^{m=\dim(M)}$. Let $(x_1, \ldots, x_m)$ denote the Euclidean coordinate functions on $\mathbb{R}^m$, and recall from Chapter 3.9 that the map φ_U can be used to identify the basis of sections $\{dx_i\}_{1\leq i\leq m}$ of $T^*\mathbb{R}^m$ with a corresponding basis of sections of $T^*M|_U$. Use the notation $\{dx_i^U\}_{1\leq i\leq m}$ to denote the latter. Let $V \subset N$ denote a coordinate chart whose intersection with $\psi(U)$ is not empty, and let $\varphi_V : V \to \mathbb{R}^{n=\dim(N)}$ denote the coordinate map. Use $(y_1, \ldots, y_n)$ to denote the Euclidean coordinates of $\mathbb{R}^n$ and use ϕ_V to define the basis of sections $\{dy_i^V\}_{1\leq i\leq n}$ for $T^*N|_V$. The pull-back via ψ of a given dy_k^V is given by

$$(\psi^* dy_k^V)|_x = \Sigma_{1\leq i\leq m} \left(\frac{\partial \phi_k}{\partial x_i}\right)\bigg|_{\varphi_U(x)} (dx_i^U)|_x$$

where ϕ here denotes $\varphi_U \circ \psi \circ (\varphi_V)^{-1}$, this is a map from a domain in $\mathbb{R}^m$ to one in $\mathbb{R}^n$.

By the way, this pull-back map ψ^* is closely related to the previously defined notion of the pull-back bundle. Indeed, ψ^* can be defined as the composition of two linear maps. The first map pulls back a section of $\wedge^k T^*N$ to give a section of $\psi^*(\wedge^k T^*N)$ over M. This as described in Chapter 5.1 above. The second is induced by a certain canonical vector bundle homomorphism from $\psi^*(\wedge^k T^*N)$ to $\wedge^k T^*M$. To say more about the latter, let $V \subset N$ and $\varphi_V: V \to \mathbb{R}^{n=\dim(N)}$ be as described in the preceding paragraph. Then the sections $\{(dy_i^V) \circ \psi\}_{1\leq i\leq n}$ give a basis of sections for $(\psi^* T^*N)|_V$. Now suppose that $U \subset M$ and $\varphi_U: U \to \mathbb{R}^{m=\dim(M)}$ is also as described in the preceding paragraph. Then the aforementioned bundle homomorphism from $\psi^* T^*N$ to T^*M restricts over U so as to send any given $((dy_i^V) \circ \psi)|_x$ to what is written on the left-hand side of the equation in the preceding paragraph.

There is a dual of sorts to the pull-back map $\psi^*: C^\infty(N; T^*N) \to C^\infty(M; T^*M)$, this canonical *push-forward* map from TM to TN that restricts to any given $x \in M$ version of $TM|_x$ as a linear map to $TN|_{\psi(x)}$. This push-forward is denoted by ψ_*. To elaborate, suppose that $\mathfrak{v} \in TM$ and ω is a section of T^*N. The pairing between ω and $\psi_*\mathfrak{v}$ is, by definition, equal to that between the pull-back $\psi^*\omega$ and $\mathfrak{v}$. Note that this push-forward can be viewed as the composition of a bundle homomorphism from TM to ψ^*TN followed by the map $\hat{\psi}: \psi^*TN \to TN$. The former is the dual homomorphism to that described in the preceding paragraph from $\psi^* T^*N$ to T^*M.

The next section of this chapter describes some important examples of pull-back and push-forward.

5.4 Invariant forms and vector fields on Lie groups

Let G denote a Lie group. Then any given element $g \in G$ defines two maps from G to itself, these being the maps that sent $\mathfrak{m} \in G$ to $\mathfrak{l}_g(\mathfrak{m}) = g\mathfrak{m}$ and to $\mathfrak{r}_g(\mathfrak{m}) = \mathfrak{m}g$. The first is called *left translation by* g, and the second is called *right translation by* g. Both are diffeomorphism with inverses given respectively by left and right translation by g^{-1}.

A 1-form ω on G is said to be *left invariant* if $\mathfrak{l}_{g_*}\omega = \omega$ for all $g \in G$. The 1-form ω is said to be *right invariant* if $\mathfrak{r}_{g_*}\omega = \omega$ for all $g \in G$. By the same token, a vector field $\mathfrak{v}$ on G is *left invariant* or *right invariant* if $\mathfrak{l}_{g_*}\mathfrak{v} = \mathfrak{v}$ or $\mathfrak{r}_{g_*}\mathfrak{v} = \mathfrak{v}$ for all $g \in G$ as the case may be.

Suppose that ω is a left invariant 1-form. If ω is nonzero on $TG|_\iota$, then ω is nowhere zero. This follows by virtue of the fact that $\mathfrak{l}_g$ is a diffeomorphism. This implies that the restriction map from $C^\infty(G; T^*G)$ to $T^*G|_\iota$ identifies the vector space of left-invariant 1-forms with $T^*G|_\iota$. This said, then a basis for $T^*G|_\iota$ gives a basis for the left-invariant 1-forms on G, and this basis gives a basis of sections for T^*G over the whole of G. This then supplies a vector bundle isomorphism from the product bundle $G \times T^*G|_\iota$ to T^*G. By the same token, the vector space of right-invariant 1-forms on G is also isomorphic to $T^*G|_\iota$ and so a basis for $T^*G|_\iota$ extends to the whole of G as a basis of right-invariant 1-forms on G, which defines a basis of sections for T^*G. This then supplies a second isomorphism between $G \times T^*G|_\iota$ and T^*G. These two isomorphisms will not, in general, agree.

Of course, a similar thing can be said with regards to TG and the vector spaces of left- and right-invariant vector fields. A basis for $TG|_\iota$ extends as a basis of left-invariant vector fields, and also as a basis of right-invariant vector fields. Either basis gives a basis of sections of TG over the whole of G and thus an isomorphism from the product vector bundle $G \times TG|_\iota$ to TG.

Example 5.5 Introduce the n^2-dimensional Euclidean space $\mathbb{M}(n; \mathbb{R})$ of $n \times n$ real matrices. In what follows, I use notation whereby the entries $\{\mathfrak{m}_{ij}\}_{1 \le i,j \le n}$ of a given matrix define the Euclidean coordinates on $\mathbb{M}(n; \mathbb{R})$. This understood, the n^2 differentials $\{d\mathfrak{m}_{ij}\}_{1 \le i,j \le n}$ give a basis of sections of $T^*\mathbb{M}(n; \mathbb{R})$. Let i: $Gl(n\ \mathbb{R}) \to \mathbb{M}(n; \mathbb{R})$ denote the tautological embedding as the group of matrices with nonzero determinant. Fix a nonzero matrix $\mathfrak{q}$ and define the 1-form $\omega_\mathfrak{q}$ on $Gl(n; \mathbb{R})$ by the rule

$$\omega_\mathfrak{q}|_\mathfrak{m} = \mathrm{tr}(\mathfrak{q}\mathfrak{m}^{-1}d\mathfrak{m}).$$

To review the notation, what is written above is $\omega_\mathfrak{q} = \sum_{1 \le i,j,k \le n} (\mathfrak{q}_{ij}(\mathfrak{m}^{-1})_{jk}d\mathfrak{m}_{ki}.)$ This 1-form is left invariant: Indeed, let $g \in Gl(n; \mathbb{R})$. Then $\mathfrak{l}_{g_*}(\mathfrak{m}_{ij}) = \sum_{1 \le k \le n} g_{ik}\mathfrak{m}_{ki}$ and so $\mathfrak{l}_{g_*}(d\mathfrak{m}_{ij}) = \sum_{1 \le k \le n} g_{ik}d\mathfrak{m}_{kj}$. Thus,

$(\mathfrak{l}_{g*}\omega_q)|_m = \text{tr}(qm^{-1}g^{-1}gdm) = \omega_q|_m$. The form ω_q is not right invariant as can be seen from the formula $(\mathfrak{r}_{g*}\omega)|_m = \text{tr}(gqg^{-1}m^{-1}dm)$, this being $\omega_{g^{-1}qg}$. A similar calculation finds that the form $\text{tr}(m^{-1}qdm)$ is right invariant but not left invariant.

The form ω_q at the identity $\iota \in Gl(n; \mathbb{R})$ is nonzero since q is nonzero, and so the form ω_q is nowhere zero on $Gl(n; \mathbb{R})$.

To see what a left-invariant vector field looks like, fix again some matrix q. This done, introduce

$$v_q|_m = -\text{tr}\left(qm^T\frac{\partial}{\partial m}\right) = -\sum_{1\le i,j,k\le n} q_{ji}m_{kj}\frac{\partial}{\partial m_{ki}}.$$

Any such vector field is left invariant, and any left-invariant vector field can be written in this way.

5.5 The exponential map on a matrix group

Recall that the space $\mathbb{M}(n; \mathbb{R})$ of $n\times n$ real valued matrices is a Euclidean space with the coordinate functions given by the n^2 entries of a matrix. Fix an $n\times n$ matrix $m \in \mathbb{M}(n; \mathbb{R})$. Use this matrix to define the *exponential* map $\mathfrak{e}_m\colon \mathbb{M}(n; \mathbb{R}) \to \mathbb{M}(n; \mathbb{R})$ by the rule $\mathfrak{e}_m(a) = m\, e^a$, where

$$e^a = \iota + a + \frac{1}{2}a^2 + \frac{1}{3!}a^3 + \cdots$$

is the power series with the coefficient of a^k equal to 1/k!. This sequence converges absolutely and so defines a smooth map from $\mathbb{M}(n; \mathbb{R})$ to itself. Note that the matrix $\mathfrak{e}_\iota(a)$ is invertible, this by virtue of the fact that $\mathfrak{e}_\iota(a)^{-1} = \mathfrak{e}_\iota(-a)$. Thus $\mathfrak{e}_m(a) \in Gl(n; \mathbb{R})$ if m is. In particular, the map $\mathfrak{e}_\iota$ maps $\mathbb{M}(n; \mathbb{R})$ into $Gl(n; \mathbb{R})$. As the differential of the map $a \to e^a$ at $a = 0$ is the identity map on $\mathbb{M}(n; \mathbb{R})$, it follows as a consequence of the inverse function theorem that this restricts to some ball in $\mathbb{M}(n; \mathbb{R})$ as a diffeomorphism onto some neighborhood of ι in $Gl(n; \mathbb{R})$.

With the preceding understood, fix $q \in \mathbb{M}(n; \mathbb{R})$ and reintroduce the left-invariant 1-form $\omega_q = \text{tr}(q\, m^{-1}dm)$ on $Gl(n; \mathbb{R})$. The pull-back of this form by e_ι is the 1-form

$$\mathfrak{e}_\iota^*\,\omega_q = \int_0^1 ds\ \text{tr}(e^{-sa}qe^{sa}da).$$

To elaborate, view $Gl(n; \mathbb{R})$ as an open subset of $\mathbb{M}(n; \mathbb{R})$ so as to identify $TGl(n; \mathbb{R})|_\iota$ with $\mathbb{M}(n; \mathbb{R})$. Granted this identification, then $\mathfrak{e}_\iota^*\,\omega_q$ is neither more nor less than a linear functional on $\mathbb{M}(n; \mathbb{R})$. This linear functional maps any given matrix $\mathfrak{c}$ to

$$\int_0^1 ds\, tr(e^{-s\mathfrak{a}}\mathfrak{q}e^{s\mathfrak{a}}\mathfrak{c}).$$

What follows considers the exponential map in the context of the groups $Sl(n; \mathbb{R})$ and $SO(n)$.

5.5.1 The group Sl(n; ℝ)

Recall that group $Sl(n; \mathbb{R})$ sits in $Gl(n; \mathbb{R})$ as the subgroup of matrices with determinant 1. As it turns out, $\mathfrak{e}_\iota$ maps the vector space of trace zero matrices in $\mathbb{M}(n; \mathbb{R})$ into $Sl(n; \mathbb{R})$. Indeed, this follows from the fact that

$$\frac{d}{ds}\ln(\det(\mathfrak{e}_\iota(s\mathfrak{a}))) = tr(\mathfrak{a}).$$

As noted in Chapter 3.5, the tangent space to $Sl(n; \mathbb{R})$ at the identity is the vector space of trace zero, $n \times n$ matrices. Given that the differential of $\mathfrak{e}_\iota$ at 0 is the identity map on $\mathbb{M}(n; \mathbb{R})$, it follows directly from the inverse function theorem that $\mathfrak{e}_\iota$ maps a ball about the origin in the vector space of trace zero, $n \times n$ matrices diffeomorphically onto an open neighborhood of the identity in $Sl(n; \mathbb{R})$. Likewise, $\mathfrak{e}_\mathfrak{m}$ for $\mathfrak{m} \in Sl(n; \mathbb{R})$ maps this same ball diffeomorphically onto an open neighborhood of $\mathfrak{m}$ in $Sl(n; \mathbb{R})$.

Let $i\colon Sl(n; \mathbb{R}) \to Gl(n; \mathbb{R})$ denote the tautological inclusion map, and let $\mathfrak{q}$ denote a given $n \times n$ matrix. Then $i^*\omega_\mathfrak{q}$ is nonzero unless $\mathfrak{q}$ is a multiple of the identity, in which case $i^*\omega_\mathfrak{q} = 0$. The fact that the pull-back of the $\mathfrak{q} = \iota$ version of $\omega_\mathfrak{q}$ is zero on $TSl(n; \mathbb{R})|_\iota$ can be seen using the formula given above for its pull-back via $\mathfrak{e}_\iota$. As $i^*\omega_\iota$ is a left-invariant 1-form on $Sl(n; \mathbb{R})$, its vanishing on $TSl(n; \mathbb{R})|_\iota$ implies that it must vanish everywhere.

5.5.2 The group SO(n)

Let $\mathbb{A}(n; \mathbb{R})$ again denote the vector space of $n \times n$ antisymmetric matrices. This is to say that $\mathfrak{a} \in \mathbb{A}(n; \mathbb{R})$ when $\mathfrak{a}$'s transpose obeys $\mathfrak{a}^T = -\mathfrak{a}$. The map $\mathfrak{e}_\iota$ as depicted above obeys $\mathfrak{e}_\iota(\mathfrak{a})^T = \mathfrak{e}_\iota(\mathfrak{a}^T)$ and so $\mathfrak{e}_\iota{}^T\mathfrak{e}_\iota = 1$ when $\mathfrak{a} \in \mathbb{A}(n; \mathbb{R})$. It follows from this that $\mathfrak{e}_\iota$ restricts to $\mathbb{A}(n; \mathbb{R})$ so as to map the letter subspace into $SO(n)$. To say more, recall that Chapter 3.5 identifies $TSO(n)|_\iota$ with $\mathbb{A}(n; \mathbb{R})$. Granted this identification, and granted that the differential at $\mathfrak{a} = 0$ of $\mathfrak{e}_\iota$ is the identity map, it follows from the implicit function theorem that $\mathfrak{e}_\iota$ maps a ball in $\mathbb{A}(n; \mathbb{R})$ diffeomorphically onto a neighborhood of $\iota \in SO(n)$. By the same token if $\mathfrak{m}$ is any given matrix in SO(n), then the map $\mathfrak{e}_{\mathfrak{m}} = \mathfrak{m}\mathfrak{e}_{\iota}$ restricts to this same ball as a diffeomorphism onto a neighborhood of $\mathfrak{m}$ in SO(n).

Let $\mathfrak{q}$ again be a matrix in $\mathbb{M}(n; \mathbb{R})$. Let i: SO(n) $\to$ Gl(n; $\mathbb{R}$) denote the tautological inclusion map. As explained next, the form $i^*\omega_{\mathfrak{q}}$ is zero if $\mathfrak{q}^T = \mathfrak{q}$ and never zero if $\mathfrak{q}^T = -\mathfrak{q}$ unless $\mathfrak{q} = 0$. To see that $i^*\omega_{\mathfrak{q}} = 0$ when $\mathfrak{q}$ is symmetric, consider its pull-back via the exponential map $\mathfrak{e}_{\iota}$ as depicted above. For any given $\mathfrak{a} \in \mathbb{A}(n; \mathbb{R})$, this defines a linear form on $\mathbb{A}(n; \mathbb{R}) = TSO(n)|_{\iota}$ whose pairing with $\mathfrak{c} \in \mathbb{A}(n; \mathbb{R})$ is $\int_0^1 ds\, \mathrm{tr}(e^{-s\mathfrak{a}}\mathfrak{q}e^{s\mathfrak{a}}\mathfrak{c})$. The term in the integrand is, for any given value of $s \in [0, 1]$, the trace of the product of a symmetric matrix with an antisymmetric matrix. Such a trace is always zero. To see that $i^*\omega_{\mathfrak{q}} \neq 0$ when $\mathfrak{q} \in \mathbb{A}(n; \mathbb{R})$, it is enough to prove it nonzero on $TSO(n)|_{\iota}$ since this form is invariant under left multiplication. The formula above shows that $i^*\omega_{\mathfrak{q}}$ at the identity pairs with $\mathfrak{c} \in \mathbb{A}(n; \mathbb{R})$ to give $\mathrm{tr}(\mathfrak{q}\mathfrak{c})$. Taking $\mathfrak{c} = -\mathfrak{q} = \mathfrak{q}^T$ shows that $i^*\omega_{\mathfrak{q}}|_{\iota}$ is nonzero on $\mathbb{A}(n; \mathbb{R})$ unless $\mathfrak{q}$ is zero. Given that $i^*\omega_{\mathfrak{q}}$ is left invariant, this implies that any $\mathfrak{q} \neq 0$ version is nowhere zero on SO(n).

5.6 The exponential map and right/left invariance on Gl(n; ℂ) and its subgroups

The group Gl(n; ℂ) is described in Chapter 2.4. Recall that there are two descriptions. The first views Gl(n; ℂ) as the open subset in the Euclidean space $\mathbb{M}(n; \mathbb{C})$ of $n \times n$ matrices with complex entries that consists of the matrices with nonzero determinant. As a manifold, $\mathbb{M}(n; \mathbb{C})$ has (real) dimension $2n^2$, and its Euclidean coordinates are taken to be the real and imaginary parts of the entries of a given matrix $\mathfrak{m}$. This understood, view $d\mathfrak{m}$ as a matrix of ℂ-valued differential forms on $\mathbb{M}(n; \mathbb{C})$ whose real and imaginary parts supply a basis of sections for $T^*\mathbb{M}(n; \mathbb{C})$.

With the preceding understood, let $\mathfrak{q} \in \mathbb{M}(n; \mathbb{C})$. Then the ℂ-valued differential form $\omega_{\mathfrak{q}} = \mathrm{tr}(\mathfrak{q}\, \mathfrak{m}^{-1}d\mathfrak{m})$ on Gl(n; ℂ) is left invariant, and any ℂ-valued, left-invariant 1-form on the group Gl(n; ℂ) can be written in this way, and any ℝ-valued left invariant 1-form is the real part of some $\omega_{\mathfrak{q}}$. By the same token, the ℂ-valued 1-form $\mathrm{tr}(\mathfrak{m}^{-1}\mathfrak{q}\; d\mathfrak{m})$ is right-invariant, and any ℂ-valued, right-invariant 1-form on the group Gl(n; ℂ) is of this sort. Moreover, any ℝ-valued right invariant 1-form is given by the real part of some $\omega_{\mathfrak{q}}$.

The ℂ-valued, right- and left-invariant vector fields on Gl(n; ℂ) are given respectively by the rules $\mathrm{tr}(\mathfrak{q}\; \mathfrak{m}^T \frac{\partial}{\partial \mathfrak{m}})$ and $\mathrm{tr}(\mathfrak{m}^T\; \mathfrak{q}\, \frac{\partial}{\partial \mathfrak{m}})$ where the notation is as follows: What is written as $\frac{\partial}{\partial \mathfrak{m}}$ denotes here the $n \times n$ matrix whose (k, j) entry is by definition $\frac{\partial}{\partial \mathfrak{m}_{kj}} = \frac{1}{2}\left(\frac{\partial}{\partial x_{kj}} - i\frac{\partial}{\partial y_{kj}}\right)$ with x_{kj} and y_{kj} denoting the respective real and imaginary parts of the complex coordinate function $\mathfrak{m}_{kj}$.

Fix $\mathfrak{m} \in \mathbb{M}(n; \mathbb{C})$ and define the exponential map $\mathfrak{e}_{\mathfrak{m}}$: $\mathbb{M}(n; \mathbb{C}) \to \mathbb{M}(n; \mathbb{C})$ using the same formula as used to define the $\mathbb{M}(n; \mathbb{R})$ version. As in the case of $\mathbb{M}(n; \mathbb{R})$, the map $\mathfrak{e}_{\mathfrak{m}}$ sends $\mathbb{M}(n; \mathbb{C})$ into $Gl(n; \mathbb{C})$ when $\mathfrak{m} \in Gl(n; \mathbb{C})$, and it restricts to some $\mathfrak{m}$-independent ball about the origin in $\mathbb{M}(n; \mathbb{C})$ as a diffeomorphism onto an open neighborhood of $\mathfrak{m}$.

It is also the case that $\mathfrak{e}_{\mathfrak{m}}$ sends the complex, codimension 1 vector space of traceless matrices in $\mathbb{M}(n; \mathbb{C})$ into $Sl(n; \mathbb{C})$ when $\mathfrak{m} \in Sl(n; \mathbb{C})$; and it restricts to a ball about the origin in this subspace as a diffeomorphism onto an open neighborhood of $\mathfrak{m}$ in $Sl(n; \mathbb{C})$. Likewise, $\mathfrak{e}_{\mathfrak{m}}$ maps the vector subspace $\mathbb{A}(n; \mathbb{C}) \subset \mathbb{M}(n; \mathbb{C})$ of anti-Hermitian matrices into $U(n)$ when $\mathfrak{m} \in U(n)$ and it restricts to some $\mathfrak{m}$-independent ball about the origin in $\mathbb{A}(n; \mathbb{C})$ as a diffeomorphism onto a neighborhood of $\mathfrak{m}$ in $U(n)$. Finally, this same $\mathfrak{e}_{\mathfrak{m}}$ maps the vector subspace $\mathbb{A}_0(n; \mathbb{C}) \subset A(n; \mathbb{C})$ of trace zero matrices into $SU(n)$ when $\mathfrak{m} \in SU(n)$; and it restricts to some $\mathfrak{m}$-independent ball about the origin in $\mathbb{A}_0(n; \mathbb{C})$ as a diffeomorphism onto an open neighborhood of $\mathfrak{m}$ in $SU(n)$. The proofs of all of these assertions are identical to the proofs of their analogs given in the preceding section of this chapter.

Let i denote the inclusion map from any of $Sl(n; \mathbb{C})$, $U(n)$ or $SU(n)$ into $Gl(n; \mathbb{C})$. In the case of $Sl(n; \mathbb{C})$, the pull-back $i^*\omega_{\mathfrak{q}}$ is zero if $\mathfrak{q} = c\iota$ with $c \in \mathbb{C}$ and nowhere zero otherwise. The form $i^*\omega_{\mathfrak{q}}$ on $Sl(n; \mathbb{C})$ is a left-invariant, $\mathbb{C}$-valued 1-form, and any such 1-form can be written as $i^*\omega_{\mathfrak{q}}$ with $tr(\mathfrak{q}) = 0$. Any $\mathbb{R}$-valued, left-invariant 1-form is the real part of some $i^*\omega_{\mathfrak{q}}$. In the case of $U(n)$, the 1-form $i^*\omega_{\mathfrak{q}}$ is zero if $\mathfrak{q}^\dagger = \mathfrak{q}$, and it is nowhere zero if $\mathfrak{q}^\dagger = -\mathfrak{q}$. Any such form is left-invariant, and any $\mathbb{C}$-valued, left-invariant 1-form on $U(n)$ is of this sort. Likewise, an $\mathbb{R}$-valued, left-invariant 1-form can be written as the real part of some $\omega_{\mathfrak{q}}$. In the case of $SU(n)$, the pull-back $i^*\omega_{\mathfrak{q}}$ is also zero if $\mathfrak{q}$ is Hermitian, or if $\mathfrak{q} = ir\iota$ with $r \in \mathbb{R}$; and it is nowhere zero if $\mathfrak{q}^\dagger = i\mathfrak{q}$ and $tr(\mathfrak{q}) = 0$. Any real valued, left-invariant 1-form is the real part of some $i^*\omega_{\mathfrak{q}}$ of this sort. The proofs of all the preceding statements are essentially identical to the proofs of the analogous statements in the preceding section of this chapter.

As noted at the outset, Chapter 2 provides two views of $Gl(n; \mathbb{C})$. The second view regards $\mathbb{M}(n; \mathbb{C}) \subset \mathbb{M}(n; 2\mathbb{R})$ as the subvector space over $\mathbb{R}$ of matrices that commute with a certain matrix, j_0, whose square is $-\iota$. The latter subvector space is denoted by $\mathbb{M}_j$. $Gl_j \subset Gl(2n; \mathbb{R})$ denote the latter's intersection with $\mathbb{M}_j$. Let $\mathfrak{a} \to \mathfrak{e}_\iota(\mathfrak{a})$ denote the exponential map defined on $\mathbb{M}(n; 2\mathbb{R})$. As can be seen from the formula, this map has the property that $j_0\, \mathfrak{e}_\iota(\mathfrak{a})\, j_0 = \mathfrak{e}_\iota(j_0\mathfrak{a}j_0)$. As a consequence, the $\mathbb{M}(n; 2\mathbb{R})$ version of $\mathfrak{e}_{\mathfrak{m}}$ restricts to $\mathbb{M}_j$ so as to map the latter vector space to the subgroup Gl_j when $\mathfrak{m} \in \mathbb{M}_j$. The identification of $\mathbb{M}_j$ with $\mathbb{M}(n; \mathbb{C})$ that is obtained by writing $\mathfrak{m} \in \mathbb{M}_j$ as a linear combination of the eigenvalue $+i$ eigenvectors of j_0 and their complex conjugates identifies the restriction to $\mathbb{M}_j$ of this $\mathbb{M}(2n; \mathbb{R})$ version $\mathfrak{e}_{\mathfrak{m}}$ with the $\mathbb{M}(n; \mathbb{C})$ version of $\mathfrak{e}_{\mathfrak{m}}$. This understood, all of the assertions given so far in this section that concern the exponential map on $Gl(n; \mathbb{C})$ and its subgroups can be reproved without recourse to complex numbers by considering the $\mathbb{M}(2n; \mathbb{R})$ version of $\mathfrak{e}_m$ on $\mathbb{M}_j$. The identification between $\mathbb{M}_j$ and $\mathbb{M}(n; \mathbb{C})$ and between Gl_j and $Gl(n; \mathbb{C})$ can also be used to prove the assertions about left- and right-invariant 1-forms and vector fields without recourse to complex numbers.

5.7 Immersion, submersion and transversality

This section of the chapter uses the notion of vector push-forward to reinterpret and then elaborate on what is said in Chapter 1.4. To set the stage, suppose that M and Y are smooth manifolds, and suppose that ψ: Y → M is a smooth map. Chapter 1.4 introduced the descriptives *immersion* and *submersion* as applied to ψ. These notions can be rephrased in terms of the push-forward ψ_* as follows:

Definition 5.6 *The map* ψ *is respectively an immersion or submersion when the vector bundle homomorphism* ψ_*: TY → ψ^*TM *is injective on each fiber or surjective on each fiber.*

A useful notion that is intermediate to the immersion and submersion cases is described next. To set this stage, suppose that Y, M and ψ are as described above. Suppose in addition that $Z \subset Y$ is a submanifold. The map ψ is said to be *transversal* to Z when the following condition is met: Let $y \in Y$ denote any given point with $\psi(y) \in Z$. Then any vector $v \in TM|_{\psi(y)}$ can be written as $\psi_* v_Y + v_Z$ with $v_Y \in TY|_y$ and $v_Z \in TZ|_{\psi(y)}$.

The proposition that follows applies the notion of transversality to construct a submanifold of Y. This proposition uses n, m and d for the respective dimensions of Y, M and Z.

Proposition 5.7 *Let* M, Y *and* $Z \subset M$ *be as just described, and suppose that* ψ: Y → M *is transversal to* Z. *Then* $\psi^{-1}(Z)$ *is a smooth submanifold of* Y *of dimension* $n + d - m$.

The proof of this proposition is given momentarily. Note in the meantime that when Z is a point, then ψ is transversal to Z if and only if ψ_* is surjective at each point in $\psi^{-1}(z)$.

***Proof of Proposition** 5.7* Let y denote a given point in $\psi^{-1}(Z)$. Fix an open neighborhood, $U \subset Y$, of y with a coordinate chart map $\varphi: U \to \mathbb{R}^n$ that sends y to the origin. Fix, in addition, an open neighborhood, $V \subset M$, of $\psi(y)$ with a coordinate chart map $\varphi': V \to \mathbb{R}^m$ that sends $\psi(y)$ to the origin and sends $Z \cap V$ to the d-dimensional subspace of points $(u_1, \ldots, u_m) \in \mathbb{R}^m$ with $u_{d+1} = \cdots = u_m = 0$.

Chapter 1.3 says something about such a coordinate chart map. Let $f = (f_1, \ldots, f_m)$ denote the map from $\mathbb{R}^n$ to $\mathbb{R}^m$ given by $\varphi' \circ \psi \circ \varphi^{-1}$.

Let $\mathfrak{f}$ denote the map from a neighborhood of the origin in $\mathbb{R}^n$ to $\mathbb{R}^{m-d}$ given by

$$x \to \mathfrak{f}(x) = (f_{d+1}(y), \ldots, f_m(y)).$$

The transversality assumption guarantees that $\mathfrak{f}$ is a submersion near the origin in $\mathbb{R}^n$. As a consequence, $\mathfrak{f}^{-1}(0)$ is a submanifold near the origin in $\mathbb{R}^n$. This submanifold is, by construction, $\varphi(\psi^{-1}(Z))$.

Additional reading

- *Differential Forms in Algebraic Topology*, Raoul Bott and Loring W. Tu, Springer, 1982.
- *Manifolds and Differential Geometry*, Jeffrey M. Lee, American Mathematical Society, 2009.
- *Geometry of Differential Forms*, Shigeyuki Morita, American Mathematical Society, 2001.
- *An Introduction to Manifolds*, Loring W. Tu, Springer, 2008.
- *Introduction to Smooth Manifolds*, John M. Lee, Springer, 2003.
- *Differential Topology*, Victor Guillemin and Alan Pollack, Prentice-Hall, 1974.
- *A Geometric Approach to Differential Forms*, David Bachman, Birkhauser, 2006.

6 Vector bundles with $\mathbb{C}^n$ as fiber

Just as there are vector spaces over $\mathbb{C}$, there are vector bundles whose fibers can be consistently viewed as $\mathbb{C}^n$ for some n. This chapter first defines these objects, and then supplies a number of examples.

6.1 Definitions

What follows directly are three equivalent definitions of a *complex vector bundle*. The first definition highlights the fact that a complex vector bundle is a vector bundle of the sort introduced in Chapter 3.1 with some extra structure.

Definition 6.1 A complex vector bundle of rank n is a vector bundle, $\pi\colon E \to M$, in the sense defined in Chapter 3.1 with fiber dimension 2n, but equipped with a bundle endomorphism $j\colon E \to E$ such that $j^2 = -1$. This j allows one to define an action of $\mathbb{C}$ on E such that the real numbers in $\mathbb{C}$ act as $\mathbb{R}$ on the underlying real bundle of fiber dimension 2n; and such that the number i acts as j. Doing this identifies each fiber of E with $\mathbb{C}^n$ up to the action of $Gl(n; \mathbb{C})$. The endomorphism j is said to be an *almost complex structure* for E. The underlying real bundle with fiber $\mathbb{R}^{2n}$ is denoted by $E_{\mathbb{R}}$ when a distinction between the two is germaine at any given time.

The preceding definition highlights the underlying real bundle structure. There are two equivalent definitions that put the complex numbers to the fore.

Definition 6.2 This definition gives the $\mathbb{C}$ analog of what is given in Chapter 3.1: A complex vector bundle E over M of fiber dimension n is a smooth manifold with the following additional structure:

- *A smooth map* $\pi\colon E \to M$.
- *A smooth map* $\hat{o}\colon M \to E$ *such that* $\pi \circ \hat{o}$ *is the identity.*
- *A smooth map* $\mu\colon \mathbb{C} \times E \to E$ *such that*

a) $\pi(\mu(c, v)) = \pi(v)$
b) $\mu(c, \mu(c', v)) = \mu(cc', v)$
c) $\mu(1, v) = v$
d) $\mu(c, v) = v$ *for* $c \neq 1$ *if and only if* $v \in \text{im}(\hat{o})$.

- *Let* $p \in M$. *There is a neighborhood*, $U \subset M$, *of* p *and a map* $\lambda_U: \pi^{-1}(U) \to \mathbb{C}^n$ *such that* $\lambda_U: \pi^{-1}(x) \to \mathbb{C}^n$ *for each* $x \in U$ *is a diffeomorphism obeying* $\lambda_U(\mu(c, v)) = c\, \lambda_U(v)$.

Definition 6.3 This definition gives the $\mathbb{C}$ analog of the cocycle definition from Chapter 3.2: A complex vector bundle of rank $n \geq 1$ is given by a locally finite open cover $\mathfrak{U}$ of M with the following additional data: An assignment to each pair $U, V \in \mathfrak{U}$ a map $g_{U,V}: U \cap V \to Gl(n; \mathbb{C})$ such that $g_{U,V} = g_{V,U}^{-1}$ with the constraint that if $U, V, W \in \mathfrak{U}$, then $g_{U,V}\, g_{V,W}\, g_{W,U} = \iota$ on $U \cap V \cap W$. Given this data, define the bundle E to be the quotient of the disjoint union $\cup_{U \in \mathfrak{U}} (U \times \mathbb{C}^n)$ by the equivalence relation that puts $(p', v') \in U' \times \mathbb{C}^n$ equivalent to $(p, v) \in U \times \mathbb{C}^n$ if and only if $p = p'$ and $v' = g_{U',U}(p)v$. The constraint involving the overlap of three sets guarantees that this does, indeed, specify an equivalence relation.

6.2 Comparing definitions

The argument to equate the Definitions 6.2 and 6.3 is almost verbatim what is given in Chapter 3.2 that equates the latter's cocycle definition of a vector bundle with the definition given in Chapter 3.1. Indeed, one need only change $\mathbb{R}$ to $\mathbb{C}$ in Chapter 3.2's argument.

To compare Definitions 6.1 and 6.2, suppose first that E is a complex vector bundle in the sense of Definition 6.2. The needed data from Chapter 3.1 for the underlying real bundle $E_{\mathbb{R}}$ consist of a projection map $\pi: E_{\mathbb{R}} \to M$, a zero section $\hat{o}: M \to E_{\mathbb{R}}$, a multiplication map $\mu: \mathbb{R} \times E \to E$. In addition, each $p \in M$ should be contained in an open set U which comes with a certain map $\lambda_U: E_{\mathbb{R}}|_U \to \mathbb{R}^{2n}$.

The $E_{\mathbb{R}}$ is, first of all, declared equal to E. Chapter 3.1's projection π is the projection given in Definition 6.2. Likewise, Chapter 3.1's zero section $\hat{o}$ is that given in Definition 6.2. Meanwhile, Chapter 3.1's map μ from $\mathbb{R} \times E_{\mathbb{R}}$ to $E_{\mathbb{R}}$ is obtained from Definition 6.2's map μ by restricting the latter to the real numbers $\mathbb{R} \subset \mathbb{C}$. Granted the preceding, any given version of Chapter 3.1's map λ_U can be taken equal to the corresponding version of Definition 6.2's map λ_U after identifying $\mathbb{C}^n$ as $\mathbb{R}^{2n}$ by writing the complex coordinates $(z_1, \ldots, z_n)$ of $\mathbb{C}^n$ in terms of their real and imaginary parts to obtain the Euclidean coordinates for $\times_n \mathbb{R}^2 = \mathbb{R}^{2n}$. With $E_{\mathbb{R}}$ so defined, Definition 6.1 requires only the endomorphism j. The latter is defined by the rule $j(v) = \mu(i, v)$ with μ here given by Definition 6.2.

Now suppose that E is a complex bundle according to the criteria from Definition 6.1. The data needed for Definition 6.2 are obtained as follows: First, Definition 6.2's projection π and Definition 6.2's zero section $\hat{o}$ are those from the underlying real bundle $E_{\mathbb{R}}$. Definition 6.2's map μ is defined as follows: Write a given complex number z in terms of its real and imaginary parts as $z = a + ib$. Let $v \in E$. Then $\mu(a + ib, v)$ is defined to be $av + bjv$. It remains only to specify, for each $p \in M$, an open set, U, containing p with a suitable map $\lambda_U\colon E|_U \to \mathbb{C}^n$. This last task requires a preliminary lemma.

To set the stage for the required lemma, reintroduce the standard almost complex structure j_0 from Chapter 2.4.

Lemma 6.4 *Let* $\pi\colon E_{\mathbb{R}} \to M$ *denote a real vector bundle with fiber dimension* 2n *with an almost complex structure* $j\colon E \to E$. *There is a locally finite cover* $\mathfrak{U}$ *for* M *of the following sort: Each set* $U \in \mathfrak{U}$ *comes with an isomorphism* $\varphi_U\colon E_{\mathbb{R}}|_U \to U \times \mathbb{R}^{2n}$ *that obeys* $\varphi_U \circ j \circ \varphi_U^{-1} = j_0$ *at each point of* U.

This lemma is proved momentarily. It has the following consequence: Let $\mathfrak{U}$ denote a cover of the sort given by the lemma. Let $U \subset \mathfrak{U}$, and set $\lambda_{U;\mathbb{R}}\colon E|_U \to \mathbb{R}^{2n}$ equal to the composition of φ_U followed by the projection to the $\mathbb{R}^{2n}$ factor. Define the complex coordinates $(z_1, \ldots, z_n)$ for $\mathbb{C}^n$ from the real coordinates $(x_1, \ldots, x_{2n})$ of $\mathbb{R}^{2n}$ by writing $z_k = x_{2k-1} + ix_{2k}$. Doing so allows $\lambda_{U;\mathbb{R}}$ to be viewed as a map, λ_U, from $E|_U$ to $\mathbb{C}^n$. Because φ_U intertwines j with j_0, the latter map is equivariant with respect to the action of $\mathbb{C}$ on E and $\mathbb{C}$ on $\mathbb{C}^n$. And, it restricts to any given fiber as an isomorphism between complex vector spaces.

Proof of Lemma 6.4 Fix $p \in M$ and an open set $V \subset M$ that contains p with an isomorphism $\psi_V\colon E_{\mathbb{R}}|_V \to V \times \mathbb{R}^{2n}$. Use j_V to denote $\psi_V \circ j \circ \psi_V^{-1}$. At each point of V, this endomorphism of $\mathbb{R}^{2n}$ defines an almost complex structure on $\mathbb{R}^{2n}$. The plan in what follows is to find a smooth map $\mathfrak{h}$, from a neighborhood of p to $Gl(2n; \mathbb{R})$ such that $\mathfrak{h} \circ j_V \circ \mathfrak{h}^{-1} = j_0$. Given such a map, set U to be this same neighborhood and define φ_U at any given point in U to be $\mathfrak{h} \circ \psi_V$.

To obtain $\mathfrak{h}$, digress momentarily and suppose that j_1 is any given almost complex structure. Then there exists $\mathfrak{g} \in Gl(2n; \mathbb{R})$ such that $\mathfrak{g} \circ j_1 \circ \mathfrak{g}^{-1} = j_0$. Indeed, this follows by virtue of the fact that j_1 and j_0 are diagonalizable and have eigenvalues only $\pm i$, each with multiplicity n. To elaborate, fix a basis of eigenvectors for j_1 with eigenvalue $+i$ and likewise a basis for j_0. Define $\mathfrak{g}$ by the rule that sends the respective real and imaginary parts of the k'th basis vector for j_1 to those of the corresponding basis vector for j_0. Granted the preceding, it follows that ψ_V can be changed by $\psi_V \to \mathfrak{g} \circ \psi_V$ with $\mathfrak{g} \in Gl(2n; \mathbb{R})$ so that the resulting version of j_V is equal to j_0 at p.

With the preceding understood, reintroduce the exponential map $\mathfrak{a} \to e^{\mathfrak{a}}$ as defined in Chapter 5.5, and recall that the latter restricts to some ball about the origin in $\mathbb{M}(2n; \mathbb{R})$ as a diffeomorphism onto a neighborhood of ι in $Gl(2n; \mathbb{R})$.

Granted the latter fact, it then follows that j_V near p can be written as $j_0 e^{\mathfrak{a}}$ where $\mathfrak{a}$ is a smooth map from a neighborhood of p in V to $\mathbb{M}(2n; \mathbb{R})$. Moreover, $\mathfrak{a}$ must be such that

$$e^{-j_0 \mathfrak{a} j_0} e^{\mathfrak{a}} = \iota,$$

for this condition is necessary and sufficient for j_V^2 to equal $-\iota$. Because $(e^{\mathfrak{a}})^{-1} = e^{-\mathfrak{a}}$, this last condition requires that $-j_0\, \mathfrak{a}\, j_0 = -\mathfrak{a}$ on some neighborhood of p. With this last point in mind, use the power series definition of $e^{\mathfrak{a}}$ to write the latter $e^{\mathfrak{a}} = e^{\mathfrak{a}/2} e^{\mathfrak{a}/2}$, and then write

$$j_V = j_0 e^{\mathfrak{a}/2} e^{\mathfrak{a}/2} = e^{-\mathfrak{a}/2} j_0 e^{\mathfrak{a}/2}.$$

Then take $\mathfrak{h} = e^{\mathfrak{a}/2}$.

6.3 Examples: The complexification

The simplest example of a complex vector bundle is the product bundle $M \times \mathbb{C}^n$. The underlying rank 2n real bundle, $E_{\mathbb{R}}$, is $M \times \mathbb{R}^{2n}$; the almost complex structure is j_0. The examples given next are not much more complicated. These examples start with some given vector bundle $\pi: E \to M$ with fiber $\mathbb{R}^n$. The resulting complex bundle has fiber $\mathbb{C}^n$; it is denoted $E_{\mathbb{C}}$ and is said to be the *complexification* of E. The bundle $E_{\mathbb{C}}$ is defined so that its fiber of $E_{\mathbb{C}}$ at any given $p \in M$ is canonically isomorphic to $\mathbb{C} \otimes_{\mathbb{R}} E|_p$ as a complex vector space. What follows describes $E_{\mathbb{C}}$ in the context of Definition 6.1.

To start, introduce the tensor product bundle $(M \times \mathbb{R}^2) \otimes E$, this a bundle with fiber $\mathbb{R}^{2n}$. Define j on this bundle as follows: Any element in the latter can be written as a finite sum of reducible elements, $\sum_{1 \le k \le n} z_k \otimes \mathfrak{e}_k$, where $z_k \in \mathbb{R}^2$ and $\mathfrak{e}_k \in E$. This understood, define an almost complex structure on this tensor product bundle so as to send $\sum_{1 \le k \le n} z_k \otimes \mathfrak{e}_k$ to $\sum_{1 \le k \le n} (j_0 z_k) \otimes \mathfrak{e}_k$, where

$$j_0 = \begin{pmatrix} 0 & -1 \\ 1 & 0 \end{pmatrix}.$$

The resulting complex bundle is $E_{\mathbb{C}}$. As advertised, $E_{\mathbb{C}}$ is a rank n complex bundle. The $\mathbb{C}$-action on $E_{\mathbb{C}}$ can be seen by writing $\mathbb{R}^2$ as $\mathbb{C}$ so as to view $E_{\mathbb{C}}$ as $(M \times \mathbb{C}) \otimes_{\mathbb{R}} E$. Write a typical element as $\mathfrak{v} = \sum_k z_k \otimes \mathfrak{e}_k$ but now view $z_k \in \mathbb{C}$. Then element $\sigma \in \mathbb{C}$ acts to send this element to $\sigma \mathfrak{v} = \sum_k \sigma z_k \otimes \mathfrak{e}_k$.

Suppose that E is a complex bundle already. This is to say that it can be viewed as a real bundle with an almost complex structure j. Then E sits inside its complexification, $E_{\mathbb{C}}$, as a complex, rank n subbundle. In fact, $E_{\mathbb{C}}$ has a direct sum decomposition, linear over $\mathbb{C}$, as $E \oplus \bar{E}$, where $\bar{E}$ is defined from the same real bundle as was E, but with j replaced by $-$j. The bundle $\bar{E}$ is called the *complex conjugate* bundle to E.

To see E inside $E_{\mathbb{C}}$, observe that the endomorphism j acts as an endomorphism of $E_{\mathbb{C}}$ sending any given $\mathfrak{v} = \sum_k z_k \otimes \mathfrak{e}_k$ to $j\mathfrak{v} = \sum_k z_k \otimes j\mathfrak{e}_k$. As an endomorphism now of $E_{\mathbb{C}}$, this j obeys $j^2\mathfrak{v} = -\mathfrak{v}$, and it also commutes with multiplication by elements in $\mathbb{C}$. This understood, the bundle E sits inside the bundle $E_{\mathbb{C}}$ as the set of vectors $\mathfrak{v}$ such that $j\mathfrak{v} = i\mathfrak{v}$. Meanwhile, the conjugate bundle $\bar{E}$ sits in $E_{\mathbb{C}}$ as the set of vectors $\mathfrak{v}$ such that $j\mathfrak{v} = -i\mathfrak{v}$. The inclusion of E into $E_{\mathbb{C}}$ as a direct summand sends any given vector $e \in E$ to the vector $\frac{1}{2}(1\otimes \mathfrak{e} - i \otimes \mathfrak{e})$.

In general, $\bar{E}$ and E are not isomorphic as complex bundles even though their underlying real bundles (i.e., $E_{\mathbb{R}}$ and $\bar{E}_{\mathbb{R}}$) are identical.

6.4 Complex bundles over surfaces in $\mathbb{R}^3$

Introduce the Pauli matrices

$$\tau_1 = \begin{pmatrix} 0 & i \\ i & 0 \end{pmatrix}, \ \tau_2 = \begin{pmatrix} 0 & -1 \\ 1 & 0 \end{pmatrix}, \ \tau_3 = \begin{pmatrix} i & 0 \\ 0 & -i \end{pmatrix}.$$

These generate the quaternion algebra as

$$\tau_1^2 = \tau_2^2 = \tau_3^3 = -\iota\,, \ \tau_1\tau_2 = -\tau_2\tau_1 = -\tau_3, \ \ \tau_2\tau_3 = -\tau_3\tau_2 = -\tau_1, \\ \tau_3\tau_1 = -\tau_1\tau_3 = -\tau_2.$$

Here, ι denotes the 2×2 identity matrix. Let $\Sigma \subset \mathbb{R}^3$ denote a given embedded or immersed surface. Assign to each point $p \in \Sigma$ its normal vector $n(p) = (n_1, n_2, n_3)$. This assignment is used to define the map $\mathfrak{n}\colon \Sigma \to \mathbb{M}(2; \mathbb{C})$ given by $\mathfrak{n} = n_1\tau_1 + n_2\tau_2 + n_3\tau_3$. Note that $\mathfrak{n}^2 = -\iota$. Define a complex vector subbundle $E_\Sigma \subset \Sigma \times \mathbb{C}^2$ to be the set of points (p, v) such that $\mathfrak{n}(p)v = iv$. As explained momentarily, this is a complex vector bundle of rank 1 over Σ.

In general, if M is a smooth manifold and $f = (f_1, f_2, f_3)\colon M \to \mathbb{R}^3$ with image in S^2 is a smooth map, then there is the corresponding subbundle $E \subset M \times \mathbb{C}^2$ given by the set $\{(p, v) : f(p)v = iv\}$ where $f = \sum_j f_j\tau_j$. What follows uses Definition 6.2 to verify that E does indeed define a complex vector bundle. Use for π the restriction of the projection from $M \times \mathbb{C}^2$ to M. Likewise, use for μ the restriction of the multiplication on $\mathbb{C}^2$ by elements in $\mathbb{C}$. Take $\hat{o}$ to be the map $x \to (x, 0) \in M \times \mathbb{C}^2$. To see about the final item, fix $p \in M$ and fix a unit length vector $v_p \in E|_p \subset \mathbb{C}^2$. This done, write the two components of v_p as (a_p, b_p) and write those of any other vector v as (a, b). Now define the $\mathbb{C}$-linear map $\eta_p\colon \mathbb{C}^2 \to \mathbb{C}$ by the rule that sends $(a, b) \in \mathbb{C}^2$ to the complex number

$\bar{a}_p a + \bar{b}_p b$. As η_p restricts to $E|_p$ as an isomorphism (over $\mathbb{C}$), so it restricts to a neighborhood $U \subset M$ of p so as to map each fiber of $E|_U$ isomorphically to $\mathbb{C}$. Define λ_U to be this restriction.

Here is another example of a map f from a compact surface to in $\mathbb{R}^3$ to S^2: Let Σ denote the surface in question. Fix a point, $x \in \mathbb{R}^3$, that lies in the interior of the region bounded by Σ. A map f: $\Sigma \to S^2$ is defined by taking for f(p) the unit vector $(p - x)/|p - x|$. Alternately, one can take x to lie outside of Σ to define a different version of E.

6.5 The tangent bundle to a surface in $\mathbb{R}^3$

The tangent bundle to a surface in $\mathbb{R}^3$ has an almost complex structure that is defined as follows: Let $\Sigma \subset \mathbb{R}^3$ denote the surface and let $n = (n_1, n_2, n_3)$ again denote its normal bundle. As explained in Chapters 3.4 and 3.5, the tangent bundle $T\Sigma$ can be identified as the set of pair $(p, v) \in \Sigma \times \mathbb{R}^3$ such that the dot product $v \cdot n = 0$. Define j: $T\Sigma \to T\Sigma$ by the rule $j(p, v) = (p, n \times v)$, where $\times$ is the cross product on $\mathbb{R}^3$. Since $n \times n \times v = -v$, it follows that j is an almost complex structure. As a consequence, we may view $T\Sigma$ as a complex vector bundle of rank 1 over Σ. This complex bundle is denoted often by $T_{1,0}\Sigma$.

6.6 Bundles over 4-dimensional submanifolds in $\mathbb{R}^5$

Introduce a set of five matrices in $\mathbb{M}(4; \mathbb{C})$ as follows: Written in 2×2 block diagonal form, they are:

$$\textit{For } j = 1, 2, 3\colon\ \gamma_j = \begin{pmatrix} 0 & \tau_j \\ \tau_j & 0 \end{pmatrix},\ \gamma_4 = \begin{pmatrix} 0 & -\iota \\ \iota & 0 \end{pmatrix},\ \gamma_5 = i\begin{pmatrix} \iota & 0 \\ 0 & -\iota \end{pmatrix}.$$

Here, $\{\tau_a\}_{1 \le a \le 3}$ again denote the Pauli matrices. Note that for each j, one has $\gamma_j^2 = -\iota$. Note also that $\gamma_i\gamma_j + \gamma_j\gamma_i = 0$ for $i \neq j$. Let $M \subset \mathbb{R}^5$ denote a dimension 4 submanifold; the sphere S^4 is an example. Let $n = (n_1, \ldots, n_5)$ denote the normal vector to M at any given point. Define the map n: $M \to \mathbb{M}(4, \mathbb{C})$ by sending p to $n(p) = \sum_j n_j\gamma_j$. Then $n^2 = -\iota$. As a consequence, at each point in M, the matrix n has a 2-dimensional (over $\mathbb{C}$) space of eigenvectors with eigenvalue $+i$. Define the bundle $E \subset M \times \mathbb{C}^4$ to be the set $\{(p, v) : n(p)v = iv.\}$ This is a vector bundle over M with fiber $\mathbb{C}^2$.

6.7 Complex bundles over 4-dimensional manifolds

What follows uses the cocycle definition of a complex vector bundle to construct some rank 2 complex bundles over a given 4-dimensional manifold. Let M denote a 4-dimensional manifold. Fix a set finite set $\Lambda \subset M$ of distinct points. Assign to each $p \in \Lambda$ an integer $m(p) \in \mathbb{Z}$. Also assign to each $p \in M$ a coordinate chart $U_p \subset M$ that contains p with diffeomorphism $\varphi_p \colon U_p \to \mathbb{R}^4$ that sends p to 0. Make these assignments so that the charts for distinct pairs p and p′ are disjoint.

To define a vector bundle from this data, introduce $\mathfrak{U}$ to denote the open cover of M given by the sets $U_0 = M-\Lambda$ and the collection $\{U_p\}_{p\in\Lambda}$. As the only nonempty intersections are between U_0 and the sets from $\{U_p\}_{p\in\Lambda}$, a vector bundle over M with fiber $\mathbb{C}^2$ is defined by specifying a set of maps $\{\mathfrak{g}_{0p} \colon U_0 \cap U_p \to Gl(2; \mathbb{C})\}_{p\in\Lambda}$.

In order to specify these maps, first define a map $\mathfrak{g} \colon \mathbb{C}^2 \to \mathbb{M}(2; \mathbb{C})$ by the rule that sends

$$z = (z_1, z_2) \to \mathfrak{g}(z) = \begin{pmatrix} z_1 & -\bar{z}_2 \\ z_2 & \bar{z}_1 \end{pmatrix}.$$

Note that $\det(\mathfrak{g}(z)) = |z|^2$ so $\mathfrak{g}$ maps $\mathbb{C}^2-\{0\}$ to $Gl(2; \mathbb{C})$. Use j_0 to identify $\mathbb{R}^4$ with $\mathbb{C}^2$ so as to view $\mathfrak{g}$ as a map from $\mathbb{R}^4$ to $\mathbb{M}(2; \mathbb{C})$ that sends $\mathbb{R}^4-\{0\}$ to $Gl(2; \mathbb{C})$.

Now let $p \in \Lambda$. Since $\varphi_p(U_0 \cap U_p) = \mathbb{R}^4-\{0\}$, the composition $x \to \mathfrak{g}(\varphi_p(x))^{m(p)}$ for $x \in U_0 \cap U_p$ defines a map from $U_0 \cap U_p$ to $Gl(2; \mathbb{C})$. The latter map is defined to be $\mathfrak{g}_{0p}$.

6.8 Complex Grassmannians

Important examples of complex vector bundles involve the $\mathbb{C}^n$ analogs of the Grassmannians that are described in Chapter 1.6. To say more, fix integers $m > 1$ and $n \in \{1, \ldots, m-1\}$; then set $Gr_{\mathbb{C}}(m; n)$ to denote the set of m-dimensional complex subvector spaces in $\mathbb{C}^n$. As is explained momentarily, $Gr_{\mathbb{C}}(m; n)$ is a smooth manifold of dimension $2n(m-n)$. Moreover, $Gr_{\mathbb{C}}(m; n)$ has over it a very natural, rank n, complex vector bundle. The latter sits in $Gr_{\mathbb{C}}(m; n) \times \mathbb{C}^m$ as the set

$$E = \{(V, v) : V \textit{ is an n-dimensional subspace of } \mathbb{C}^m \textit{ and } v \in V\}.$$

The bundle E is the $\mathbb{C}$-analog of what is described in Chapter 3.3. This $Gr_{\mathbb{C}}(m; n)$ is called the *Grassmannian of n-planes in* $\mathbb{C}^m$, and the bundle E is called the *tautological n-plane bundle.*

The argument to prove that $Gr_{\mathbb{C}}(m; n)$ has the structure of a smooth manifold is much the same as the argument used in Chapter 1.6 to prove that the $\mathbb{R}$ versions of the Grassmannians are smooth manifolds. Likewise, the argument that E is a complex vector bundle over $Gr_{\mathbb{C}}(m; n)$ is much the same as that given in Chapter 3.3 for the $\mathbb{R}$ analog of E. Even so, it is worth summarizing both arguments because of a point that is referred to in a later chapter.

What follows directly is a brief interlude to set some notation. To start, suppose that v and w are vectors in $\mathbb{C}^m$. Their Hermitian inner product is denoted by $\bar{v} \cdot w$; it is defined using their respective components by

$$\bar{v} \cdot w = \bar{v}_1 w_1 + \cdots + \bar{v}_m w_m.$$

Thus, the norm of a vector $v \in \mathbb{C}^n$ is $|v| = (\bar{v} \cdot v)^{1/2}$. A basis $(e_1, \ldots, e_n)$ for an n-dimensional subspace $V \subset \mathbb{C}^m$ is orthonormal when $\bar{e}_i . e_j = 1$ or 0 depending on whether $i = j$ or $i \neq j$. The orthogonal projection $\Pi_V\colon \mathbb{C}^m \to V$ can be defined as follows: Fix an orthonormal basis, $(e_1, \ldots, e_n)$ for V. Then

$$\Pi_V(w) = \sum\nolimits_{1 \le j \le n} (\bar{e}_j \cdot w)\, e_j.$$

To continue setting notation, introduce $\mathbb{M}_{\mathbb{C}}(m; n)$ to denote the space of $\mathbb{C}$-linear homomorphisms from $\mathbb{C}^n$ to $\mathbb{C}^{m-n}$. This is to say that each point in $\mathbb{M}_{\mathbb{C}}(m; n)$ is a matrix with n columns and $m-n$ rows whose entries are complex numbers. The entries serve as coordinates and so identify $\mathbb{M}_{\mathbb{C}}(m; n)$ with $\mathbb{C}^{n(m-n)}$.

With this notation set, what follows next is a description of the smooth manifold structure on $Gr_{\mathbb{C}}(m; n)$. To start, define a topology on $Gr_{\mathbb{C}}(m; n)$ as follows: A basis of open neighborhoods for any given n-dimensional subspace V is labeled by the positive numbers, with the neighborhood labeled by a given $\varepsilon > 0$ consists of the set of subspace V' such that $|\Pi_V(v') - v| < \varepsilon$.

A coordinate chart for a neighborhood of a given $V \subset Gr_{\mathbb{C}}(m; n)$ is the set $\mathcal{O}_V$ of n-planes V' such that Π_V maps V' isomorphically to V. A coordinate map, φ_V, from $\mathcal{O}_V$ to $\mathbb{R}^{2n(m-n)} = \mathbb{C}^{n(m-n)} = \mathbb{M}_{\mathbb{C}}(m; n)$ is defined as follows: Fix an orthonormal basis $\{e_k\}_{1 \le k \le n}$ for V so as to identify this vector space with $\mathbb{C}^n$. Let $V^{\perp} \subset \mathbb{C}^m$ denote the Hermitian conjugate space, thus the kernel of Π_V. Fix an orthonormal basis $\{u_a\}_{1 \le a \le m-n}$ for $V^{\perp}$ to make a $\mathbb{C}$-linear identification of the latter with $\mathbb{C}^{m-n}$. Now let $\mathfrak{p}$ denote a given point in $\mathbb{C}^{n(m-n)}$. Viewed as a matrix in $\mathbb{M}_{\mathbb{C}}(m; n)$, it has components $\{\mathfrak{p}_{ia}\}_{1 \le i \le n, 1 \le a \le n-m}$. The map φ_V^{-1} sends $\mathfrak{p}$ to the n-dimensional subspace spanned by the set $\{e_i + \sum_{1 \le a \le n} \mathfrak{p}_{ia} u_a\}_{1 \le i \le n}$.

Let V and V' denote any two points in $Gr_{\mathbb{C}}(m; n)$. Fix orthonormal bases $\{e_i'\}_{1 \le i \le n}$ and $\{u_a'\}_{1 \le a \le m-n}$ for V' and $V'^{\perp}$ to define the coordinate chart map $\varphi_{V'}\colon \mathcal{O}_{V'} \to \mathbb{M}_{\mathbb{C}}(m; n) = \mathbb{C}^{n(m-n)}$. This done, then the corresponding transition function $\varphi_{V'} \circ \varphi_V^{-1}$ sends a given point $\mathfrak{p} \in \varphi_V(\mathcal{O}_V \cap \mathcal{O}_{V'})$ to the point $\mathfrak{p}' \in \varphi_{V'}(\mathcal{O}_V \cap \mathcal{O}_{V'})$ with the latter determined by $\mathfrak{p}$ using the rule

$$\mathfrak{p}'_{ia} = \sum\nolimits_{1 \le k \le n} (G^{-1})_{ik} \left(\bar{u}'_a \cdot e_k + \sum\nolimits_{1 \le b \le m-n} \mathfrak{p}_{kb}\, \bar{u}'_a \cdot u_b \right)$$

where the notation is as follows: First, any given $i, k \in \{1, \ldots, n\}$ version of G_{ik} is given by

$$G_{ik} = \bar{e}'_i \cdot e_k + \sum\nolimits_{1 \le a \le n} \mathfrak{p}_{ia}\, \bar{e}'_k \cdot u_a .$$

Second, G is the $n \times n$ matrix with (i, k) entry equal to G_{ik}. Note that $G \subset Gl(n; \mathbb{C})$ because $\varphi_V^{-1}(\mathfrak{p}) \in \mathcal{O}_V \cap \mathcal{O}_{V'}$.

The simplest example is that of $Gr_{\mathbb{C}}(m; 1)$, the set of 1-dimensional complex subspaces in $\mathbb{C}^m$. This manifold is usually called the $(m-1)$-dimensional *complex projective space*; and it is denoted by $\mathbb{CP}^{m-1}$. This manifold has an open cover given by the collection of sets $\{\mathcal{O}_1, \ldots, \mathcal{O}_m\}$ where $\mathcal{O}_k$ is the version of what is denoted above by $\mathcal{O}_V$ with V the complex line spanned by the vector in $\mathbb{C}^m$ with k'th entry 1 and all other entries 0. The complex span of a vector $v \in \mathbb{C}^m$ is in $\mathcal{O}_k$ if and only if the k'th entry of v is nonzero. Noting that $\mathbb{M}_{\mathbb{C}}(m; 1) = \mathbb{C}^{m-1}$, the $\mathcal{O}_k$ version of the coordinate map φ_V^{-1} sends $(p_1, \ldots, p_{m-1}) \in \mathbb{C}^{m-1}$ to the line spanned by the vector $(p_1, \ldots, p_{k-1}, 1, p_k, \ldots, p_{m-1})$. Thus, the i'th entry of this vector is equal to p_i when $i < k$, it is equal to 1 when $i = k$, and it is equal to p_{i-1} when $k < i$. For $k' < k$, the matrix G for the intersection of $\mathcal{O}_k$ with $\mathcal{O}_{k'}$ is a 1–1 matrix, this the nonzero, $\mathbb{C}$-valued function on the set $\mathcal{O}_k \cap \mathcal{O}_{k'}$ given by $p_{k'} \in \mathbb{C}-\{0\}$. The transition function writes $(p'_1, \ldots, p'_{m-1})$ as the function of $(p_1, \ldots, p_{m-1})$ given as follows:

- If $i < k'$ *then* $p'_i = p_i/p_{k'}$.
- If $k' \le i < k$, *then* $p'_i = p_{i+1}/p_{k'}$.
- $p'_k = 1/p_{k'}$.
- *If* $k < i$, *then* $p'_i = p_i/p_{k'}$.

What follows next is a brief description of the tautological bundle $E \to Gr_{\mathbb{C}}(m; n)$. The data needed for Definition 6.2 are a projection map to $Gr_{\mathbb{C}}(m; n)$, a zero section, a multiplication by $\mathbb{C}$ and a suitable map defined over a neighborhood of any given point from E to $\mathbb{C}^n$. The bundle projection map π: $E \to Gr_{\mathbb{C}}(m; n)$ sends any given pair (V, v) with $v \in V$ and V an n-dimensional subspace of $\mathbb{C}^m$ to V. The zero section $\hat{o}$ sends this same $V \in Gr_{\mathbb{C}}(m; n)$ to the pair $(V, 0)$. Multiplication by $\mathbb{C}$ on E by any given $c \in \mathbb{C}$ sends (V, v) to (V, cv). For the final requirement, fix the basis $\{e_i\}_{1 \le i \le n}$ for V as above. Let $W \in \mathcal{O}_V$. Then the map $\lambda_{\mathcal{O}_V}$ sends $(W, w) \in E|_W$ to the vector in $\mathbb{C}^n$ with j'th entry equal to $\bar{e}_j \cdot w$.

The map $\lambda_{\mathcal{O}_V}$ just defined gives a $\mathbb{C}$-linear map from $E|_{\mathcal{O}_V}$ to $\mathcal{O}_V \times \mathbb{C}^n$ that respects the vector bundle structure of each fiber. These maps are used next to give a cocycle definition of E in the manner of Definition 6.3. A suitable chart for this consists of a finite collection of sets of the form $\mathcal{O}_V$ for $V \in Gr_{\mathbb{C}}(m; n)$.

To define the transition functions, choose for each point V represented a basis, $\{e_i\}_{1\le i\le n}$, for V. This done, suppose that V and V′ are two points in $Gr_{\mathbb{C}}(m; n)$ that are represented. The corresponding bundle transition function: $g_{\mathcal{O}_{V'}, \mathcal{O}_{V'}}: \mathcal{O}_V \cap \mathcal{O}_{V'} \to Gl(n; \mathbb{C})$ is the map G with entries $\{G_{ij}\}_{1\le i,j\le n}$ as depicted above.

6.9 The exterior product construction

The notion of the exterior product for $\mathbb{C}^n$ can be used to define a complex vector bundle of rank 2^n over any given 2n-dimensional manifold.

To set the stage, let $\wedge^*\mathbb{C}^n = \bigoplus_{k=0,1,\ldots,n} (\wedge^k\mathbb{C}^n)$. Here, $\wedge^0\mathbb{C}^n$ is defined to be $\mathbb{C}$. Meanwhile, $\wedge^1\mathbb{C}^n = \mathbb{C}^n$ and any given $k > 1$ version of $\wedge^k\mathbb{C}^n$ is the k'th exterior power of $\mathbb{C}^n$. Fix a basis $\{e_1, \ldots, e_n\}$ for $\mathbb{C}^n$ and the vector space $\wedge^k\mathbb{C}^n$ has a corresponding basis whose elements are denoted by

$$\{e_{i_1} \wedge \cdots \wedge e_{i_k}\}_{1\le i_1 < \ldots < i_k \le n}.$$

Note that $\wedge^k\mathbb{C}^n$ has dimension $\frac{n!}{k!(n-k)!}$ and so $\dim(\wedge^*\mathbb{C}^n) = 2^n$.

There are two actions of $\mathbb{C}^n$ on this vector space. The first is exterior multiplication. This action has $z \in \mathbb{C}^n$ sending any given element $\omega \in \wedge^*\mathbb{C}^n$ to $z \wedge \omega$. By way of reminder, the action is such as to send $\wedge^k\mathbb{C}^n$ to $\wedge^{k+1}\mathbb{C}^n$. It can be defined by as follows: First, $z \wedge (\cdot)$ sends the element $1 \in \wedge^0\mathbb{C}^n$ to $z \in \mathbb{C}^n = \wedge^1\mathbb{C}^n$. It is defined on the $k > 0$ versions of $\wedge^k\mathbb{C}^n$ by its action on the basis vectors given above. To define the latter, write $z = z_1e_1 + \cdots + z_ne_n$. Then

$$z \wedge e_{i_1} \wedge \cdots \wedge e_{i_k} = \sum\nolimits_{1\le i\le n} z_i(e_i \wedge (e_{i_1} \wedge \cdots \wedge e_{i_k}))$$

where $e_i \wedge (e_{i_1} \wedge \ldots \wedge e_{i_k})$ is zero if some $i_k = i$, and it is ± 1 times the basis vector obtained by adjoining i to the set $\{i_1, \ldots, i_k\}$ otherwise. The coefficient here is +1 if the ordering $(i, i_1, \ldots, i_k)$ requires an even number of interchanges to make an increasing sequence, and −1 otherwise. Note that this ±1 business guarantees that $z \wedge (z \wedge (\cdot)) = 0$.

The second action of z is denoted in what follows by ι_z. This homomorphism is zero on $\wedge^0\mathbb{C}^n$ and maps any $k > 0$ version of $\wedge^k\mathbb{C}^n$ to $\wedge^{k-1}\mathbb{C}^n$. It is defined inductively as follows: First, $\iota_z W = \bar{Z} \cdot W \in \wedge^0\mathbb{C}^n = \mathbb{C}$. Now suppose that ι_z is defined on $\wedge^k\mathbb{C}^n$ for $k \ge 1$. To define it on $\wedge^{k+1}\mathbb{C}^n$ it is enough to specify the action on elements of the form $w \wedge \omega$ with $w \in \mathbb{C}^n$ and $\omega \in \wedge^k\mathbb{C}^n$. This understood, the homomorphism ι_z sends $w \wedge \omega$ to $(\bar{z} \cdot w)\, \omega - w \wedge \iota_z(\omega)$.

Use $\iota_{(\cdot)}$ and the exterior product to define the map $\mathfrak{g}: \mathbb{C}^n \to End(\wedge^*\mathbb{C}^n) = \mathbb{M}(2^n; \mathbb{C})$ by the rule

$$\mathfrak{g}(z) = \iota_z + z\wedge.$$

Noting that $\mathfrak{g}^2 = |z|^2\,\iota$, it follows that $\mathfrak{g}$ maps $\mathbb{C}^n - \{0\}$ to $Gl(2^n; \mathbb{C}))$.

To continue with the preliminaries, suppose now that M is a given 2n-dimensional manifold. Fix a finite set, Λ, of distinct points in M. For each $p \in \Lambda$, fix a coordinate chart $U_p \subset M$ centered at p with its map $\varphi_p\colon U_p \to \mathbb{R}^{2n}$. Choose these charts so that they are pair-wise disjoint. Use the almost complex structure j_0 to identify $\mathbb{R}^{2n}$ with $\mathbb{C}^n$ so as to view any given $p \in \Lambda$ version of φ_p as a map to $\mathbb{C}^n$.

The data just given will now be used to give a cocycle definition of a complex vector bundle with fiber dimension 2^n. Take for the open cover the sets $\{U_p\}_{p\in\Lambda}$ with the extra set $U_0 = M - \Lambda$. The only relevant bundle transition functions are those for the sets $\{U_0 \cap U_p\}_{p\in\Lambda}$. For a given point $p \in \Lambda$, take the transition function g_{U_0,U_p} to be the map from $U_0 \cap U_p$ to $Gl(2^n; \mathbb{C})$ that sends any given point x to $\mathfrak{g}(\varphi_p(x))$.

6.10 Algebraic operations

All of the algebraic notions from Chapter 4, such as bundle isomorphisms, homomorphisms, subbundles, quotient bundles, tensor products, symmetric products, antisymmetric projects and direct sums, have their analogs for complex vector bundles. The definitions are identical to those in Chapter 4 save that all maps are now required to be $\mathbb{C}$-linear.

There are two points that warrant an additional remark with regards to Definition 6.1. The first involves the notion of a bundle homomorphism. In the context of Definition 6.1, a complex bundle homomorphism is a homomorphism between the underlying real bundles that intertwines the respective almost complex structures. The second point involves the notion of tensor product. To set the stage, suppose that E and E' are complex vector bundles over M of rank n and n' respectively. Their tensor product is a complex vector bundle over M of rank nn'. Note, however, that the tensor product $E_{\mathbb{R}} \otimes E'_{\mathbb{R}}$ of their underlying real bundles has dimension $4nn'$ which is twice that of $(E \otimes E')_{\mathbb{R}}$. The latter sits in $E_{\mathbb{R}} \otimes E'_{\mathbb{R}}$ as the kernel of the homomorphism $j \otimes \iota' - \iota \otimes j'$ where j and j' denote the respective almost complex structures on E and E' while ι and ι' denote the respective identity homomorphisms on E and E'.

A third comment is warranted with regards to the exterior product. Recall that if $E \to M$ is a real bundle of fiber dimension n, then $\wedge^n E$ is a vector bundle of fiber dimension 1 over M. In the case when E is a complex vector bundle, then $\wedge^n E$ is a vector bundle with fiber $\mathbb{C}$. The latter is often denoted by det(E). The reason for this appelation is as follows: Suppose that E is given by the cocycle definition from Chapter 6.1. This is to say that a locally finite open cover, $\mathfrak{U}$, of M is given with suitable bundle transition functions from overlapping sets to Gl(n; $\mathbb{C}$) with n the fiber dimension of E. Then $\wedge^n E$ is defined by using the same open cover and with the transition function given by the determinant of the corresponding transition function for E. Note in this regard that any transition function for $\wedge^n E$ must map to Gl(1; $\mathbb{C}$) $= \mathbb{C}-\{0\}$.

6.11 Pull-back

Let M and N denote smooth manifolds and let ψ: M $\to$ N denote a smooth map. Chapter 5.1 explains how ψ is used to construct a real vector bundle over M from a given real vector bundle over N. If π: E $\to$ N is the given bundle over N, then the corresponding bundle over M is denoted by ψ^*E. If E has an almost complex structure, j, then there is a unique almost complex structure for ψ^*E which is intertwined with j by the map $\hat{\psi}: \psi^*E \to E$. Indeed, with ($\psi^*E$) viewed in M $\times$ E as the set of pairs (x, v) with $\psi(x) = \pi(v)$, this almost complex structure acts to send (x, v) to (x, jv).

Granted the preceding observation, it follows that the pull-back of any given complex vector bundle π: E $\to$ N can be viewed in a canonical way as a complex vector bundle over M; and having done so, the covering map $\hat{\psi}: \psi^*E \to E$ is $\mathbb{C}$-linear on each fiber.

By way of an example, return to the milieu of Chapter 6.4. Let $\Sigma \subset \mathbb{R}^3$ denote a given embedded surface. As explained in Chapter 6.4, the association to each point in Σ of its normal vector can be used to define a map n: $\Sigma \to \mathbb{M}(2; \mathbb{C})$ and the vector subbundle $E \subset \Sigma \times \mathbb{C}^2$ given by the set of points (p, v) with $\mathfrak{n}(p)v = iv$. With this as background, remark that the assignment to p of its normal vector also defines a map from Σ to S^2. The latter, n(·), is called the Gauss map. Let $E_1 \to S^2$ denote the subbundle of the product bundle $S^2 \times \mathbb{C}^2$ given by the set of points $((x_1, x_2, x_3), v)$ with $(\Sigma_{1 \le i \le 3} x_i \tau_i)\, v = iv$. The bundle E is the pull-back n^*E_1.

By way of a second example, suppose that $\Sigma \subset \mathbb{R}^3$ is a embedded surface. Chapter 6.5 defines the complex rank 1 bundle $T_{1,0}\Sigma \to \Sigma$. The latter is $n^*T_{1,0}S^2$.

The following proposition gives a third application of pull-back. But for some notation, its proof is the same as that of its real bundle analog in Chapter 5.2.

Proposition 6.5 *Let* M *denote a smooth manifold, let* n *denote a positive integer, and let* π: E $\to$ M *denote a given rank n, complex vector bundle. If* m *is sufficiently large, there exists a map* ψ_m: M $\to$ $Gr_{\mathbb{C}}(m; n)$ *and an isomorphism between* E *and the pull-back via* ψ_m *of the tautological bundle over* $Gr_{\mathbb{C}}(m; n)$.

Additional reading

- *Differential Geometry of Complex Vector Bundles*, Shoshichi Kobayashi, Princeton University Press, 1987.
- *Characteristic Classes*, John Milnor and James D. Stasheff, Princeton University Press, 1974.
- *Basic Bundle Theory and K-Cohomology Invariants*, Dale Husemoller, Michael Joachim, Branislaw Jurco and Marten Schottenloher, Springer, 2009.
- *Algebraic Geometry: A First Course*, Joe Harris, Springer, 2010.

7 Metrics on vector bundles

A metric on $\mathbb{R}^n$ is a positive definite symmetric bilinear form. This is to say that if g is a metric, then $g(v, w) = g(w, v)$ and $g(v, v) > 0$ if $v \neq 0$. A metric on a real vector bundle $E \to M$ is a section, g, of $Sym^2(E^*)$ that restricts to each fiber as symmetric, positive definite quadratic form.

Metrics always exist. To see this, let $m = \dim(M)$ and let $\{\varphi_\alpha: U_\alpha \to \mathbb{R}^m\}$ denote a locally finite cover of M by coordinate charts such that $\{\varphi_\alpha^{-1}\{x : |x| < ½\}\}$ is also a cover of M and such that $E|_{U_\alpha}$ over U_α has an isomorphism, $\upsilon_\alpha: E|_{U_\alpha} \to U_\alpha \times \mathbb{R}^n$. Use the function χ from Appendix 1.2 to define $\chi_\alpha: M \to [0, 1)$ so as to be zero on $M - U_\alpha$ and equal to $\chi(|\varphi_\alpha|)$ on U_α. Let δ denote a given metric on $\mathbb{R}^n$, and view δ as a metric on the product bundle $U_\alpha \times \mathbb{R}^n$ for each α. A metric, g, on E is $g(v, w) = \sum_\alpha \chi_\alpha\, \delta(\upsilon_\alpha v, \upsilon_\alpha w)$.

Note in this regard that if g and g′ are metrics on E, then so is $ug + u'g'$ for any functions u and v with both u and u′ nonnegative functions such that $u + u' > 0$ at each point.

A Hermitian metric on $\mathbb{C}^n$ is a bilinear form, g, on $\mathbb{C}^n$ with the following properties: First, $g(u, cu') = cg(u, u')$ and $g(cu, u') = \bar{c}g(u, u')$ for all $u, u' \in \mathbb{C}^n$ and all $c \in \mathbb{C}$. Second, $g(u, u) > 0$ for all $u \in \mathbb{C}^n \neq 0$. Finally, $\overline{g(u,u')} = g(u', u)$ for all $u, u' \in \mathbb{C}^n$. A Hermitian metric on a complex vector bundle $\pi: E \to M$ is a section of $\overline{E}^* \otimes E^*$ that restricts to each fiber as a Hermitian metric. Here is an equivalent definition: Let $E_\mathbb{R}$ denote the underlying real bundle and let j denote its almost complex structure. A Hermitian on E is defined by a metric, $g_\mathbb{R}$, on $E_\mathbb{R}$ if $g_\mathbb{R}(\cdot, j(\cdot)) = -g_\mathbb{R}(j(\cdot),(\cdot))$. This condition asserts that j must define a skew-symmetric endomorphism on each fiber of $E_\mathbb{R}$. To obtain g from $g_\mathbb{R}$, view $E \subset E_\mathbb{R}$ by writing $E_\mathbb{R} \otimes_\mathbb{R} \mathbb{C}$ and then identifying E as the set of vectors of the form $u = e - i(je)$ with $e \in E_\mathbb{R}$. Then inner product $g(u, u')$ is defined to be

$$g(u, u') = g_\mathbb{R}(e, e') - ig_\mathbb{R}(e, je').$$

Note that if $c = c_1 + ic_2$ is any given element in $\mathbb{C}$, then $cu = (c_1 e + c_2 je) - ij(c_1 u + c_2 ju)$. As a consequence, the preceding formula finds

$$g(cu, u') = c_1 g_\mathbb{R}(e, e') + c_2 g(je, e') - ic_1 g_\mathbb{R}(e, je') - ic_2 g_\mathbb{R}(je, je'),$$

which is equal to $\bar{c}g(u, u')$ by virtue of the fact that j is skew-symmetric. A similar computation finds that $g(u, cu') = c\, g(u, u')$ and also $\overline{g(u,u')} = g(u', u)$.

Meanwhile, a given Hermitian metric, g, on E defines the metric $g_{\mathbb{R}}$ on $E_{\mathbb{R}}$ by setting $g_{\mathbb{R}}(e, e')$ to be the real part of $g(u, u')$.

With only notational changes, the construction given above of a metric on a real bundle will give a Hermitian metric on a complex bundle. If g and g′ are Hermitian metrics on a complex vector bundle, then so is $u\, g + u'\, g'$ when u and u′ are functions with $u + u' > 0$.

7.1 Metrics and transition functions for real vector bundles

A real bundle $E \to M$ of fiber dimension n with a metric can be used to construct a locally finite, open cover $\mathfrak{U}$ of M with two salient features: First, each $U \in \mathfrak{U}$ comes with an isomorphism from $E|_U$ to $U \times \mathbb{R}^n$. Second, the corresponding bundle transition functions map nonempty intersections of sets from $\mathfrak{U}$ into O(n). To see why this is true, fix a locally finite cover, $\mathfrak{U}$, for E such that each $U \in \mathfrak{U}$ comes with an isomorphism from $\varphi_U : E|_U \to U \times \mathbb{R}^n$. Use φ_U^{-1} to define a basis of sections of $E|_U$ over U, and then use the Gram–Schmid procedure to construct a basis of sections of $E|_U$ over U which is orthonormal at each point as defined by the fiber metric. Let $\{\mathfrak{s}_{Uj}\}_{1 \le j \le n}$ denote these sections. Define a new isomorphism $E|_U$ to $U \times \mathbb{R}^n$ by declaring its inverse to be the map that sends any given point $(x, (v_1, \ldots, v_n)) \in U \times \mathbb{R}^n$ to $\sum_j v_j \mathfrak{s}_{Uj}|_x$. The corresponding bundle transition functions map any orthonormal vector in $\mathbb{R}^n$ to an orthonormal vector. As a consequence, all transition functions map to O(n).

The question arises as to whether a trivializing chart can be found whose corresponding transition functions map to SO(n).

Definition 7.1 *A real vector bundle* $E \to M$ *is said to be* <u>*orientable*</u> *if it has a trivializing cover such that the corresponding vector bundle transition functions on the overlaps have positive determinant.*

If a vector bundle is orientable, and if it has a fiber metric, then its transition functions can be chosen to map to SO(n). The following lemma says something about when a bundle is orientable:

Lemma 7.2 *A real vector bundle* $E \to M$ *with some given fiber dimension* $n \ge 1$ *is orientable if and only if the real line bundle* $\det(E) = \wedge^n E$ *is isomorphic to the product bundle* $M \times \mathbb{R}$.

Proof of Lemma 7.2 Let $\mathfrak{U}$ denote a locally finite cover of M such that each chart $U \in \mathfrak{U}$ comes with an isomorphism $\varphi_U : E|_U \to U \times \mathbb{R}^n$. As noted above, these isomorphisms can be chosen so that the corresponding bundle transition functions map to O(n). The isomorphism φ_U induces an isomorphism ψ_U: $\det(E)|_U \to U \times \mathbb{R}$. If U and U′ are two intersecting sets from $\mathfrak{U}$, then the corresponding vector bundle transition function for E is a map $g_{U',U}$: $U \cap U' \to O(n)$. The corresponding bundle transition function for det(E) is the map from, $U \cap U'$ to $\{\pm 1\}$ given by $\det(g_{U',U})$. This understood, if E is orientable, one can take $\mathfrak{U}$ and the corresponding isomorphisms $\{\varphi_U\}_{U\in\mathfrak{U}}$ such that $\det(g_{U',U}) = 1$ for all pairs U, $U' \in \mathfrak{U}$. As a consequence, an isomorphism ψ: $\det(E) \to M \times \mathbb{R}$ is defined by declaring ψ on $\det(E)|_U$ for any given $U \in \mathfrak{U}$ to be ψ_U.

To prove the converse, suppose that η: $\det(E) \to M \times \mathbb{R}$ is an isomorphism. If $U \in \mathfrak{U}$, then $\eta \circ \psi_U{}^{-1}$ can be written as $(x, r) \to (x, \eta_U r)$ where $\eta_U \in \{\pm 1\}$. The latter is such that

$$\eta_{U'} \det(g_{U',U})\eta_U^{-1} = 1$$

on $U \cap U'$ when U and U′ are any two sets from $\mathfrak{U}$ with nonempty overlap. This understood, define a new isomorphism, φ'_U: $E|_U \to U \times \mathbb{R}^n$ by $\varphi'_U = \sigma_U \circ \varphi_{\mathfrak{U}}$ where σ_U at any given point in U is a diagonal matrix with entries $(\eta_U, 1, \ldots, 1)$ on the diagonal. It then follows from the preceding equation that the corresponding transition functions for the data $\{\varphi'_U\}_{U\in\mathfrak{U}}$ map to SO(n).

What follows are some examples.

7.1.1 The Mobius bundle

This bundle over S^1 is defined in Chapter 1.3. It is not orientable. To see why, suppose to the contrary. This bundle would admit a section that has no intersections with the zero section. By definition, a section is a map v: $S^1 \to \mathbb{R}^2$ whose coordinate entries (v_1, v_2) obey the condition $\cos(\theta)v_1 + \sin(\theta)v_2 = v_1$ and $\sin(\theta)v_1 - \cos(\theta)v_2 = v_2$ at each $\theta \in S^1$. This requires that v_1 and v_2 can be written as $v_1 = r(\theta)\cos(\frac{1}{2}\theta)$ and $v_2 = r(\theta)\sin(\frac{1}{2}\theta)$ with r: $S^1 \to \mathbb{R}-\{0\}$. Such a pair is not continuous on the whole of S^1.

7.1.2 The tautological bundle over $\mathbb{RP}^n$

This bundle is also not orientable. To see why, consider the map from S^1 into $\mathbb{RP}^n$ that sends the angle coordinate $\theta \in \mathbb{R}/(2\pi\mathbb{Z})$ to the line with tangent $(\cos(\frac{1}{2}\theta), \sin(\frac{1}{2}\theta), 0, \ldots, 0) \in \mathbb{R}^{n+1}$. The pull-back via this map of the tautological bundle is the Mobius bundle. Given that the Mobius bundle is not orientable, it follows that neither is the tautological bundle over $\mathbb{RP}^n$.

7.1.3 A bundle with an almost complex structure

As is explained next, a bundle with an almost complex structure is orientable. Indeed, let π: E $\to$ M denote such a bundle and let 2n denote its fiber dimension. Recall from Chapter 6.2 that there is a locally finite cover $\mathfrak{U}$ for M such that each U $\in \mathfrak{U}$ comes with an isomorphism to the product bundle, and such that all transition functions map to the group $G_j \subset Gl(2n; \mathbb{R})$. As noted at the end of Chapter 2.4, matrices in G_j have positive determinant.

7.2 Metrics and transition functions for complex vector bundles

Let π: E $\to$ M denote now a complex vector bundle with fiber dimension n (over $\mathbb{C}$). Suppose that g is a Hermitian metric on E. Let U $\subset$ M denote an open set and suppose that $(\mathfrak{s}_1, \ldots, \mathfrak{s}_n)$ is a basis for $E|_U$ is said to be orthonormal if $g(\mathfrak{s}_i, \mathfrak{s}_j) = 1$ when $i = j$ and zero otherwise. As the Gram–Schmid algorithm works just as well over $\mathbb{C}$ as over $\mathbb{R}$, it follows that M has a locally finite cover such that E's restriction to each set from the cover has an orthonormal basis. The following is a consequence: There is a locally finite cover, $\mathfrak{U}$, of M with two salient features: First, each U $\in \mathfrak{U}$ comes with a bundle isomorphism from $E|_U$ to $U \times \mathbb{C}^n$. Second, all transition functions map to the group U(n).

Recall from Chapter 6.10 the definition of the complex bundle $\det(E) = \wedge^n E$. This bundle has fiber $\mathbb{C}$.

Lemma 7.3 *The bundle* det(E) *is isomorphic to the product bundle* $E \times \mathbb{C}$ *if and only if the following is true: There is a locally finite cover,* $\mathfrak{U}$, *of* M *such that each* U $\in \mathfrak{U}$ *comes with an isomorphism from* $E|_U$ *to* $U \times \mathbb{C}^n$ *and such that all transition functions map to* SU(n).

The proof of this lemma is very much like that of Lemma 7.2 and so left to the reader.

7.3 Metrics, algebra and maps

Suppose that E is a vector bundle over a given manifold M with fiber $\mathbb{R}^n$. Assume that E has a given fiber metric. The latter endows bundles the various algebraically related bundles such as E^*, $\bigoplus_m E$, $\bigotimes_m E$, $\wedge^m E$ and $\mathrm{Sym}^m E$ with fiber metrics.

Consider, for example, the case of E^*. To obtain an orthonormal basis for the fiber of the latter at a given point in M, choose an orthonormal basis $\{e_1, \ldots, e_n\}$ for E at the point in question. A corresponding basis for E^* at the given point is given by the elements $\{\upsilon_1, \ldots, \upsilon_n\}$ with the property that $\upsilon_i(e_k) = 0$ if $i \neq k$ and 1 if $i = k$. Note that the metric on E at this point is given by the bilinear form $\sum_{1 \leq i \leq n} \upsilon_i \otimes \upsilon_i$. With these corresponding metrics on E and E^*, the metric on E can be viewed as an isometric isomorphism between E and E^*.

Consider for a second example the bundle $\wedge^m E$. To obtain an orthonormal basis for the fiber of the latter at a given point in M, fix an orthonormal basis $\{e_1, \ldots, e_n\}$ for E at the point in question. The set $\{e_{i_1} \wedge \cdots \wedge e_{i_m}\}_{1 \leq i_1 < \cdots < i_m \leq n}$ defines an orthonormal basis for $\wedge^m E$. Note that the metric so defined does not depend on the orthonormal basis.

By the same token, if $E \to M$ is a complex vector bundle with fiber $\mathbb{C}^n$, a Hermitian metric on E endows all of the analogous algebraically related bundles with Hermitian metrics. Note that with the given metric on E and the corresponding metric on E^*, the metric on E can be viewed as an isometry $\tau \colon E \to E^*$ with the property that $\tau(ce) = \bar{c}\tau(v)$ for any given complex number c. Alternately, the metric can be viewed as giving $\mathbb{C}$-linear isometry between the complex conjugate bundle $\overline{E}$ and E^*.

Suppose next that E and E' are vector bundles over M, either both real or both complex. Suppose in addition that each has a fiber metric. Then $E \otimes E'$ and $E \oplus E'$ inherit natural fiber metrics. If only E' has a fiber metric, then an injective vector bundle homomorphism $f \colon E \to E'$ can be used to give E a fiber metric. Indeed, the metric on E is defined as follows: Let g' denote the metric on E'. Then the induced metric, g, on E is such that $g(e, e') = g'(f(e), f(e'))$. For example, this induced metric construction gives a fiber metric to any subbundle of $M \times \mathbb{R}^n$ or $M \times \mathbb{C}^n$ if one agrees to give the latter their obvious base point independent fiber metrics. Given that any vector bundle is isomorphic to a subbundle of a product bundle, this metric induction procedure gives another proof of the assertion that all vector bundles have fiber metrics.

If $f \colon E \to E'$ is an injective vector bundle homomorphism then the quotient bundle E'/E also inherits a metric from a metric on E'. This comes about by writing E' as $f(E) \oplus E^{\perp}$ where $E^{\perp} \subset E$ is the subbundle of vectors that are orthogonal to the image of f. As the map from E' to E'/E restricts to $E^{\perp}$ as an isomorphism, a metric on $E^{\perp}$ gives a metric on E'/E. Meanwhile, $E^{\perp}$ has its fiber metric as a subbundle of E'.

The final remark in this section concerns pull-backs: Suppose that M and N are smooth manifolds and $\psi \colon M \to N$ is a smooth map. If $\pi \colon E \to N$ is a vector bundle with fiber metric, then the pull-back bundle ψ^*E inherits what is perhaps an obvious fiber metric. To be specific, the latter is defined by viewing $\psi^*E \subset M \times E$ as the set of pairs of the form (p, v) with $\psi(v) = \pi(p)$. The metric assigns to pairs (p, v) and (p, w) the inner-product of v with w.

7.4 Metrics on TM

A metric on TM is called a *Riemannian metric*. To see what such a metric looks like, suppose that $U \subset M$ is an open set with a diffeomorphism $\varphi: U \to \mathbb{R}^n$. Let g denote a metric on TM. Then $(\varphi^{-1})^*g$ defines a metric on $T\mathbb{R}^n$. Now, $T\mathbb{R}^n$ has its standard Euclidean basis of sections, this the basis given by the vector fields $\{\frac{\partial}{\partial x_i}\}_{i=1,2,\ldots,n}$. This understood, the metric $(\varphi^{-1})^*g$ can be viewed as a map from $\mathbb{R}^n$ to the space of $n \times n$, symmetric, positive definite matrices, this the map with entries

$$g_{ik} = g\left((\varphi^{-1})_*\left(\frac{\partial}{\partial x_i}\right), (\varphi^{-1})_*\left(\frac{\partial}{\partial x_k}\right)\right).$$

Thus, $(\varphi^{-1})^*g$ is the section of $\mathrm{Sym}^2(T^*\mathbb{R}^n)$ given by $\sum_{1\le i\le k\le n} g_{ik}\, dx_i \otimes dx_k$.

A manifold with a given metric on its tangent bundle is said to be a *Riemannian* manifold.

Here is an often used way to obtain a Riemannian metric on a given manifold M: Suppose that X is a Riemannian manifold (for example $\mathbb{R}^N$ with its Euclidean metric), and suppose that M is given as a submanifold of X. Then the Riemannian metric on X induces a Riemannian metric on M because the tangent bundle to M sits in $TX|_M$ as a subbundle. Said explicitly, the inner product between two given vectors in TM is obtained by viewing the latter as vectors in $TX|_M$ and taking their inner product using the metric on TX.

Additional reading

- *Riemannian Geometry*, Sylvestre Gallot, Dominque Hulin and Jacques Lafontaine, Springer, 2004.
- *An Introduction to Differentiable Manifolds and Riemannian Geometry*, William M. Boothby, Academic Press, 2002.
- *Metric Structures in Differential Geometry*, Gerald Walschap, Springer, 2004.
- *Fiber Bundles*, Dale Husemöller, Springer, 1993.

8 Geodesics

Let M denote a smooth manifold. A metric on TM can be used to define a notion of the distance between any two points in M and the distance traveled along any given path in M. What follows in this chapter first explains how this is done. The subsequent parts of this chapter concern the distance minimizing paths.

8.1 Riemannian metrics and distance

A Riemannian metric on M gives a distance function on M which is defined as follows: Let γ denote a smooth map from an interval $I \subset \mathbb{R}$ into M. Let t denote the Euclidean coordinate on $\mathbb{R}$. Define $\dot{\gamma}$ to be the section of γ^*TM that is defined by $\gamma_* \frac{\partial}{\partial t}$, this the push-forward of the tangent vector to $\mathbb{R}$. This has norm $|\dot{\gamma}| = g(\dot{\gamma}, \dot{\gamma})^{1/2}$ where g here denotes the given Riemannian metric. The norm of $\dot{\gamma}$ is said to be the *speed* of γ. The length of γ is defined to be

$$\ell_\gamma = \int_I g(\dot{\gamma}, \dot{\gamma})^{1/2} dt.$$

Note that this does not depend on the parametrization of γ, so it is intrinsic to the image of γ in M. In the case $M = \mathbb{R}^n$ with its Euclidean metric, what is defined here is what is usually deemed to be the length of γ.

Let p and q denote points in M. A *path* from p to q is defined to be the image in M of a continuous, piece-wise smooth map from an interval in $\mathbb{R}$ to M that maps the smaller endpoint of the interval to p and the larger endpoint to q. It follows from what was just said that each path in M has a well-defined length. Define

$$\text{dist}(p, q) = \inf_\gamma \ell_\gamma$$

where the infimum is over all paths in M from p to q. Note that dist(p, q) = dist(q, p) and that the triangle inequality is obeyed:

$$\text{dist}(p, q) \leq \text{dist}(p, x) + \text{dist}(x, q) \quad \textit{for any } x \in M.$$

Moreover, dist(p, q) > 0 unless p = q. To prove this last claim, go to a coordinate chart that contains p. Let $(x_1, \ldots, x_n)$ denote the coordinates in this chart, and view p as a point, $z \in \mathbb{R}^n$. Since $q \neq p$, there is a ball in $\mathbb{R}^n$ of some radius $r > 0$ that has z in its center, and does not contain the image of q via the coordinate chart. (This is certainly the case if q is not in the coordinate chart at all.) Then any path from p to q has image via the coordinate chart that enter this ball on its boundary and ends at the origin. The Euclidean distance of this image is no less than r. Meanwhile, because g is a metric, when written in the coordinates via the coordinate chart, g appears as a symmetric, bilinear form with entries g_{ij}. As it is positive definite, there is a constant, c_0, such that $\sum_{i,j} g_{ij}(x)v^iv^j \geq c_0^{-1}|v|^2$ for any vector $v \in \mathbb{R}^n$ and any point x in the radius r ball about z. This implies that the length of any path from q to p is at least $c_0^{-1/2}r > 0$.

The function dist(·, ·) can be used to define a topology on M whereby the open neighborhoods of a given point $p \in M$ are generated by sets of the form $\{\mathcal{O}_{p,\varepsilon}\}_{\varepsilon>0}$ where the set $\mathcal{O}_{p,\varepsilon} = \{q \in M : \text{dist}(p, q) < \varepsilon)$. The argument given in the preceding paragraph (with the fact that $\sum_{i,j} g_{ij}(x)\, v^iv^j \leq c_0\, |v|^2$) implies that this metric topology on M is identical to the given topology.

8.2 Length minimizing curves

The following is the fundamental theorem that underlies much of the subject of Riemannian geometry. This theorem invokes two standard items of notation: First, repeated indices come with an implicit sum. For example, $g_{ij}v^iv^j$ means $\sum_{1\leq i\leq j\leq n} g_{ij}v^iv^j$. Second, g^{ij} denotes the components of the inverse, g^{-1}, to a matrix g with components g_{ij}.

Theorem 8.1 (the geodesic theorem) *Suppose that* M *is a compact, manifold and that* g *is a Riemannian metric on* M. *Fix any points* p *and* q *in* M.

- *There is a smooth curve from* p *to* q *whose length is the distance between* p *and* q.
- *Any length minimizing curve is an embedded, 1-dimensional submanifold that can be reparametrized to have constant speed; thus* $g(\dot\gamma, \dot\gamma)$ *is constant.*
- *A length minimizing, constant speed curve is characterized as follows: Let* $U \subset M$ *denote an open set with a diffeomorphism* $\varphi: U \to \mathbb{R}^n$. *Introduce the Euclidean coordinates* $(x^1, \ldots, x^n)$ *for* $\mathbb{R}^n$ *and let* g_{ij} *denote the components of the metric in these coordinates. This is to say that the metric on* U *is given by* $g|_U = \varphi^*(g_{ij}\, dx^i \otimes dx^j)$. *Denote the coordinates of the* φ*-image of a length*

minimizing curve by $\gamma = (\gamma^1, \ldots, \gamma^n)$. *Then the latter curve in* $\mathbb{R}^n$ *obeys the equation:*

$$\ddot{\gamma}^i + \Gamma^j_{km}\dot{\gamma}^k\dot{\gamma}^m = 0$$

where $\Gamma^j_{km} = \frac{1}{2}g^{jp}(\partial_m g_{pk} + \partial_k g_{pm} - \partial_p g_{km})$. *Here,* $\dot{\gamma}^i = \frac{d}{dt}\gamma^i$ *and* $\ddot{\gamma}^i = \frac{d^2}{dt^2}\gamma^i$. *(Here, as above, repeated indices are implicitly summed.)*

- *Any curve that obeys this equation in coordinate charts is locally length minimizing. This is to say that there exists* $c_0 > 1$ *such that when* p *and* q *are two points on such a curve with* $\text{dist}(p, q) \le c_0^{-1}$, *then there is a segment of the curve with one endpoint* p *and the other* q *whose length is* dist(p, q).

This theorem is called the *geodesic theorem* in what follows. A proof is given at the end of the next chapter. The next section of this chapter supplies an existence theorem for geodesics, and the remaining sections consider various examples where the geodesics can be described in very explicit terms. What follows directly are some additional remarks about the statement of the theorem.

The equation depicted above in the theorem is called the *geodesic equation*. This equation refers to a particular set of coordinates. Even so, the notion that a map from an interval to M obeys this equation is coordinate independent. This is to say the following: Let $U \subset M$ denote an open set with a diffeomorphism, $\varphi: U \to \mathbb{R}^n$. Now let $I \subset R$ denote an interval and let $\gamma_M: I \to U$ denote a smooth map such that $\gamma = \varphi \circ \gamma_M$ obeys the geodesic equation. Then so does the map γ' given by $\varphi' \circ \gamma_M$ if $\varphi': U \to \mathbb{R}^n$ is any other diffeomorphism. You can use the Chain rule to prove this assertion if you make sure to account for the fact that the metric coefficients g_{ij} depend on the choice of coordinates. To say more about this, write the coordinates of $\psi = \varphi \circ (\varphi')^{-1}$ as $(\psi^1, \ldots, \psi^n)$. Let g_{ij} and g'_{ij} denote the components of the metric when written using the coordinates φ and φ'. Then

$$g'_{ij}\Big|_{x'} = \left(\frac{\partial\psi^k}{\partial x_i}\frac{\partial\psi^m}{\partial x_j}g_{km}(\psi(\cdot))\right)\Bigg|_{x'}.$$

Meanwhile the respective components of γ' are given by $(\psi^{-1})^i(\gamma)$. Granted all of this, then the argument for coordinate independence amounts to a tedious exercise with the Chain rule. A *geodesic* is a map γ, from an interval $I \subset \mathbb{R}$ into M that obeys the geodesic equation in one, and thus all coordinate charts that contain points in the image of γ.

Here is a last bit of commonly used terminology: The functions

$$\left\{\Gamma^j_{km} = \frac{1}{2}g^{jp}(\partial_m g_{pk} + \partial_k g_{pm} - \partial_p g_{km})\right\}_{1 \le j,k,m \le n}$$

that appear in the theorem's statement are called the *Christoffel symbols*.

8.3 The existence of geodesics

The proof given momentarily for the existence of geodesics invokes a fundamental theorem about differential equations.

Theorem 8.2 (the vector field theorem) *Let* m *denote a positive integer and let* $\mathfrak{v}\colon \mathbb{R}^m \to \mathbb{R}^m$ *denote a given smooth map. Fix* $y_0 \in \mathbb{R}^m$ *and there exists an interval* $I \subset R$ *centered on* 0, *a ball* $B \subset \mathbb{R}^m$ *about the point* y_0, *and a smooth map* $z\colon I \times B \to \mathbb{R}^m$ *with the following property: If* $y \in B$, *then the map* $t \to z(t) = \mathfrak{z}(t, y)$ *obeys*

$$\frac{d}{dt} z = \mathfrak{v}(z).$$

with the initial condition $z|_{t=0} = y$. *Moreover, there is only one solution of this equation that equals* y *at* $t = 0$.

The name *vector field theorem* refers to the fact that $\mathfrak{v}$ can be viewed as a vector field on $\mathbb{R}^m$. When viewed in this light, the image of the map $t \to z(t)$ is said to be an integral curve of $\mathfrak{v}$.

This vector field theorem is proved in Appendix 8.1.

Proposition 8.3 below states a basic existence and uniqueness theorem for the geodesic equation.

Proposition 8.3 *Let* M *be a smooth manifold and let* g *denote a Riemannian metric on* M. *Let* $p \in M$ *and let* $v \in TM|_p$. *There exists* $\varepsilon > 0$ *and a unique map from the interval* $(-\varepsilon, \varepsilon)$ *to* M *that obeys the geodesic equation, sends* 0 *to* p, *and whose differential at* 0 *sends the vector* $\frac{\partial}{\partial t}$ *to* v.

Proof of Proposition 8.3 Fix a coordinate chart centered at p with coordinates $(x^1, \ldots, x^n)$ such that p is the origin. Let $I \subset R$ denote an open interval centered at 0, and view the geodesic equation as an equation for a map $z\colon I \to \mathbb{R}^n \times \mathbb{R}^n$ that obeys a certain system of differential equations that involve only first derivatives. To obtain this system of equations, write the coordinates of z as (γ, u). Then

$$\frac{d}{dt}(\gamma, u) = \left(u, -\Gamma^j_{jk}(\gamma)u^j u^k\right).$$

It is left as an exercise to verify that (γ, u) obeys the latter equation if and only if γ obeys the geodesic equation. This last equation has the schematic form $\frac{d}{dt}z = \mathfrak{v}(z)$ for a function $z\colon I \to \mathbb{R}^m$ where in this case $m = 2n$. This understood, the assertions of Proposition 8.3 follow directly from the vector field theorem.

The upcoming Proposition 8.4 states a stronger version of Proposition 8.3. As with Proposition 8.3, it is also a direct consequence of the preceding theorem. To set the stage, introduce π: TM $\to$ M to denote the vector bundle projection map. Suppose that O $\subset$ TM is an open set. Theorem 8.1 above assigns the following data to each point v $\in$ O: First, a neighborhood of $\pi(v)$, and second, a unique geodesic in this neighborhood that contains $\pi(v)$ and whose tangent vector at $\pi(v)$ is v. Thus, the points in O parametrize a family of geodesics in M. Among other things, the next proposition says that this family is *smoothly* parametrized by the points in O.

Proposition 8.4 *Let* M *be a smooth manifold and let* g *denote a Riemannian metric on* M. *Fix a point in* TM. *Then there exists an open neighborhood* O $\subset$ TM *of this point, a positive number* $\varepsilon > 0$ *and a smooth map* $\gamma_O: (-\varepsilon, \varepsilon) \times O \to M$ *with the following property: If* v $\in$ O, *then the map* $\gamma_O(\cdot, v): (-\varepsilon, \varepsilon) \to M$ *is a geodesic with* $\gamma_O(0, v) = \pi(v)$ *and with* $\gamma_O(\cdot, v)_*\left(\frac{\partial}{\partial t}\Big|_{t=0}\right)$ *equal to the vector* v.

Proof of Proposition 8.4 Fix a coordinate chart for a neighborhood of $\pi(x)$ to identify the latter with a neighorhood of the origin in $\mathbb{R}^n$ and to identify TM on this neighborhood with $\mathbb{R}^n \times \mathbb{R}^n$. As noted above, the geodesic equation can be viewed as an equation for a map, $z = (\gamma, u)$, from an interval in $\mathbb{R}$ about the origin to a neighborhood of $\{0\} \times \mathbb{R}^n$ in $\mathbb{R}^n \times \mathbb{R}^n$ that has the form

$$\frac{d}{dt}(\gamma, u) = \left(u, -\Gamma^j_{jk}(\gamma)u^j u^k\right).$$

As this equation has the form $\frac{d}{dt}z = \mathfrak{v}(z)$, an appeal to the theorem above proves Proposition 8.4.

8.4 First examples

What follows gives some examples of metrics with their corresponding geodesics.

8.4.1 The standard metric on $\mathbb{R}^n$

This is the Euclidean metric $g = dx_i \otimes dx_i$. As you might expect, the geodesics are the straight lines.

8.4.2 The round metric on S^n

The *round* metric is that induced by its embedding in $\mathbb{R}^{n+1}$ as the set $\{x \in \mathbb{R}^{n+1} : |x| = 1\}$. This embedding identifies TS^n as the set of pairs $(x, v) \in \mathbb{R}^{n+1} \times \mathbb{R}^{n+1}$

with $|x| = 1$ and with $x \cdot v = 0$. Here, $x \cdot v = \Sigma_{1 \le j \le n+1} x_j v_j$. The round metric is such that the inner product between elements (x, v) and (x, w) in TS^n is $v \cdot w$.

Suppose $\gamma: \mathbb{R} \to S^n \subset \mathbb{R}^{n+1}$ is a smooth curve. The geodesic equation when written in coordinates $\gamma(t) = (x^1(t), \ldots, x^{n+1}(t))$ asserts that

$$\ddot{x}^j + x^j |\dot{x}|^2 = 0.$$

To see that the solutions to this equation are the geodesics, view $\mathbb{R}^{n+1}$ as $\mathbb{R}^n \times \mathbb{R}$ and introduce the coordinate embedding $y \to (y, (1 - |y|^2)^{1/2})$ of the ball of radius 1 in $\mathbb{R}^n$ into $S^n \subset \mathbb{R}^n \times \mathbb{R}$. The round metric on S^n pulls back via this embedding as the metric on radius 1 ball in $\mathbb{R}^n$ with entries $g_{ij} = \delta_{ij} + y_i y_j (1 - |y|^2)^{1/2}$. Here, $\delta_{ij} = 1$ if $i = j$ and $\delta_{ij} = 0$ otherwise. Thus, $g_{ij} = \delta_{ij} + y_i y_j + \mathcal{O}(|y|^4)$. As a consequence, $\Gamma^i_{jk} = \delta_{jk} y_i + \mathcal{O}(|y|^2)$. The equation asserted by the theorem is $\ddot{y} + y_j |\dot{y}|^2 + \mathcal{O}(|y|^2) = 0$. This agrees with what is written above to leading order in y. Since the metric and the sphere are invariant under rotations of S^n, as is the equation for x above, this verifies the equation at all points.

Note that this equation implies that $t \to x^j(t)$ lies in a plane. Indeed, if $|\dot{x}| = 1$ at time $t = 0$, then this is true at all times t. Then the equation above finds that

$$x^j(t) = a^j \cos(t) + b^j \sin(t)$$

where $|a|^2 + |b|^2 = 1$ and $a_j b_j = 0$. In particular, the geodesics are the curves in S^n that are obtained by intersecting S^n with a plane through the origin in $\mathbb{R}^{n+1}$.

8.4.3 The induced metric on a codimension 1 submanifold in $\mathbb{R}^{n+1}$

This last example where the geodesic equation for the sphere is written as an equation for a curve in $\mathbb{R}^{n+1}$ has an analog of sorts for any n-dimensional submanifold in $\mathbb{R}^{n+1}$. Let $\Sigma \subset \mathbb{R}^{n+1}$ denote the submanifold in question. The relevant metric is defined as follows: Let $x \to n(x) \in \mathbb{R}^{n+1}$ denote a smooth assignment to a point $x \in \Sigma$ of a unit length unit normal vector to Σ at x. View $T\Sigma \subset \mathbb{R}^{n+1} \times \mathbb{R}^{n+1}$ as the set of pairs of the form (x, v) with $x \in \Sigma$ and with $v \cdot n(x) = 0$. The metric inner product between vectors (x, v) and (x, w) is declared to be $v \cdot w$. To say something about the geodesics, write a curve in Σ as a map from an interval in $\mathbb{R}$ to $\mathbb{R}^{n+1}$ as a map $t \to (x^1(t), \ldots, x^{n+1}(t))$ whose image lies in Σ. The geodesic equation asks that the curve $t \to x(t)$ in $\mathbb{R}^{n+1}$ obey

$$\ddot{x}^j + n^j|_x (\partial_k n_i)|_x \dot{x}^k \dot{x}^i = 0.$$

The proof that solutions to this equation give the geodesics in Σ is left to the reader. As a guide, remark that a proof can be had by choosing a very convenient coordinate chart about any given point, then writing this equation in the coordinate chart, and comparing it to the geodesic equation.

8.4.4 The hyperbolic metric on $\mathbb{R}^n$

Let $M \subset \mathbb{R}^{n+1}$ denote the branch of the hyperbola where $|x_{n+1}|^2 - \sum_{1\le k\le n} |x_k|^2 = 1$ and $x_{n+1} > 0$. This manifold M is diffeomorphic to $\mathbb{R}^n$ with diffeomorphism given by the map that sends $y \in \mathbb{R}^n$ to

$$\psi(y) = (y, (1 + |y|^2)) \in \mathbb{R}^{n+1} = \mathbb{R}^n \times \mathbb{R}.$$

In any event, view $TM \subset T\mathbb{R}^{n+1} = \mathbb{R}^{n+1} \times \mathbb{R}^{n+1}$ as the space of pairs (x, v) with $x \in M$ and

$$v_{n+1}x_{n+1} - \sum_{1\le k\le n} v_k x_k = 0.$$

Granted this identification of TM, define a Riemannian metric on TM by the rule that makes the inner product between pairs (x, v) and (x, w) equal to

$$\sum_{1\le k\le n} v_k w_k - v_{n+1}w_{n+1}.$$

This metric is positive definite, as can be seen by writing $v_{n+1} = x_{n+1}^{-1}\sum_{1\le k\le n} v_k x_k$ and using the fact that $x_{n+1} = (1 + \sum_{1\le k\le n} x_k^2)^{1/2}$. I use $g_{\mathbb{H}}$ in what follows to denote this metric.

The metric $g_{\mathbb{H}}$ just defined is called the *hyperbolic metric*. To justify this name, remark that the geodesics are as follows: Fix any point $(x, v) \in TM \subset \mathbb{R}^{n+1} \times \mathbb{R}^{n+1}$ with $\sum_{1\le k\le n} v_k^2 - v_{n+1}^2 = 1$. and $v \in TM|_x$. Then the curve $t \to x(t) = \cosh(t)x + \sinh(t)v$ is a geodesic curve in M. Note that these curves are the intersections between M and the planes through the origin in $\mathbb{R}^{n+1}$.

The proof that the curves just described are the geodesics can be had by using the coordinate chart given by the map ψ above. Note in this regard that it is enough to focus on the case where $x = (0, 1) \in \mathbb{R}^n \times \mathbb{R}$ by invoking symmetry arguments that are much like those used in the discussion above for the round metric on S^n. To say more about this, I need to introduce the *Lorentz group*, SO(n, 1). This is the subgroup of $Gl(n+1; \mathbb{R})$ that preserves the indefinite quadratic form $\eta(x, x) = |x_{n+1}|^2 - \sum_{1\le k\le n} |x_k|^2$. This is to say that SO(n,1) consists of the matrices $\mathfrak{m} \in Gl(n+1; \mathbb{R})$ for which $\eta(\mathfrak{m}x, \mathfrak{m}x) = \eta(x, x)$ for all $x \in \mathbb{R}^{n+1}$. As it turns out, this is a Lie group whose dimension is $\frac{1}{2}n(n + 1)$.

In any event, $\mathfrak{m}x \in M$ if $x \in M$ and $\mathfrak{m} \in SO(n, 1)$. This is to say that the linear map $\mu_m: \mathbb{R}^{n+1} \to \mathbb{R}^{n+1}$ given by $x \to \mathfrak{m}x$ restricts to define a diffeomorphism of M. Moreover, this diffeomorphism is such that $\mu_m^* g_{\mathbb{H}} = g_{\mathbb{H}}$ and so the length of a curve and the length of its μ_m image are the same. As a consequence, μ_m maps geodesics to geodesics. As explained momentarily, given $x \in M$ there exists $\mathfrak{m} \in SO(n, 1)$ such that $\mathfrak{m}x$ is the point $(0, 1) \in \mathbb{R}^n \times \mathbb{R}$. Thus, one need only identify the geodesics through the latter point.

What follows is a matrix $\mathfrak{m} \in SO(n, 1)$ that maps a given point $x \in M$ to the point (0, 1): Write $x = (\hat{x}, x_{n+1}) \in \mathbb{R}^{n+1} = \mathbb{R}^n \times \mathbb{R}$. Let $o \in SO(n)$ denote a matrix which is such that $o\hat{x} = (|\hat{x}|, 0, \ldots, 0)$. Let $\mathfrak{o} \in SO(n, 1)$ denote the matrix that acts on any given $v = (\hat{v}, v_{n+1}) \in \mathbb{R}^{n+1} = \mathbb{R}^n \times \mathbb{R}$ to give $\mathfrak{o}v = (o\hat{v}, v_{n+1})$. Now set $\mathfrak{p} \in SO(n+1)$ to be the matrix that acts on $(v_1, v_2, \ldots, v_{n+1}) \in \mathbb{R}^{n+1}$ to give

$$(x_{n+1}v_1 - |\hat{x}|v_{n+1}, v_2, \ldots, v_n, x_{n+1}v_{n+1} + |\hat{x}|v_1).$$

The matrix $\mathfrak{m} = \mathfrak{p}\mathfrak{o}$ does the job.

8.5 Geodesics on SO(n)

To define the relevant metric, first define an inner product on the space $\mathbb{M}(n; \mathbb{R})$ of $n \times n$ matrices by $\langle \mathfrak{a}, \mathfrak{a}' \rangle = \mathrm{tr}(\mathfrak{a}^T\mathfrak{a}')$. This is a Euclidean metric. Fix an orthonormal basis. As I will be talking momentarily about SO(n) also, it is convenient to fix one so that each basis element is either a symmetric matrix or an antisymmetric one. (Note that the symmetric matrices are orthogonal to the antisymmetric ones.) Let $\{\mathfrak{a}_j\}$ denote the basis elements. Here, j runs from 1 to n^2. Assume that the first $n(n-1)/2$ basis elements are antisymmetric.

For each $j \in \{1, \ldots, n^2\}$, let ω^j denote the 1-form on $Gl(n; \mathbb{R})$ given by

$$\omega^j|_m = \mathrm{tr}(\mathfrak{a}_j m^{-1} dm).$$

As noted in Chapter 5.3, these forms are left-invariant 1-forms. Since the collection spans $T^*Gl(n; \mathbb{R})$ at the identity, so these forms span $TGL(n; \mathbb{R})$ at any give point. Thus, they give an isomorphism of vector bundles $TGl(n; \mathbb{R}) \to Gl(n; \mathbb{R}) \times \mathbb{M}(n; \mathbb{R})$ by the rule that associates to a pair $m \in Gl(n; \mathbb{R})$ and $v \in TGl(n; \mathbb{R})|_m$ the pair $(m, \sum_i \mathfrak{a}_i \langle \omega^i|_m, v \rangle)$ in the product $Gl(n; \mathbb{R}) \times \mathbb{M}(n; \mathbb{R})$. This basis $\{\omega^i\}$ gives the metric

$$g = \sum_{1 \leq i \leq n^2} \omega^i \otimes \omega^i$$

on $Gl(n; \mathbb{R})$. Note that this metric does not depend on the chosen orthonormal basis since any one basis can be obtained from any other by the action of the group $SO(n^2)$ on the Euclidean space $\mathbb{M}(n; \mathbb{R})$.

The metric g is invariant under the action on $Gl(n; \mathbb{R})$ of left translation by elements in $Gl(n; \mathbb{R})$. This means the following: If $u \in Gl(n; \mathbb{R})$, then the diffeomorphism $\mathfrak{l}_u: Gl(n; \mathbb{R}) \to Gl(n; \mathbb{R})$ that sends m to um is such that $\mathfrak{l}_u^* g = g$. Such is the case because $\mathfrak{l}_u^* \omega^i = \omega^i$. (In general, metrics on a manifold M are sections of $\mathrm{Sym}^2(T^*M)$ and so pull-back under diffeomorphisms.)

As noted in Chapter 5.5, the form ω^j restricts as zero on $TSO(n) \subset TGL(n; \mathbb{R})|_{SO(n)}$ when $\mathfrak{a}_j = \mathfrak{a}_j^T$. Meanwhile, the collection $\{\omega^j : \mathfrak{a}_j^T = -\mathfrak{a}_j\}$

span TSO(n) along SO(n). This understood, introduce again $\mathbb{A}(n;\mathbb{R}) \subset \mathbb{M}(n;\mathbb{R})$ to denote the vector space of anti-symmetric matrices. The set $\{\omega^j\}_{1\le j\le n(n-1)/2}$ restrict to SO(n) to give a bundle isomorphism $TSO(n) \to SO(n) \times \mathbb{A}(n;\mathbb{R})$. It follows as a consequence that the restriction to $TSO(n) \subset TGl(n;\mathbb{R})|_{SO(n)}$ of the metric defined above on $TGl(n;\mathbb{R})$ defines the metric

$$g = \sum_{1\le i\le n(n-1)/2} \omega^i \otimes \omega^i$$

on TSO(n). This metric is left invariant with respect to the left multiplication map on SO(n) because it is left invariant on $Gl(n;\mathbb{R})$ and, in addition, because left multiplication on $Gl(n;\mathbb{R})$ by an element in SO(n) maps SO(n) to itself.

As an aside, note that this metric $g = \sum_{1\le i\le n(n-1)/2} \omega^i \otimes \omega^i$ on TSO(n) is also invariant with respect to the right multiplication map on SO(n). This is to say that it pulls back to itself under the right multiplication map $\mathfrak{r}_u$ on SO(n) that sends $m \to mu^{-1}$ with u any given element in SO(n). To see why this is so, remark first that

$$\mathfrak{r}_u^* \omega^i = \mathrm{tr}(u^{-1}\mathfrak{a}_i\, u\, m^{-1}dm).$$

Note next that the map $\mathfrak{a} \to u^{-1}\mathfrak{a}u$ maps $\mathbb{A}(n;\mathbb{R})$ to itself when $u \in SO(n)$. Moreover, $\mathrm{tr}((u^{-1}\mathfrak{a}u)^T(u^{-1}\mathfrak{a}'u)) = \mathrm{tr}(\mathfrak{a}^T\mathfrak{a}')$ for any $\mathfrak{a} \in \mathbb{M}(n;\mathbb{R})$. Thus, the map $\mathfrak{a} \to u^{-1}\mathfrak{a}u$ acts as an isometry on $\mathbb{A}(n;\mathbb{R})$ when $u \in SO(n)$. This understood, it follows that

$$u^{-1}\mathfrak{a}_i u = \sum_{1\le j\le n(n-1)/2} S_i^j \mathfrak{a}_j$$

when $\mathfrak{a}_i \in \mathbb{A}(n;\mathbb{R})$ where S_i^j are the entries of a $(n(n-1)/2)\times(n(n-1)/2)$ orthogonal matrix, S. This implies that

$$\mathfrak{r}_u^* g = \sum_i \sum_{jk} S_i^j S_i^k\, \omega^j \otimes \omega^k$$

where the sums range from 1 to $n(n-1)/2$. Since S is orthogonal, the sum on the right over i of $S_i^j S_i^k$ is δ_{jk} where δ_{jk} here denotes the $(n(n-1)/2) \times (n(n-1)/2)$ identity matrix. This is to say that $\mathfrak{r}_u^* g = \sum_{1\le j,k\le n(n-1)/2} \delta_{jk}\, \omega^j \otimes \omega^k = g$. The metric g is said to be *bi-invariant* because it is both left and right invariant. As it turns out, any bi-invariant metric on SO(n) has the form cg with $c > 0$ a constant. Thus, the bi-invariant metrics are rather special.

Return now to the issue of the geodesics for this metric g for TSO(n). The proposition that follows describes these geodesics. The statement of the theorem refers to the exponential map from $\mathbb{M}(n;\mathbb{R})$ to $Gl(n;\mathbb{R})$ that sends a given matrix $\mathfrak{a} \in \mathbb{M}(n;\mathbb{R})$ to $e^{\mathfrak{a}} = \iota + \mathfrak{a} + \frac{1}{2}\mathfrak{a}^2 + \frac{1}{3!}\mathfrak{a}^3 + \cdots$. As noted in Chapter 5.5, this map sends $\mathbb{A}(n;\mathbb{R})$ into SO(n) and it restricts to some small radius ball about the origin in $\mathbb{A}(n;\mathbb{R})$ as a diffeomorphism onto a neighborhood in SO(n) of ι.

Proposition 8.5 *Identify* TSO(n) *with* SO(n) × $\mathbb{A}$(n; $\mathbb{R}$) *as just described. With the metric* g *as given above, any solution to the geodesic equation on* SO(n) *and has the form* t → m $e^{t\mathfrak{a}}$ *where* m ∈ SO(n) *and* $\mathfrak{a}$ ∈ $\mathbb{A}$(n; $\mathbb{R}$). *Conversely, any such map from* $\mathbb{R}$ *into* SO(n) *parametrizes a geodesic.*

The proof of this proposition (and of others to come) make use of an important observation about geodesics for induced metrics. To set the stage, suppose that X is a given Riemannian manifold and that M is a submanifold of X. Give M the induced metric. A geodesic in M need not be a geodesic in X. Witness, for example, the case of the round sphere in $\mathbb{R}^n$ as described in the previous part of this chapter. By contrast, a geodesic in X that lies entirely in M is a priori also a geodesic for the induced metric on M. This follows from the fact that geodesics are locally length minimizing.

With the preceding understood, the proposition is proved by demonstrating that any path of the form m → m$e^{t\mathfrak{a}}$ for m ∈ Gl(n; $\mathbb{R}$) and $\mathfrak{a}$ ∈ $\mathbb{A}$(n; $\mathbb{R}$) is a priori a geodesic in Gl(n; $\mathbb{R}$). Note that this last result implies the assertion made by the proposition for the following reason: If every path through a given matrix m ∈ SO(n) of the form t → m$e^{t\mathfrak{a}}$ with $\mathfrak{a}$ ∈ $\mathbb{A}$(n; $\mathbb{R}$) is a geodesic, then Proposition 8.3 guarantees that such paths account for all of the geodesics that pass through m.

By way of notation, a submanifold M in a Riemannian manifold X is said to be *totally geodesic* when all geodesics in M as defined by the induced metric from X are also geodesics for the metric on X. The proof given here of the proposition demonstrates that SO(n) is a totally geodesic submanifold in Gl(n; $\mathbb{R}$).

8.5.1 Geodesics on Gl(n; $\mathbb{R}$)

The proposition that follows describes certain of the geodesics on Gl(n; $\mathbb{R}$).

Proposition 8.6 *Identify* TGl(n; $\mathbb{R}$) *with* Gl(n; $\mathbb{R}$) × $\mathbb{M}$(n; $\mathbb{R}$) *as described above. With the metric* g *as above, paths of the form* t → m $e^{t\mathfrak{a}}$ *with* m ∈ Gl(n; $\mathbb{R}$) *and* $\mathfrak{a}$ ∈ $\mathbb{M}$(n; $\mathbb{R}$) *obeying* $\mathfrak{a}\,\mathfrak{a}^T - \mathfrak{a}^T\mathfrak{a} = 0$ *are geodesics. In particular, paths of the form* t → m $e^{t\mathfrak{a}}$ *are geodesics in* Gl(n; $\mathbb{R}$) *if* $\mathfrak{a} = \pm\mathfrak{a}^T$.

By way of notation, if $\mathfrak{a}$ and $\mathfrak{a}'$ are any two n × n matrices, the matrix $\mathfrak{a}\mathfrak{a}' - \mathfrak{a}'\mathfrak{a}$ is said to be their commutator and is written $[\mathfrak{a}, \mathfrak{a}']$.

Proof of Proposition 8.6 The metric is left invariant and this has the following consequence: If I is an interval containing 0 and γ: I → Gl(n; $\mathbb{R}$) is a solution to the geodesic with γ(0) = m, then the map t → $\mathfrak{L}_{m^{-1}}(\gamma(t))$ is a solution to the geodesic equation that hits the identity matrix ι at t = 0. This understood, no generality is lost by showing that a solution, γ, to the geodesic equation with γ(0) = ι has the form t → $e^{t\mathfrak{a}}$ if $[\mathfrak{a}, \mathfrak{a}^T] = 0$.

To see this, remember that the map from $\mathbb{M}(n;\mathbb{R}) \to Gl(n;\mathbb{R})$ that sends a to e^{a} gives a diffeomorphism of some neighborhood of ι in $Gl(n;\mathbb{R})$ with a ball about the origin in $\mathbb{M}(n;\mathbb{R})$. This understood, it is enough to prove that any line through the origin in the vector space $\mathbb{M}(n;\mathbb{R})$ of the form $t \to ta$ with $[a, a^{T}] = 0$ is a geodesic for the metric that is obtained by pulling back the metric g on $Gl(n;\mathbb{R})$. As a line obeys $\ddot{\gamma}^{i} = 0$, so it is sufficient to prove that the term $\Gamma^{i}{}_{jk}\dot{\gamma}^{j}\dot{\gamma}^{k}$ is also zero when γ is the line $t \to t\,a$ with $[a, a^{T}] = 0$.

To show this, consider first the pull-back of g via the map $a \to e^{a}$. Recall from Chapter 5.5 that the pull-back of the 1-form ω_q for $q \in \mathbb{M}(n;\mathbb{R})$ via the exponential map is the 1-form

$$\upsilon_q = \int_0^1 \mathrm{tr}(e^{sa} q e^{-sa} da)ds.$$

It follows as a consequence that the metric g appears with respect to this coordinate chart as

$$g|_a = \int_0^1\int_0^1 \mathrm{tr}(e^{-sa^T} da^T e^{sa^T} e^{s'a} da e^{-s'a})ds ds' .$$

This is to say that

$$g_{ij}|_a = \int_0^1\int_0^1 \mathrm{tr}(e^{-sa^T} a_i^T e^{sa^T} e^{s'a} a_j e^{-s'a})ds ds' .$$

Granted the preceding, what follows explains why $\Gamma^{i}_{jk}\dot{\gamma}^{i}\dot{\gamma}^{j}$ is zero when $t \to \gamma(t) = ta$ with a obeying $[a, a^{T}] = 0$. To show this, digress momentarily and fix elements $q_1, q_2 \in \mathbb{M}(n,\mathbb{R})$. Consider the function on $\mathbb{M}(n;\mathbb{R})$ given by

$$a \to \mathrm{tr}(e^{-sa^T} q_1^T e^{sa^T} e^{s'a} q_2 e^{-s'a}) .$$

Note that the integrand above that defines the pull-back at a of the metric g involves this sort of expression. The directional derivative of the function just described in the direction of $q \in Gl(n;\mathbb{R})$ is given by

$$s\int_0^1 d\tau\, \mathrm{tr}((-e^{-(1-\tau)sa^T} q^T e^{-\tau sa^T} q_1^T e^{sa^T} + e^{-sa^T} q_1^T e^{(1-\tau)sa^T} q^T e^{\tau sa^T}) e^{s'a} q_2 e^{-s'a})$$

$$+ s' \int_0^1 d\tau\, \mathrm{tr}(e^{-sa^T} q_1^T e^{sa^T} (e^{(1-\tau)s'a} q e^{\tau s'a} q_2 e^{-s'a} - e^{s'a} q_2 e^{-(1-\tau)s'a} q e^{-\tau s'a})).$$

This last observation is relevant because the evaluation $\Gamma^i_{jk}\dot\gamma^j\dot\gamma^k$ requires evaluating expressions of this sort when two of $\mathfrak{q}_1$, $\mathfrak{q}_2$ or $\mathfrak{q}$ have the form $\sum_i \dot\gamma^i \mathfrak{a}_i$. Note in particular, that this last sum is $\mathfrak{a}$ when γ is the line $t \to t\mathfrak{a}$.

Granted the preceding consider the expression above first when $\mathfrak{q}$ and either of $\mathfrak{q}_1$, $\mathfrak{q}_2$ are multiples of $\mathfrak{a}$. If this is the case, then the cyclic property of the trace implies that the integrand above is zero. Meanwhile, if $\mathfrak{q}_1$ and $\mathfrak{q}_2$ are multiples of $\mathfrak{a}$, then what is written above is proportional to

$$\int_0^1 d\tau\, \mathrm{tr}((s e^{-(1-\tau)s\mathfrak{a}^T}\mathfrak{q}^T e^{(1-\tau)s\mathfrak{a}^T} + s' e^{(1-\tau)s'\mathfrak{a}}\mathfrak{q} e^{-(1-\tau)s'\mathfrak{a}})[\mathfrak{a},\mathfrak{a}^T]) .$$

In particular, the integral of this last expression with respect to s and s′ is zero since the result can be written as $\mathrm{tr}(\mathfrak{b}\,[\mathfrak{a}, \mathfrak{a}^T])$. Thus $\Gamma^i_{jk}\,\dot\gamma^j\,\dot\gamma^k = 0$ for the line $t \to t\mathfrak{a}$ when $[\mathfrak{a}, \mathfrak{a}^T] = 0$.

8.6 Geodesics on U(n) and SU(n)

The story for these groups is very much the $\mathbb{C}$ analog of what was done just now for $Gl(n; \mathbb{R})$ and $SO(n)$. To start, define a Hermitian inner product on the vector space $\mathbb{M}(n; \mathbb{C})$ of $n \times n$ matrices with complex entries by the rule $(\mathfrak{a}, \mathfrak{a}') \to \mathrm{tr}(\mathfrak{a}^\dagger\mathfrak{a}')$. Fix an orthonormal basis $\{\mathfrak{a}_i\}_{1\le i\le n^2}$ for $\mathbb{M}(n; \mathbb{C})$ with respect to this Hermitian inner product. For each integer $i \in \{1, \ldots, n^2\}$, let ω^i denote the $\mathbb{C}$-valued 1-form on $Gl(n; \mathbb{C})$ given by

$$\omega^i|_m = \mathrm{tr}(\mathfrak{a}_i m^{-1} dm).$$

As noted in Chapter 5.6, each of these forms is a left invariant 1-form on $Gl(n; \mathbb{C})$. As their real and imaginary parts span $T^*Gl(n; \mathbb{C})|_\iota$, so their real and imaginary parts span $T^*Gl(n; \mathbb{C})$ at each point. This understood, their real and imaginary parts define an isomorphism from $TGl(n; \mathbb{C})$ to the product bundle $Gl(n; \mathbb{C}) \times \mathbb{M}(n; \mathbb{C})$. The isomorphism in question sends a given tangent vector $\mathfrak{v}$ at a given point $m \in Gl(n; \mathbb{C})$ to the pair $(m, \sum_{1\le i\le n^2} \mathfrak{a}_i \langle\omega^i, \mathfrak{v}\rangle)$. Meanwhile, the collection $\{\omega^i\}$ defines the metric

$$g = \textstyle\sum_{1\le i\le n^2} \mathrm{re}(\bar\omega^i \otimes \omega^i)$$

on $TGl(n; \mathbb{C})$. Note that g does not depend on the chosen Hermitian basis because any such basis is obtained from any other by the action of some element in $U(n^2)$.

As an aside, note that $T\,\mathbb{M}(n; \mathbb{C})$ has an almost complex structure such that the resulting complex, rank n^2 vector bundle is spanned at each point by the entries of the matrix $\frac{\partial}{\partial \mathfrak{m}}$. As an open set in $\mathbb{M}(n; \mathbb{C})$, the tangent bundle to $Gl(n; \mathbb{C})$ inherits this almost complex structure. This understood, the isomorphism defined by the forms $\{\omega^i\}$ from $TGl(n; \mathbb{C})$ to $Gl(n; \mathbb{C}) \times \mathbb{M}(n; \mathbb{C})$ is $\mathbb{C}$-linear, and thus an isomorphism of complex vector bundles. The metric g just described is defined on the underlying real tangent bundle of $Gl(n; \mathbb{C})$ by the Hermitian metric $\sum_{1 \le i \le n^2} \bar{\omega}^i \otimes \omega^i$ on the complex bundle.

To say more about the metric g, introduce from Chapter 2.4 the almost complex structure $j_0 \in \mathbb{M}(2n; \mathbb{R})$ so as to view $\mathbb{M}(n; \mathbb{C})$ as the subvector space $\mathbb{M}_j \subset \mathbb{M}(n; 2\mathbb{R})$ of matrices that obey $\mathfrak{m} j_0 - j_0 \mathfrak{m} = 0$. In particular, recall that $\mathfrak{m} \in \mathbb{M}_j$ if and only if its entries obey $\mathfrak{m}_{2k,2i} = \mathfrak{m}_{2k-1,2i-1}$ and $\mathfrak{m}_{2k,2i-1} = -\mathfrak{m}_{2k-1,2i}$ for any pair of i, k from $\{1, \ldots, n\}$. This understood, the identification between $\mathbb{M}(n; \mathbb{C})$ and $\mathbb{M}_j$ associates to a given $\mathfrak{a} \in \mathbb{M}(n; \mathbb{C})$ the matrix $\mathfrak{m} \in \mathbb{M}_j$ with $\mathfrak{m}_{2k-1,2i-1}$ and $\mathfrak{m}_{2k,2i-1}$ given by the respective real and imaginary parts of $\mathfrak{a}$. This identification sends $Gl(n; \mathbb{C}) \subset \mathbb{M}(n; \mathbb{C})$ to the subspace $\mathbb{G}_j \subset \mathbb{M}_j$ of invertible elements. In particular, it follows directly from all of this that the metric g defined above on $Gl(n; \mathbb{C})$ is the restriction to $\mathbb{G}_j \subset Gl(2n; \mathbb{R})$ of the metric that was defined above in the preceding part of this chapter.

8.6.1 The geodesics on Gl(n; ℂ)

The next proposition describes some of the geodesics for the metric just described.

Proposition 8.7 *Identify* $TGl(n; \mathbb{C})$ *with* $Gl(n; \mathbb{C}) \times \mathbb{M}(n; \mathbb{C})$ *as just described. With the metric* g *as above, any path of the form* $t \to \mathfrak{m}\, e^{t\mathfrak{a}}$ *with* $\mathfrak{m} \in Gl(n; \mathbb{C})$ *and* $\mathfrak{a} \in \mathbb{M}(n; \mathbb{C})$ *obeying* $[\mathfrak{a}, \mathfrak{a}^\dagger] = 0$ *is a geodesic.*

***Proof of Proposition** 8.7* There are two ways to argue. Here is the first: Suppose that $\mathfrak{a} \in \mathbb{M}_j$ and $\mathfrak{m} \in \mathbb{G}_j$. The path $t \to \mathfrak{m}\, e^{t\mathfrak{a}}$ in $Gl(2n; \mathbb{R})$ lies entirely in $\mathbb{G}_j$, and so if it is a geodesic in $Gl(2n; \mathbb{R})$, then it is a geodesic in $\mathbb{G}_j$ as defined using the metric on the latter that is induced by the inclusion $\mathbb{G}_j \subset Gl(2n; \mathbb{R})$. It is a geodesic in $Gl(2n; \mathbb{R})$, if it is the case that $[\mathfrak{a}, \mathfrak{a}^T] = 0$. This understood, let $\mathfrak{a}_\mathbb{C}$ denote $\mathfrak{a}$'s counterpart in $\mathbb{M}(n; \mathbb{C})$. As explained momentarily, the commutator $[\mathfrak{a}, \mathfrak{a}^T] = 0$ in $Gl(2n; \mathbb{R})$ if and only if $[\mathfrak{a}_\mathbb{C}, \mathfrak{a}_\mathbb{C}^\dagger] = 0$ in $Gl(n; \mathbb{C})$. As noted in Chapter 2.4, the product $\mathfrak{b}\mathfrak{b}'$ of matrices in $\mathbb{M}_j$ is in $\mathbb{M}_j$ and obeys $(\mathfrak{b}\mathfrak{b}')_\mathbb{C} = \mathfrak{b}_\mathbb{C}\mathfrak{b}'_\mathbb{C}$. Thus $(\mathfrak{m}e^{t\mathfrak{a}})_\mathbb{C} = \mathfrak{m}_\mathbb{C} e^{t\mathfrak{a}_\mathbb{C}}$, and so the paths in $Gl(n; \mathbb{C})$ as described by the proposition are geodesics.

To see that $[\mathfrak{a}, \mathfrak{a}^T] = 0$ in $Gl(2n; \mathbb{R})$ if and only if $[\mathfrak{a}_\mathbb{C}, \mathfrak{a}_\mathbb{C}^\dagger] = 0$ in $Gl(n; \mathbb{C})$, note first that the matrix j_0 as defined in Chapter 2.4 obeys $j_0^T = -j_0$. This implies that $\mathfrak{a}^T \in \mathbb{M}_j$ if $\mathfrak{a} \in \mathbb{M}_j$. It is left for the reader to verify that $(\mathfrak{a}^T)_\mathbb{C} = \mathfrak{a}_\mathbb{C}^\dagger$. Granted these last points, it then follows $[\mathfrak{a}, \mathfrak{a}^T] \in \mathbb{M}_j$ and $([\mathfrak{a}, \mathfrak{a}^T])_\mathbb{C} = [\mathfrak{a}_\mathbb{C}, \mathfrak{a}_\mathbb{C}^\dagger]$.

The second proof is the $\mathbb{C}$ analog of the argument for the Gl(n; $\mathbb{R}$) version of this proposition. The only substantive difference is that the expression for the metric in the case of Gl(n; $\mathbb{C}$) replaces each occurrence of the transpose of a matrix with the Hermitian conjugate of the matrix. To elaborate just a little, as with Gl(n; $\mathbb{R}$), the fact that the metric is invariant with respect to the left translation diffeomorphisms implies that it is sufficient to prove that the paths $t \to e^{t\mathfrak{a}}$ are geodesics if $\mathfrak{a} \in \mathbb{M}(n; \mathbb{C})$ obeys $[\mathfrak{a}, \mathfrak{a}^\dagger] = 0$. To see if this is the case, recall from Chapter 5.6 that the exponential map $\mathfrak{a} \to e^{\mathfrak{a}}$ from the vector space $\mathbb{M}(n; \mathbb{C})$ into Gl(n; $\mathbb{C}$) restricts to a small radius ball about 0 in $\mathbb{M}(n; \mathbb{C})$ as a diffeomorphism from said ball onto a neighborhood of the identity matrix in Gl(n; $\mathbb{C}$). The metric pulls back via this map to

$$g|_{\mathfrak{a}} = -\mathrm{re}\int_0^1\int_0^1 \mathrm{tr}(e^{-s\mathfrak{a}^\dagger}\, d\mathfrak{a}^\dagger e^{s\mathfrak{a}^\dagger} e^{s'\mathfrak{a}} d\mathfrak{a} e^{-s'\mathfrak{a}})ds ds'.$$

Note in particular that Hermitian adjoints appear in this formula, not transposes. But for this replacement here and subsequently the argument for the proposition is identical to that used to prove the Gl(n; $\mathbb{R}$) version of the proposition. As such, the details are left to the reader.

8.6.2 The geodesics on U(n) and SU(n)

View U(n) as the subgroup in Gl(n; $\mathbb{C}$) of matrices m with $m^{-1} = m^\dagger$ and view SU(n) $\subset$ U(n) as the subgroup of matrices with determinant 1. As such, the metric on Gl(n; $\mathbb{C}$) described above induces metrics on both U(n) and SU(n). Note in this regard that the 1-form $\mathrm{tr}(\mathfrak{a}\, m^{-1}dm)$ restricts to zero on TU(n) if $\mathfrak{a}^\dagger = \mathfrak{a}$, and it is nowhere zero and real valued on TU(n) if $\mathfrak{a} \neq 0$ and $\mathfrak{a}^\dagger = -\mathfrak{a}$. Such a form is nonzero on TSU(n) if and only if $\mathfrak{a}$ is not a multiple of the identity.

Keeping this in mind, let $\{\mathfrak{a}_i\}_{1\le i\le n^2}$ denote an orthonormal basis for the vector space $\mathbb{A}(n; \mathbb{C}) \subset \mathbb{M}(n; \mathbb{C})$, this the vector subspace (over $\mathbb{R}$) of skew Hermitian matrices. Thus $\mathfrak{a}^\dagger = -\mathfrak{a}$ when $\mathfrak{a} \in \mathbb{A}(n; \mathbb{C})$. Agree to take $\mathfrak{a}_i$ to have trace zero when $i < n^2$ and to take $\mathfrak{a}_{n^2} = \frac{i}{\sqrt{n}}\iota$. The induced metric on U(n) is $g = \sum_{1\le i\le n^2} \omega^i \otimes \omega^i$, and that on SU(n) is $g = \sum_{1\le i\le n^2-1} \omega^i \otimes \omega^i$.

Let $m \in$ U(n) and let $\mathfrak{a} \in \mathbb{A}(n; \mathbb{C})$. As noted in Chapter 5.6, the matrix $me^{\mathfrak{a}}$ is in U(n). Moreover, if $m \in$ SU(n) and $\mathfrak{a}$ has zero trace, then $m\, e^{\mathfrak{a}}$ is in SU(n). This is background for the proposition that follows.

Proposition 8.8 *The geodesics for the metric defined above on* U(n) *are of the form* $t \to m\, e^{t\mathfrak{a}}$ *where* $m \in$ U(n) *and* $\mathfrak{a} \in \mathbb{A}(n; \mathbb{C})$. *The geodesic for the metric just defined on* SU(n) *are of this form with* $m \in$ SU(n) *and* $\mathfrak{a}$ *a traceless matrix* in $\mathbb{A}(n; \mathbb{C})$.

Proof of Proposition 8.8 Because the curves described are geodesics in Gl(n; $\mathbb{C}$), they are a priori geodesics in the subgroup in question. The first proposition in Chapter 8.3 above guarantees that these account for all of the geodesics.

8.7 Geodesics and matrix groups

A Lie group G is said in what follows to be a *compact matrix* group if it appears as a subgroup of some $n \geq 1$ version of either SO(n) or U(n). Thus, G is a submanifold with the property that $\mathfrak{m}^{-1} \in G$ if $\mathfrak{m}$ is a matrix in G, and also $\mathfrak{m}\mathfrak{h} \in G$ when both $\mathfrak{m}$ and $\mathfrak{h}$ are in G. What was said above about the geodesics in SO(n), U(n) and SU(n) has an analog for any compact matrix group. The metric here is assumed to be the one induced by the metric on SO(n) or U(n) that is discussed in Chapters 8.5 and 8.6. This is to say that a tangent vector to G is a priori a tangent vector to the linear group, and so the metric used in Chapters 8.5 and 8.6 can be used to define its norm.

To elaborate, remark first that $TG|_\iota$ appears as a subvector space (over $\mathbb{R}$) in the vector space of rank $n \times n$ real, antisymmetric matrices, or complex anti-Hermitian matrices, as the case may be. It is customary to use $\mathfrak{lie}(G)$ to denote this vector space. The following proposition describes the geodesics in the general case:

Proposition 8.9 *Suppose that* G *is a compact matrix group, and let* $\mathfrak{m} \in G$.

- *The map* $\mathfrak{a} \to \mathfrak{m}\, e^{\mathfrak{a}}$ *from* $\mathfrak{lie}(G)$ *to the general linear group has image in* G.
- *This map restricts to some ball about the origin in* $\mathfrak{lie}(G)$ *as a diffeomorphism of the latter onto a neighborhood of* $\mathfrak{m}$ *in* G.
- *For any given* $\mathfrak{a} \in \mathfrak{lie}(G)$, *the map* $t \to \mathfrak{m}\, e^{t\mathfrak{a}}$ *from* $\mathbb{R}$ *to* G *is a geodesic; and all geodesics through* $\mathfrak{m} \in G$ *are of this sort.*

***Proof of Proposition** 8.9* The second and third bullets of Proposition 8.9 follow with the verification of the first bullet using the same arguments given above for the $G = SO(n)$ or U(n) or SU(n) cases. To see about the first bullet, note first that it is enough to consider the case when $\mathfrak{m}$ is the identity. To prove the assertion in the latter case, it is enough to prove that the map $\mathfrak{a} \to e^{\mathfrak{a}}$ maps $\mathfrak{lie}(G)$ into G. Indeed, as the differential of this map at the origin is the identity on $\mathfrak{lie}(G)$, it then follows from the inverse function theorem (Theorem 1.1) that this map restricts to a ball about the origin in $\mathfrak{lie}(G)$ as a diffeomorphism onto a neighborhood in G of ι.

Granted the preceding, fix some very large integer, N, and write $e^{\mathfrak{a}}$ as $(e^{\mathfrak{a}/N})^N$. Given the definition of $\mathfrak{lie}(G)$ as $TG|_\iota$, it follows that $e^{\mathfrak{a}/N}$ can be written as $\mathfrak{g}_N + \mathfrak{r}_N$ where $\mathfrak{g}_N \in G$ and where $|\mathfrak{r}_N| \leq c_0 N^{-2}$. As a consequence, $e^{\mathfrak{a}} = (\mathfrak{g}_N)^N + \mathfrak{w}_N$

where $|\mathfrak{w}_N| \leq c_0 N^{-1}$. As each $\mathfrak{g}_N \in G$, so $(\mathfrak{g}_N)^N \in G$. Taking $N \to \infty$ shows that $e^{\mathfrak{a}} \in G$ also.

Appendix 8.1 The proof of the vector field theorem

The primary purpose of this appendix is to prove the vector field theorem (Theorem 8.2) in Chapter 8.3. By way of reminder, this theorem concerns solutions to an equation for a map from a neighorhood of the origin in $\mathbb{R}$ to $\mathbb{R}^m$ of the following sort: Let $\mathfrak{v}: \mathbb{R}^m \to \mathbb{R}^m$ denote a given smooth map, and let z denote a map from a neighborhood of the origin in $\mathbb{R}$ to $\mathbb{R}^m$. The map z is asked to obey the equation

$$\frac{d}{dt} z = \mathfrak{v}(z).$$

Proof of the vector field theorem (Theorem 8.2) Let $B \subset \mathbb{R}^n$ denote the radius 2 ball. Fix $\varepsilon > 0$ and let $I = [-\varepsilon, \varepsilon]$. Let $C^0_*(I; \mathbb{R}^m)$ denote the space of continuous maps from I to $\mathbb{R}^m$ that equal $0 \in \mathbb{R}^m$ at $0 \in I$. Keep in mind that this is a complete metric space where the topology is defined by the distance function that assigns to maps u and w the distance $\|u - w\| = \sup_{t \in I} |u(t) - w(t)|$. Define a map, $T: B \times C^0_*(I; \mathbb{R}^m) \to C^0_*(I; \mathbb{R}^m)$ by the rule

$$T(y, u)|_t = \int_{[0,t]} \mathfrak{v}(y + u(s))ds\ .$$

If $T(y, u) = u$ and if u is a smooth map, then the map $t \to z(t) = y + u(t)$ obeys the desired equation with initial condition $z(0) = y$. Conversely, a solution to the desired equation with $z(0) = y$ is a fixed point of the map $u \to T_y(u) = T(y; u)$.

Existence of a fixed point: To see that there is a fixed point, let $c_0 \geq 1$ denote a constant such that $|\mathfrak{v}(z)| \leq c_0$ when $|z| \leq 1$. Let $\mathcal{B}_1$ denote the ball in $C^0_*(I; \mathbb{R}^m)$ with radius 1 centered about the constant map. On this ball,

$$\left|T_y(u)|_t\right| \leq |t|\ c_0 \leq \varepsilon\ c_0.$$

Thus T_y maps $\mathcal{B}_1$ to itself if $\varepsilon < c_0^{-1}$. To see that T_y is a contraction for small ε, let $c_0 \geq 1$ be such that $|\mathfrak{v}(z) - \mathfrak{v}(z')| \leq c_0\ |z - z'|$ when $|z|$ and $|z'|$ are both less than 1. Then

$$\left|(T_y(u) - T_y(u'))|_t\right| \leq c_0\ t\ \sup_{|s| \leq t} |u(s) - u'(s)| \leq c_0\ t\ \|u - u'\|.$$

Thus if $\varepsilon < c_0^{-1}$, then $\|T_y(u) - T_y(u')\| \leq \|u - u'\|$ and T_y is uniformly contracting. As a consequence, it has a unique fixed point on the ball $\mathcal{B}_1$.

Fixed points are smooth maps: To see that a fixed point is smooth, note that what is written on the right-hand side (8.2) is a C^1 function. So, if u is equal to what is written on the right-hand side of (8.2), then u is a C^1 function. This implies that the right-hand side of (8.2) is a C^2 function. But then u is a C^2 function and so the right-hand side of (8.2) is a C^3 function. Continuing in this vein shows that u is infinitely differentiable. It follows, in particular, that any fixed point of the map $U \to T_y(U)$ obeys the desired equation.

There is a unique solution with $z(0) = y$: To prove this, suppose that $t \to z(t)$ is a solution to the equation for $t \in I$. I need to prove that $u = z - y$ is in $\mathcal{B}_1$, for then the result follows from the contraction mapping theorem (Theorem 1.6). To see that u is in $\mathcal{B}_1$, I need to prove that $|u(t)|$ is small for $t \in I$. To see that such is the case, let $s \in (0, \infty)$ denote the smallest value of t such that $|u(t)| = 1$. Since u is a fixed point of the map T_y, it follows that

$$|u(\pm s)| \le c_0|s|$$

and so if the constant ε that defines the interval I is chosen so that $\varepsilon < c_0^{-1}$, then this shows that $s > \varepsilon$ and so $u \in \mathcal{B}_1$.

The dependence on $y \in B$: Let $y \to u_y$ now denote the mapping that associates to each $y \in B$ the unique fixed point of the map T_y in $\mathcal{B}_1$. It is a consequence of U being a fixed point and the fact that $|\mathfrak{v}(z) - \mathfrak{v}(z')| \le c_0 |z - z'|$ when both $|z|$ and $|z'|$ are less than 1 that

$$\|u_y - u_{y'}\| \le c_0 \left(|y - y'| + \varepsilon \|u_y - u_{y'}\|\right)$$

thus, given that $\varepsilon < c_0^{-1}$, this tells us that the assignment $y \to u_y$ is Lipschitz as a map to the metric space $C^0_*(I, \mathbb{R}^m)$. In particular, for each $t \in I$, one has $|u_y(t) - u_{y'}(t)| \le c_0 |y - y'|$.

To say something about higher-order derivatives, one takes successively more involved difference quotients. This is straightforward but tedious, and I leave it to you.

Additional reading

- *Introduction to Smooth Manifolds*, John M. Lee, Springer, 2002.
- *An Introduction to Riemannian Geometry and the Tensor Calculus*, C. E. Weatherburn, Cambridge University Press, 2008.
- *Elementary Differential Geometry*, A. N. Pressley, Springer, 2010.
- *Introduction to Differentiable Manifolds*, Louis Auslander and Robert F. MacKenzie, Dover Publications, 2009.
- *Three Dimensional Geometry and Topology*, William Thurston and Silvio Levy, Princeton University Press, 1997.

- *The Shape of Space*, Jeffrey R. Weeks, CRC Press, 2001.
- *Differential Geometry, Lie Groups and Symmetric Spaces*, Sigurdur Helgason, American Mathematical Society, 2001.
- *A Course in Metric Geometry*, Dmitri Burago, Yuri Burago and Sergei Ivanov, American Mathematical Society, 2001.

9 Properties of geodesics

This chapter discusses various notions that involve geodesics. The final parts of this chapter use some of these to prove the geodesic theorem (Theorem 8.1).

9.1 The maximal extension of a geodesic

Let $I \subset \mathbb{R}$ denote a closed interval that is not the whole of $\mathbb{R}$, and let s denote an end point of I, for the sake of the discussion, the maximum point. Let $\gamma: I \to M$ denote a geodesic. It follows from Proposition 8.3 that there exists an extension of γ to an interval with maximal endpoint $s + \varepsilon$ for some $\varepsilon > 0$. To see this, take a point $q = \gamma(s')$ for s' nearly s. Then q is very close to $p = \gamma(s)$. Proposition 8.3 tells us that the geodesic $t \to \gamma_q(t) = \gamma(s' + t)$ is defined for $|t| \leq \varepsilon$ for some ε that depends only on p. This extends γ. It follows as a consequence that each geodesic in M has a maximum extension. In particular, if M is compact, then each geodesic extends as a geodesic mapping $\mathbb{R}$ into M.

A manifold M is called *geodesically complete* if each geodesic can be extended so as to map the whole real line into M. All compact manifolds are geodesically complete. To see a manifold with metric that is not geodesically complete, take M to be the complement in $\mathbb{R}^n$ of any nonempty set but take the Euclidean metric from $\mathbb{R}^n$. In general, the complement of any point in a geodesically complete manifold is not geodesically complete.

9.2 The exponential map

The following proposition holds when M is geodesically complete:

Proposition 9.1 *There exists a smooth map,* $\mathfrak{e}: \mathbb{R} \times TM \to M$ *with the following property: Fix* $v \in TM$. *Then the corresponding map* $\mathfrak{e}_v = \mathfrak{e}(\cdot, v): \mathbb{R} \to M$ *is the* *unique solution to the geodesic equation with the property that* $\mathfrak{e}_v(0) = \pi(v)$ *and* $(\mathfrak{e}_v)_*\left(\frac{\partial}{\partial t}\Big|_{t=0}\right) = 1$.

Proof of Proposition 9.1 The map $\mathfrak{e}$ is well defined because M is geodesically complete. The issue is whether $\mathfrak{e}$ is smooth. This is a local question in the following sense: The map $\mathfrak{e}$ is smooth if and only if it is smooth near any given point in its domain. To see that such is the case, fix a point $v \in TM$, a neighborhood $O \subset TM$ of v and $\varepsilon > 0$ such that $\mathfrak{e}$ is smooth on $(-\varepsilon, \varepsilon) \times O$. Proposition 8.4 can be used to obtain ε. Let $T \geq \varepsilon$ denote either ∞ or the smallest number such that $\mathfrak{e}$'s restriction to $[-T, T] \times \hat{O}$ is smooth for any open set $\hat{O} \subset TM$ that contains v. The claim made by the proposition follows with a proof that $T = \infty$.

To see that $T = \infty$, suppose to the contrary that T is finite so as to derive some nonsense. Let $p = \mathfrak{e}_V(T) = \mathfrak{e}(T, v)$ and let $v_T \in TM|_p$ denote the vector $(\mathfrak{e}_v)_* \frac{\partial}{\partial t}$ at $t = T$. Fix an open set $O_V \subset TM|_p$ containing v_T. According to Proposition 8.4, there exists $\delta > 0$ and a smooth map, γ_{O_V}, from $(-\delta, \delta) \times O_V$ into M such that if $v' \in O_V$, then $\gamma_{O_V}(\cdot, v')$ is a geodesic that starts at $\pi(v') \in M$ and has initial vector v'.

With the preceding understood, fix an open neighborhood $O' \subset TM$ of v such that $\mathfrak{e}$ is smooth on $(T - \delta, T - \frac{1}{2}\delta) \times O'$. Define the map $\hat{e}$ to map this same domain into TM by the rule $\hat{e}(t, v') = \mathfrak{e}_* \frac{\partial}{\partial t}$. Thus, $\pi \circ \hat{e} = \mathfrak{e}_*$. There is then a neighborhood $O'' \subset O'$ such that $\hat{\mathfrak{e}}$ maps $\{T - \frac{3}{4}\delta\} \times O''$ into O_v. This understood, the uniqueness assertion in Proposition 8.3 implies that

$$\mathfrak{e}\left(T + s - \frac{3}{4}\delta, v''\right) = \gamma_{O_V}\left(s, \hat{e}\left(T - \frac{3}{4}\delta, v'\right)\right) \textit{ when } v'' \in O'' \textit{ and } s \in \left(-\frac{1}{4}\delta, \delta\right).$$

This exhibits $\hat{e}$ on $(0, T + \frac{1}{4}\delta) \times O''$ as the composition of two smooth maps, and hence smooth. This is nonsense given the definition of T.

Given the map $\mathfrak{e}$, define the *exponential* map,

$$\exp\colon TM \to M$$

by the rule $\exp(v) = \mathfrak{e}(1, v)$. This is to say that exp(v) is the point in M that is obtained from $\pi(v)$ by traveling for time $t = 1$ from $\pi(v)$ along the geodesic that starts at $\pi(v)$ at time 0, has initial direction v at $\pi(p)$ and has speed $|v|$. When $p \in M$ has been specified, the restriction of exp to $TM|_p$ is denoted in what follows by $\exp_p$.

The map exp is called the *exponential* map because this map is given by the exponential of a matrix when M is SO(n) or U(n) or one of their subgroups. This identification of maps follows from what is said about the geodesics on such groups in Chapters 8.5 to 8.7.

9.3 Gaussian coordinates

The exponential map can be used to obtain a preferred coordinate chart for a neighborhood of any given point in M. This is because the exponential map restricts to a small radius ball in the fiber of TM over any given point as a diffeomorphism. To see that such is the case, fix a point $p \in M$. Then it is enough to prove that the differential of $\exp_p$ at the $v = 0$ in $TM|_p$ is an isomorphism. In fact, it is an isometry. To elaborate, remark that $T(TM|_p)$ is canonically isomorphic to $\pi^*(TM|_p)$ where π: $TM \to M$ is the projection map. This follows from the fact that the pull-back $\pi^*(TM|_p) \to TM|_p$ consists of pairs $(v, v') \in TM|_p$.

The differential at $0 \in \exp_p$ sends $v \in TM|_p$ to the push-forward of the vector $\frac{\partial}{\partial t}$ at $s = 0$ via the map $s \to \exp_p(sv)$. At any given s, the point $\exp_p(sv) \in M$ is the point given by $\mathfrak{e}(1, sv)$ with $\mathfrak{e}$ as defined in the preceding section. This is the same point as $\mathfrak{e}(s, v)$. As a consequence, $((\exp_p)_* v)|_{0\in TM|_p}$ is $(\mathfrak{e}(\cdot, v)_* \frac{\partial}{\partial t}|_{t=0}) = v$.

Fix an isometric identification between $TM|_p$ and $\mathbb{R}^n$. The image via a small radius ball in $\mathbb{R}^n$ via the exponential map gives a diffeomorphism between this ball and an open neigborhood of p in M. The inverse of this diffeomorphism gives what are called *Gaussian* or *normal* coordinates centered at p.

You may see the term *injectivity radius* at p used with regards to Gaussian coordinates. This is the least upper bound on the radii of a ball in $\mathbb{R}^n$ about the origin which is mapped diffeomorphically onto its image by $\exp_p$.

As is demonstrated next, the metric g_{ij} in these coordinates and the geodesics have a particularly nice form.

Proposition 9.2 Let $(x^1, \ldots, x^n)$ *denote Gaussian coordinates centered at any given point in* M. *Then the metric in these coordinates has the form* $g_{ij} = \delta_{ij} + K_{ij}(x)$ *where* $K_{ij}(x)x^j = 0$ *and also* $|K_{ij}| \le c_0|x|^2$. *Moreover, the geodesics through the given point appear in these coordinates as the straight lines through the origin.*

***Proof of Proposition** 9.2* The fact that the straight lines through the origin are the geodesics through the given point follows directly from the definition of the exponential map. As noted above, the differential of the exponential map at the origin in $TM|_p$ is the identity, and this implies that $g_{ij} = \delta_{ij} + K_{ij}$ where $|K| \le c_0|x|$. To see that it is $\mathcal{O}(|x|^2)$, suppose that $K_{ij} = K_{ij,k}x^k + \mathcal{O}(|x|^2)$. Note that $K_{ij,k} = K_{ji,k}$. Then $\Gamma^i_{jk} = \frac{1}{2}(K_{ij,k} + K_{ik,j} - K_{jk,i}) + \mathcal{O}(|x|)$. This being the case, then the line $t \to t\, v^i$ with v a constant vector can be a geodesic only if $K_{ij,k}v^jv^k - \frac{1}{2}K_{jk,i}v^jv^k = 0$. Taking all possible v, this says that $K_{ij,k} + K_{ik,j} - K_{jk,i} = 0$. A little linear algebra will show that this equation can be satisfied for all i, j and k only if $K_{ij,k} = 0$ for all i, j, k. Indeed, first take $i = j = k$ shows that $K_{11,1} = 0$, and so all cases where $i = j = k$ are zero. Next take $i = 1$

and $j = 1$ and $k = 2$ to see that $K_{11,2} = 0$. With the latter zero, taking $i = 2$ and $j = k = 1$ finds that $K_{21,1} = 0$. By symmetry, $K_{12,1} = 0$. Thus, all cases where two of i, j and k agree are zero. Now consider the case where where i, j, and k are distinct, say 1, 2, 3. Then $i = 1, j = 2$ and $k = 3$ finds that $K_{12,3} + K_{13,2} - K_{23,1} = 0$, while $i = 2$, $j = 3$ and $k = 1$ finds $K_{23,1} + K_{21,3} - K_{13,2} = 0$. Adding these equations and using the symmetry $K_{ij,k} = K_{ji,k}$ tells us that $K_{21,3} = 0$. Thus, all cases where i, j and k are distinct are zero.

To see that $K_{ij}(x)x^i = 0$, fix $\varepsilon > 0$ but with $\varepsilon < c_0^{-1}$. Fix $v \in \mathbb{R}^n$ with small norm; so that the point $x = v$ lies in the Gaussian coordinate chart. Let $t \to \eta(t)$ denote a smooth map from $[0, \infty)$ to $\mathbb{R}^n$ such that $\eta(0) = 0$ and $\eta = 0$ for $t \geq \varepsilon$. For each $s \in (-\varepsilon, \varepsilon)$ consider the path $\gamma[s]$ given by $t \to tv + s\eta(t)$. If ε is small, then this will be a path in the coordinate chart for all t with $|t| \leq 1$. The $s = 0$ version is a geodesic, and as a consequence, the function

$$S \to \ell_{\gamma[s]} = \int_0^1 \left(g_{ij}(\gamma[s])\,\dot\gamma\,[s]^i\,\dot\gamma\,[s]^j\right)^{1/2} dt$$

has derivative zero at $s = 0$. The derivative of the integrand at $s = 0$ is

$$|v|^{-1}\left(K_{ij}(tv)v^i\dot\eta^j + \frac{1}{2}\eta^k\partial_k K_{ij}v^jv^i\right).$$

Integrating this expression, and then integrating by parts finds that

$$\left(\partial_k K_{ij} - \frac{1}{2}\partial_i K_{jk}\right)\Big|_{tv} v^k v^j w^i = 0$$

for all $t \in [0, 1]$ and for all vectors v and w in $\mathbb{R}^n$. To see what this implies, consider first taking $v = w$. Then this says that $\partial_t\,(K_{ij}(tv)v^iv^j) = 0$. Since $K_{ij}(0) = 0$, this implies that $K_{ij}(tv)v^iv^j = 0$ for all t. This understood, take w so that $w^iv^i = 0$. (Thus, w is Euclidean orthogonal to v.) Then the preceding equation asserts that

$$\partial_t K_{ij}(tv)v^iw^j = \frac{1}{2}w^k(\partial_k K_{ij})|_{tv}v^iv^j.$$

To see what to make of this equation, differentiate with respect to s the equation

$$K_{ij}\left(t(v + sw)\right)(v^i + sw^i)(v^j + sw^j) = 0$$

and set $s = 0$ to see that $tw^k(\partial_k K_{ij})|_{tv}v^iv^j = -2K_{ij}(tv)v^iw^j$. Using this fact tells us that

$$t\partial_t(K_{ij}(tv)v^iw^j) = -K_{ij}(tv)v^iw^j.$$

Integration of this tells us the following: If $t > t'$, then $K_{ij}(tv)v^i w^j = K_{ij}(t'v)v^i w^j\, e^{-(t-t')}$. Since $K_{ij}(0) = 0$, taking $t' \to 0$ gives what we want, that $K_{ij}(tv)v^i w^j = 0$ for all t, v and w.

This last proposition has the following consequence: Introduce "spherical" coordinates $(r, \theta^1, \ldots, \theta^{n-1})$ for $\mathbb{R}^n$. This is to say that $r = |x|$ and the coordinates $\{\theta^a\}_{1\le a\le n-1}$ are angle coordinates on S^{n-1}. The metric in Gaussian coordinates when written in terms of these spherical coordinates for $\mathbb{R}^n$ has the form

$$g = dr^2 + r^2\, k_{ab}\, d\theta^a\, d\theta^b$$

where k_{ab} is a function of r and the coordinates for S^{n-1}. Here, dr^2 is a common shorthand for $dr \otimes dr$, and $d\theta^a\, d\theta^b$ is a common shorthand of $d\theta^a \otimes d\theta^b$.

9.4 The proof of the geodesic theorem

The proof in what follows of the geodesic theorem (Theorem 8.1) is broken into various parts.

9.4.1 Part 1

A map, γ, from an interval $I \subset \mathbb{R}$ to M is said to be *locally length minimizing* when the following is true: Fix any point $t_0 \in I$ and there is a neighborhood in I of t_0 such that if t is in this neighborhood, then the length of γ's restriction to the interval between t and t_0 is equal to $\mathrm{dist}(\gamma(t_0), \gamma(t))$.

This part of the proof states and then proves a lemma to the effect that geodesics are locally length minimizing. The proof uses the Gaussian coordinates that were just introduced in the previous section.

Lemma 9.3 *Let* $p \in M$ *denote a given point. There exists* $\varepsilon > 0$ *such that if* $q \in M$ *has distance* ε *or less from* p, *then there is a unique geodesic that starts at p, ends at q and has length equal to the distance between* p *and* q. *Moreover, this is the only path in* M *between* p *and* q *with this length.*

Proof of Lemma 9.3 If q is close to p, then q will lie in a Gaussian coordinate chart centered at p. Such a chart is given by a diffeomorphism, φ, from an open neighborhood, U, of p to the interior of a ball of some radius centered at 0 in $\mathbb{R}^n$. Let ρ denote the radius of this ball. Take $x = (x^1, \ldots, x^n)$ to be the Euclidean coordinates for $\mathbb{R}^n$, thus the image of φ is the ball where $|x| < \rho$. With these coordinates understood, suppose that $\varphi(q)$ is the point $x = v$ where $0 < |v| < \rho$. The part of the geodesic between p and q is the path $t \to x(t) = tv$, parametrized by $t \in [0, 1]$. What follows explains why this path has length equal to the distance between p and q; and why it is the only such path in M.

To start, suppose that $\gamma\colon [0, 1] \to M$ is any other path between p and q. Let γ_p denote the component of $\gamma \cap U$ that starts at p as $\gamma(0)$. Write the φ image of this part of γ in polar coordinates $t \to (r(t), \theta^1(t), \ldots, \theta^n(t))$. Write the metric $g_{ij}\, dx^i \otimes dx^j$ in polar coordinates as at the end of the preceding section. Having done so, one can write the square of the norm of γ for this part of γ as

$$|\dot{\gamma}|^2 = \dot{r}^2 + r^2\, k_{ab}\, \dot{\theta}^a\, \dot{\theta}^b.$$

It follows from this that $|\dot{\gamma}| \geq |\dot{r}|$ at points where γ is in U, with equality if and only if $\gamma(t) = f(t)\, v_*$ where $v_* \in \mathbb{R}^n$ has norm I and where f is a smooth function on an $[0, t_0) \subset [0, 1]$ that contains 0. Note that this interval is a proper subset of $[0, 1)$ if γ exits the coordinate patch, in which case $f(t) \to \rho$ as $t \to t_0$. If $t_0 = 1$, then v_* is the unit vector in the direction of v, and $f(t) \to |v|$ as $t \to 1$. In any event, it follows from this that the length of γ is greater than $|v|$ unless the image of γ is the line segment between 0 and v.

9.4.2 Part 2

Lemma 9.3 has the following corollary: A path in M whose length is the distance between its endpoints parametrizes a subset of a geodesic. The next lemma asserts this and a bit more.

Lemma 9.4 *Let* p, q *be points in* M *and suppose that* $\gamma \subset M$ *is a path from* p *to* q *whose length is the distance between* p *and* q. *Then* γ *is an embedded closed subset of a geodesic.*

Proof of Lemma 9.4 It follows from Lemma 9.3 that γ is part of a geodesic. As can be seen from Proposition 9.2, a geodesic is locally embedded. Thus, γ is embedded unless it crosses itself. This is to say that there is a positive length subpath $\gamma' \subset \gamma$ that starts and ends at the same point. However, if such a subpath exists, then the complement in γ of the interior of γ' would be a path from p to q with length less than that of γ. By assumption, no such path exists.

9.4.3 Part 3

Suppose M is a manifold with a complete Riemannian metric, g. Fix any two points, p and q, in M. This part proves that there is geodesic from p to q whose length is equal to the distance between p and q.

To see why this is, suppose that L is the distance between p and q. Fix $\varepsilon > 0$ and a map $\gamma\colon [0, 1] \to M$ with length less than $L + \varepsilon$ and such that $\gamma(0) = p$ and

$\gamma(1) = q$. I can find $N \geq 1$ (depending only on p and q) such that the following is true: There is a sequence of times

$$0 = t_0 < t_1 < \cdots < t_{N-1} < t_N = 1$$

for which the points $\gamma(t_k)$ and $\gamma(t_{k+1})$ for $k \geq 0$ have distance $2L/N$ from each other and lie in a Gaussian coordinate chart centered at $\gamma(t_k)$. I can also assume that $\gamma(t_{k+2})$ also lies in this coordinate chart if $k \leq N - 2$. It follows that the shortest path from $\gamma(t_k)$ to $\gamma(t_{k+1})$ is the geodesic arc between them. This understood, I can find a continuous, piece-wise smooth path from p to q with N smooth segments, with length less than $L + \varepsilon$, and such that each segment is a geodesic arc between its endpoints, and in particular a geodesic arc that lies entirely in a single Gaussian coordinate chart.

Now, consider a sequence $\varepsilon_1 > \varepsilon_2 > \cdots > \varepsilon_m$ with limit zero, and for each m, a piece-wise smooth path from p to q of the sort just described. Let γ_m denote this path. There are N points, $P_m = \{p_{1,m}, \ldots, p_{N-1,m}\}$ in M where the segments that comprise γ_m join. This ordered set defines a point in $\times_{N-1}$ M. Thus, we have a sequence $\{P_m\}_{m=1,2,\ldots} \subset \times_{N-1}M$. Give the latter set the metric topology from the product of the Riemannian metric topologies on the factors. The sequence $\{P_m\}_{m=1,2,\ldots}$ then lies in a bounded subset, and so in a compact subset. This is so because M is geodesically complete. This sequence has a convergent subsequence because M is geodesically complete. Let $(p_1, \ldots, p_{N-1})$ denote the limit. It follows from the way things were defined that each $k \geq 0$ version of p_{k+1} is contained in a Gaussian coordinate chart about p_k. Here, $p_0 = p$ and $p_N = q$. Likewise, p_{k+2} is also contained in this chart if $k \leq N{-}2$. The sum of the distances between p_k and p_{k+1} for $k \in \{0, \ldots, N{-}1\}$ is equal to L. This understood, it follows that the union of the geodesic arcs between p_k and p_{k+1} for $k \in \{0, \ldots, N{-}1\}$ is a length minimizing, piece-wise smooth path from p to q. To see that this curve has no kinks, use the fact that p_{k+1} and p_{k+2} are both in the Gaussian coordinate chart centered at p_k. It follows that the shortest path between p_k and p_{k+2} is the arc between them, and this must then be the union of the geodesic segments between p_k and p_{k+1} and between p_{k+1} and p_{k+2}.

Note that the assumption of geodesic completeness is necessary. Consider, for example the case where $M = \mathbb{R}^2{-}\{0\}$ with its Euclidean metric. There is no shortest path between the points (1, 0) and (−1, 0) in M.

9.4.4 Part 4

Given what is said in the preceding parts of the proof, there is only one loose end left to tie so as to complete the proof of the geodesic theorem. This is the assertion that the length minimizing path can be parametrized to have constant speed. Since this path is a geodesic, it is sufficient to prove that a solution to the geodesic equation has constant speed. That such is the case follows directly from Proposition 9.2. One can also prove this by writing the function $t \to g(\dot{\gamma}, \dot{\gamma})$ with γ a geodesic in any local coordinate chart, and differentiating the resulting expression with respect to t. Some algebraic manipulations with the geodesic equation can be used to prove that this derivative is zero.

Additional reading

- *A Panoramic View of Riemannian Geometry*, Marcel Berger, Springer, 2003.
- *Riemannian Geometry*, Peter Petersen, Springer, 2006.
- *Riemannian Manifolds: An Introduction to Curvature*, John M. Lee, Springer, 1997.
- *Riemannian Geometry*, Syllvestre Gallot, Dominique Hulin and Jacques Lafontaine, Springer, 2004.
- *Variational Problems in Riemannian Geometry*, Seiki Nishikawa, American Mathematical Society, 2002.

10 Principal bundles

A principal bundle is the Lie group analog of a vector bundle, and they are, in any event, intimately related to vector bundles. The definition requires the specification of a Lie group G. This chapter contains the beginnings of the principal G-bundle discussion. As was the case with vector bundles, my favorite definition is given first; the standard definition (which defines the same object) is the second definition.

10.1 The definition

Fix a smooth manifold M, and a Lie group G. A principal G-bundle is a smooth manifold, P, with the following extra data:

- *A smooth action of* G *by diffeomorphisms; thus a map* $m: G \times P \to P$ *with the property that* $m(\iota, p) = p$ *and* $m(h, m(g, p)) = m(hg, p)$. *It is customary to write this action as* $(g, p) \to pg^{-1}$.
- *A surjective map* $\pi: P \to M$ *that is G-invariant. Thus,* $\pi(pg^{-1}) = p$. *The map* π *is called the projection from* P *to* M.
- *Any given point in* M *has an open neighborhood,* U, *with a G-equivariant diffeomorphism* $\varphi: P|_u \to U \times G$ *that intertwines the map* π *with the evident projection from* $U \times G$ *to* U. *This is to say that if* $\varphi(p) = (\pi(p), h(p))$, *with* $h(p) \in G$, *then* $\varphi(pg^{-1}) = (\pi(p), h(p)g^{-1})$.

Principal G-bundles P and P′ over M are said to be *isomorphic* when there is a G-equivariant diffeomorphism $\psi: P \to P'$ that intertwines the respective projection maps. A principal G-bundle is said to be *trivial* when it is isomorphic to $M \times G$, where the latter has the obvious projection and the obvious G-action. The examples discussed here concern solely the cases where $G = Gl(n; \mathbb{R})$, $Gl(n; \mathbb{C})$ or a matrix group such as SO(n), U(n) or SU(n).

If $N \subset M$ is any subset, then $P|_N$ is used to denote $\pi^{-1}(N) \subset P$. If N is a submanifold, then the restriction of π and the multiplication map to $P|_N$ defines $P|_N$ as a principal G-bundle over N.

Note that the third bullet of the definition asserts that any given point in M has a neighborhood, U, such that $P|_U$ is isomorphic to the product bundle.

Examples of principal bundles are given momentarily.

10.2 A cocycle definition

The cocycle definition of vector bundles given in Chapter 3.2 has a principal bundle analog. This definition requires the specification of the following data: The first thing needed is a locally finite, open cover $\mathfrak{U}$, of M. Needed next is a collection of smooth maps from the intersections of the various sets in $\mathfrak{U}$ to the group G; thus a set $\{g_{U,U'}: U \cap U' \to G\}_{U,U' \in \mathfrak{U}}$. This data is required to obey the following conditions:

- $g_{U,U}$ *is the constant map to the identity,* $\iota \in G$.
- $g_{U,U'}^{-1} = g_{U',U}$.
- *if* U, U' *and* U'' *are any three sets from* $\mathfrak{U}$ *with* $U \cap U' \cap U'' \neq \emptyset$, *then the condition* $g_{U,U'}\, g_{U',U''}\, g_{U'',U} = \iota$ *must hold.*

The map $g_{U,U'}$ is said to be a *principal bundle transition function* and the conditions listed above are called *cocycle constraints*. Given the data just described, the principal G-bundle is defined to be the quotient of the disjoint union $\cup_{U \subset \mathfrak{U}} (U \times G)$ by the equivalence relation that puts $(x, g) \in U \times G$ equivalent to $(x', g') \in U' \times G$ if and only if $x = x'$ and $g = g_{U,U'}(x)\, g'$.

To see that this definition is equivalent to the previous, consider first a principal bundle that is defined by this cocycle data. The group G acts so as to have $h \in G$ send the equivalence class of a pair (x, g) to (x, gh^{-1}). Meanwhile, the projection to M sends this same equivalence class to p. Finally, the isomorphism required by the third bullet in Chapter 10.1 is as follows: Let $x \in M$. Fix a set $U \subset \mathfrak{U}$ that contains x. The map to $U \times G$ of the equivalence class of $U \times G$ assigns to this class none other but the product $U \times G$. The cocycle conditions guarantee that all of this is well defined.

To go the other way, suppose that $P \to M$ is a given principal bundle as defined in Chapter 10.1. It is a consequence of the third bullet of the definition in Chapter 10.1 that there is a locally finite cover, $\mathfrak{U}$, for M with the following property: If $U \subset \mathfrak{U}$, then there is an isomorphism, $\varphi_U: P|_U \to U \times G$. This understood, suppose that U and U' are any two intersecting sets from $\mathfrak{U}$. Then the composition $\varphi_U \circ \varphi_{U'}^{-1}$ maps $U \cap U'$ to G. Use $g_{U,U'}$ to denote the latter. The collection $\{g_{U,U'}\}_{U,U' \in \mathfrak{U}}$ satisfies the cocycle constraints given above.

Let $P_{\mathfrak{U}}$ denote the principal G-bundle as defined using this cocycle data. Define a map Φ: $P \to P_{\mathfrak{U}}$ so as to send any given point $p \in P$ to the equivalence class of the point $\varphi_U(p)$ with $U \in \mathfrak{U}$ any set that contains $\pi(p)$. This map Φ identifies P with $P_{\mathfrak{U}}$.

10.3 Principal bundles constructed from vector bundles

What follows explains how to construct a principal bundle from a vector bundle.

10.3.1 Frame bundles

Let π: $E \to M$ denote any given, rank n-vector bundle. Introduce $P_{Gl(E)} \to M$ to denote the submanifold in $\bigoplus_n E$ that consists of the n-tuples $(e_1, \ldots, e_n)$ that give basis over their common base point in M. This manifold $P_{GL(E)}$ is a principal $Gl(n; \mathbb{R})$-bundle over M. To say more, remark that the projection to M sends a n-tuple to its base point. Meanwhile, a matrix $g \in Gl(n; \mathbb{R})$ acts to send any given n-tuple $(e_1, \ldots, e_n)$ to $(\sum_k g^{-1}_{k1} e_k, \sum_k g^{-1}_{k2} e_k, \ldots, \sum_k g^{-1}_{kn} e_k)$. To see the required isomorphisms to trivial bundles over suitable neighborhoods of points, fix attention on an open set $U \subset M$ where there is a vector bundle isomorphism φ_U: $E|_U \to U \times \mathbb{R}^n$. This isomorphism identifies $(\bigoplus_n E)|_U$ with $U \times (\bigoplus_n \mathbb{R}^n)$ and so identifies $P_{Gl(E)}|_U$ as $U \times Gl(n; \mathbb{R})$ where the embedding $Gl(n; \mathbb{R}) \to \times_n \mathbb{R}^n$ sends any given $n \times n$ matrix to the n-tuple given by its n columns.

By the way, the transition functions for $P_{Gl(E)}$ are the maps to $Gl(n; \mathbb{R})$ that are given by the transition functions for the vector bundle E. To see why this is, consider the open set U described in the previous paragraph, and a second open set $U' \subset M$ that intersects U and comes with a trivializing isomorphism $\varphi_{U'}$: $E|_{U'} \to U' \times \mathbb{R}^n$. Introduce $g_{U,U'}$: $U \cap U' \to Gl(n; \mathbb{R})$ to denote the corresponding transition function for $E|_{U \cap U'}$. Now, let $(x, g = (v_1, \ldots, v_n)) \in U' \times Gl(n; \mathbb{R}) \subset U' \times (\bigoplus_n \mathbb{R}^n)$. The transition function for E identifies this point with $(x, (g_{U,U'}(x) v_1, \ldots, g_{U,U'},(x) v_n)) = (x, g_{U,U'}(x) g) \subset U \times Gl(n; \mathbb{R})$.

The principal bundle $P_{Gl(E)}$ is called the frame bundle of E. In the case when $E = TM$, the corresponding $P_{GL(TM)}$ is called the *frame bundle of* M.

10.3.2 Orthonormal frame bundles

Suppose that $E \to M$ is a rank n-vector bundle with a fiber metric. The metric gives the principal O(n)-bundle $P_{O(E)} \to M$. This is the submanifold of $P_{Gl(E)}$ whose n-tuples supply an orthonormal frame for E over their common base point. To elaborate, recall from Chapter 7.1 that there is a locally finite, cover $\mathfrak{U}$ for M with the following property: Each set $U \in \mathfrak{U}$ comes with an isometric isomorphism φ_U: $E|_U \to U \times \mathbb{R}^n$. This isometry induces the principal bundle isomorphism φ_U: $P_{O(E)} \to U \times O(n)$ by viewing $P_{O(E)}$ as the submanifold in $\bigoplus_n E$ of n-tuples that give an orthonormal frame for E over their common base point while viewing O(n) as the subset of $\bigoplus_n \mathbb{R}^n$ of n-tuples of vectors that define an orthonormal basis for $\mathbb{R}^n$.

Recall from Chapter 7.1 that the bundle E is said to be orientable if it can be defined using transition functions that take values in SO(n). Lemma 7.2 asserts that this is the case if and only if the real line bundle $\wedge^n E$ is isomorphic to the product line bundle. In any event, if E is orientable, then there is the analogous principal SO(n)-bundle $P_{SO(E)} \subset P_{O(E)} \subset \bigoplus_n E$ that consists of the n-tuples that define an oriented frame over their common base point.

A natural case to consider arises when M has a Riemannian metric and $E = TM$. The bundle $P_{O(TM)}$ consists of the subspace in $\bigoplus_n TM$ of n-tuples that give an orthonormal basis for the tangent bundle to M over their common base point. The manifold M is said to be orientable when $\wedge^n TM$ is the trivial line bundle. One has in this case the principal SO(n)-bundle of oriented, orthonormal frames in TM.

As an example of the latter, consider the case of S^n with its round metric. The oriented orthonormal frame bundle for TS^n turns out to be the group $SO(n + 1)$ with the projection map to S^n identifying the latter with the quotient space $SO(n + 1)/SO(n)$. To see how this comes about, view S^n as the unit sphere in $\mathbb{R}^{n+1}$. A point in the oriented, orthonormal frame bundle consists of an $(n + 1)$-tuple $(x, v_1, \ldots, v_n)$ of orthonormal vectors in $\mathbb{R}^{n+1}$; where the projection to S^n gives the first element, x. The identification with $SO(n + 1)$ takes this $(n + 1)$-tuple to the matrix with these $n + 1$ vectors as its columns.

10.3.3 Complex frame bundles

Suppose that π: $E \to M$ is a complex vector bundle with fiber $\mathbb{C}^n$. Sitting inside $\bigoplus_n E$ is the principal $Gl(n; \mathbb{C})$-bundle $P_{Gl(E)} \to M$ of n-tuples that give a $\mathbb{C}$-linear basis for the fiber at each point. If E has a Hermitian metric, then one can define inside $P_{GL(E)}$ the principal U(n)-bundle $P_{U(E)} \to M$ of n-tuples that define an orthonormal frame at each point. Recall in this regard Lemma 7.4 which finds such transition functions if and only if the complex rank 1 bundle $\wedge^n E \to M$ is isomorphic to the product bundle $M \times \mathbb{C}$.

What follows is an example of $P_{U(E)}$. View S^2 as the unit radius ball about the origin in $\mathbb{R}^3$. Reintroduce the Pauli matrices $\{\tau^1, \tau^2, \tau^3\}$ from Chapter 6.4, and let $E \to S^2$ denote the subbundle in $S^2 \times \mathbb{C}^2$ of pairs (x, v) such that $x_j \tau^j v = iv$. This bundle has its Hermitian metric from $\mathbb{C}^2$. Let (z, w) denote the complex coordinates of $\mathbb{C}^2$. Let π: $S^3 \to S^2$ denote the map that sends (z, w) with norm 1 to $x = (2\mathrm{re}(z\bar{w}), 2\mathrm{im}(z\bar{w}), |w|^2 - |z|^2)$. The matrix $x_j\tau^j$ is

$$i\begin{pmatrix} |w|^2 - |z|^2 & 2\bar{z}w \\ 2z\bar{w} & -|w|^2 - |z|^2 \end{pmatrix}.$$

Note in particular that the vector $v = \begin{pmatrix} w \\ z \end{pmatrix}$ is such that $x^j\tau^j\ v = iv$. Thus, $P_{U(E)} = S^3$.

10.4 Quotients of Lie groups by subgroups

The example described above with the oriented orthonormal frame bundle of S^n is but one application of a much more general construction that is summarized in the central proposition of this section. The statement of the proposition requires introducing four new notions and some notation. These are introduced in what follows directly.

10.4.1 Quotients of Lie groups

Suppose that G is a Lie group and $H \subset G$ is a Lie subgroup. This is to say that H is a smooth submanifold of G that has the following three properties: First, the identity element is in H. Second, the inverse of an element in H is also in H. Finally, the product of any two elements in H is in H. For example, SO(n) is a subgroup of Gl(n; $\mathbb{R}$). Likewise, SU(n) is a subgroup of U(n).

If H is a subgroup of G, use G/H to denote the space of equivalence classes that are defined by the rule $g \sim g'$ if and only if $g' = gh$ for some $h \in H$. The topology here is the quotient topology.

10.4.2 The Lie algebra

The Lie algebra of the group G is the tangent space to G at the identity ι. This vector space is denoted by $\mathfrak{lie}(G)$ in what follows. For example, the Lie algebra of Gl(m; $\mathbb{R}$) is the vector space of n × n real matrices, and that of Gl(m; $\mathbb{C}$) is the vector space of n × n complex matrices. If G = SO(n), then $\mathfrak{lie}(G)$ is the vector space $\mathbb{A}(n; \mathbb{R})$ of n × n, antisymmetric matrices. If G = U(n), then $\mathfrak{lie}(G)$ is the vector space $\mathbb{A}(n; \mathbb{C})$ of n × n, complex, anti-Hermitian matrices. If G = SU(n), then $\mathfrak{lie}(G)$ is the codimension 1 subspace in $\mathbb{A}(n; \mathbb{C})$ of trace zero matrices.

The *algebra* designation for the vector space $\mathfrak{lie}(G)$ signifies the existence of a bilinear map from $\mathfrak{lie}(G) \times \mathfrak{lie}(G)$ to $\mathfrak{lie}(G)$ of a certain sort. This map is called the

Lie bracket and the Lie bracket of vectors $\sigma, \tau \in \mathfrak{lie}(G)$ is traditionally written as $[\sigma, \tau]$. It is such that $[\tau, \sigma] = -[\sigma, \tau]$ and it obeys the *Jacobi relation*

$$[\sigma, [\tau, \mu]] + [[\sigma, \tau], \mu] + [\tau, [\mu, \sigma]] = 0.$$

In the examples in this book, G is given as a subgroup of some $n \geq 1$ version of $Gl(n; \mathbb{R})$ or $Gl(n; \mathbb{C})$. If G is a subgroup of some $n \geq 1$ version of $Gl(n; \mathbb{R})$ or $Gl(n; \mathbb{C})$, then $\mathfrak{lie}(G)$ appears as a vector subspace of the corresponding vector space of real or complex $n \times n$ matrices. The Lie bracket in this case sets $[\sigma, \tau] = \sigma\tau - \tau\sigma$. This Lie bracket can be thought of as a measure of the failure of elements in G to commute because $e^{\sigma}e^{\tau}e^{-\sigma}e^{-\tau} = \iota + [\sigma, \tau] + \mathfrak{e}$ where the remainder, $\mathfrak{e}$, has norm that is the order of $|\sigma||\tau|(|\sigma| + |\tau|)$. Note in this regard what is said in Chapters 8.5 and 8.6: The map $\sigma \to e^{\sigma}$ sends $\mathfrak{lie}(G)$ into G.

If G is not given as a subgroup of some $n \geq 1$ version of $Gl(n; \mathbb{R})$ or $Gl(n; \mathbb{C})$, then the Lie bracket can still be defined in the following way: Extend a given pair $\sigma, \tau \in TG|_{\iota}$ as vector fields, v_{σ} and v_{τ}, that are defined in a neighborhood of the identity of G. View these as derivations of the algebra $C^{\infty}(G)$. The commutator of two derivations is also a derivation; in particular, this is the case for the commutator of v_{σ} with v_{τ}. View the latter derivation as a vector field defined near $\iota \in G$. Its value at ι in $TG|_{\iota}$ is $[\sigma, \tau]$.

If G appears as a subgroup of some $n \geq 1$ version of $Gl(n; \mathbb{R})$, then its Lie algebra inherits the inner product that assigns the trace of $\sigma^{T}\tau$ to a given pair σ, $\tau \in \mathfrak{lie}(G)$. If $G \subset Gl(n; \mathbb{C})$, then the inner product of a given pair σ and τ from $\mathfrak{lie}(G)$ is assumed to be given by the real part of the trace of $\sigma^{\dagger}\tau$. These inner products are used implicitly in what follows to define norms and orthogonal complements. If G is not given as a subgroup of one of these groups, agree beforehand to fix a convenient inner product on $TG|_{\iota}$.

10.4.3 Representations of Lie groups

Let V denote a vector space, either $\mathbb{R}^m$ or $\mathbb{C}^m$. Introduce $Gl(V)$ to denote either the group $Gl(m; \mathbb{R})$ or $Gl(m; \mathbb{C})$ as the case may be. Now suppose that G is a given Lie group. A *representation* of G on V is a defined to be a smooth map $\rho: G \to Gl(V)$ with the property that $\rho(\iota) = \iota$ and $\rho(gg') = \rho(g)\rho(g')$ for any pair $g, g' \in G$. Suppose that $\mathfrak{v} \in V$. The *stabilizer* of $\mathfrak{v}$ is the subset $\{g \in G : \rho(g)\mathfrak{v} = \mathfrak{v}\}$.

10.4.4 Lie algebras and representations

Suppose that $\rho: G \to Gl(V)$ is a representation. Introduce $\rho_*: \mathfrak{lie}(G) \to \mathfrak{lie}(Gl(V))$ to denote the differential of ρ at the identity. In the case when G is a matrix group, the map ρ_* is given by the rule $\rho_*(q) = (\frac{d}{dt}\rho(e^{tq}))|_{t=0}$. This ρ_* is a linear map that intertwines the Lie bracket on $\mathfrak{lie}(G)$ with the matrix Lie bracket on $\mathfrak{lie}(Gl(V))$ in the sense that $\rho_*([\sigma, \tau]) = [\rho_*\sigma, \rho_*\tau]$ for any given pair $\sigma, \tau \in \mathfrak{lie}(G)$. That such is the case follows from the fact that ρ intertwines group multiplication on G with matrix multiplication.

This intertwining property can be verified when G is a subgroup of some $n \geq 1$ version of either $Gl(n; \mathbb{R})$ or $Gl(n; \mathbb{C})$ as follows: Let $\mathfrak{a}, \mathfrak{a}' \in \mathfrak{lie}(G)$. For any given pair $t, s \in \mathbb{R}$, both $e^{t\mathfrak{a}}$ and $e^{s\mathfrak{a}'}$ are in G. As a consequence, the assignment to (t, s) of $e^{t\mathfrak{a}}\, e^{s\mathfrak{a}'} e^{-t\mathfrak{a}} e^{-s\mathfrak{a}'}$ defines a smooth map from a neighborhood of the origin in $\mathbb{R}^2$ to G. It follows from what was said above about the commutator that this map has the form $(t, s) \to \iota + ts\,[\mathfrak{a}, \mathfrak{a}'] + \mathfrak{r}$ where $|\mathfrak{r}|$ is bounded by some multiple of $|t|\ |s|\ (|t| + |s|)$. Given that ρ is a representation, it follows that

$$\rho(e^{t\mathfrak{a}} e^{s\mathfrak{a}'} e^{-t\mathfrak{a}} e^{-s\mathfrak{a}'}) = \rho(e^{t\mathfrak{a}})\rho(e^{s\mathfrak{a}'})\rho(e^{-t\mathfrak{a}})\rho(e^{-s\mathfrak{a}'}).$$

Now use Taylor's expansion with remainder to write this as

$$\iota + ts\rho_*([\mathfrak{a}, \mathfrak{a}']) = \iota + ts\,[\rho_*[\mathfrak{a}], \rho_*[\mathfrak{a}']) + \mathfrak{e}$$

where $|\mathfrak{e}|$ is also bounded by a multiple of $|t|\ |s|\ (|t| + |s|)$. Given such bound for $\mathfrak{e}$, what is written above is possible for all t, s near $0 \in \mathbb{R}^2$ only if ρ_* intertwines the Lie bracket on $\mathfrak{lie}(G)$ with the commutator bracket on the lie algebra of Gl(V).

The stage is now set. In the proposition below, G is a lie group, V is a vector space, either $\mathbb{R}^m$ or $\mathbb{C}^m$ for some $m \geq 1$ and $\mathfrak{v} \in V$ is a given nonzero element. Meanwhile, ρ: $G \to Gl(V)$ is a representation.

Proposition 10.1 *The stabilizer of* $\mathfrak{v} \in V$ *is a Lie subgroup* $H \subset G$ *whose tangent space at the identity is the subspace* $\mathfrak{H} = \{\mathfrak{q} \in \mathfrak{lie}(G) : (\rho_*(\mathfrak{q}))\mathfrak{v} = 0\}$. *If* G *is compact, then the following is also true:*

- *The subspace* $M_{\mathfrak{v}} = \{\rho(\mathfrak{g})\mathfrak{v} : \mathfrak{g} \in G\} \subset V$ *is a smooth manifold, homeomorphic to the quotient space* G/H.
- *The tangent space to* $M_{\mathfrak{v}}$ *at* $\mathfrak{v}$ *is canonically isomorphic to the orthogonal complement of* $\mathfrak{H}$ *in* $\mathfrak{lie}(G)$ *with the map that sends any given* $\mathfrak{z}$ *in this orthogonal complement to* $\rho_*(\mathfrak{z})\mathfrak{v}$.
- *The map* π: $G \to M_{\mathfrak{v}}$ *defines a principal* H*-bundle.*

A proof of this proposition is given in Appendix 10.1; but only in the special case when G is a subgroup of some $n \geq 1$ version of either $Gl(n; \mathbb{R})$ or $Gl(n; \mathbb{C})$.

10.5 Examples of Lie group quotients

This part of the chapter gives some examples of principal bundles that arise using the preceding proposition.

Example 10.2 This first case concerns SO(n). Use the inner product on $\mathbb{M}(n; \mathbb{R})$ given by $\langle \mathfrak{m}, \mathfrak{m}' \rangle = \mathrm{tr}(\mathfrak{m}^T \mathfrak{m}')$ to view $\mathbb{M}(n; \mathbb{R})$ as a Euclidean space of dimension n^2. A representation ρ: $SO(n) \to SO(n^2) \subset Gl(n^2; \mathbb{R})$ is defined by having $\mathfrak{m} \in SO(n)$ act on $a \in \mathbb{M}(n; \mathbb{R})$ via $\mathfrak{m} \to \mathfrak{m} \mathfrak{a} \mathfrak{m}^{-1}$. Given a nonzero $\mathfrak{a} \in \mathbb{M}^n(n; \mathbb{R})$, define the subset $M_\mathfrak{a} \subset \mathbb{M}(n; \mathbb{R})$ to be the set of elements of the form $\mathfrak{m} \mathfrak{a} \mathfrak{m}^{-1}$ for $\mathfrak{m} \in SO(n)$. Use $H \subset SO(n)$ to denote the subgroup of matrices $\mathfrak{m}$ such that $\mathfrak{m} \mathfrak{a} \mathfrak{m}^{-1} = \mathfrak{a}$. The proposition asserts that $M_\mathfrak{a}$ is a smooth manifold and that the map $SO(n) \to M_\mathfrak{a}$ defines a principal H-bundle.

For a concrete example, choose $\mathfrak{a}$ to be a diagonal matrix with distinct eigenvalues. The group H in this case is a finite group, being the group of diagonal matrices in SO(n). For a somewhat more interesting example, fix a set $n_1 \leq n_2 \leq \cdots \leq n_k$ of positive integers that sum to n. Take $\mathfrak{a}$ to be diagonal with m distinct eigenvalues, these of multiplicities $n_1, n_2, \ldots, n_k$. The subgroup H in this case is isomorphic to the subgroup $S(O(n_1) \times \cdots \times O(n_k)) \subset O(n_1) \times \cdots \times O(n_k)$ consisting of the m-tuples of matrices for which the corresponding product of determinants is equal to 1.

In the case when $k = 2$, the resulting manifold $M_\mathfrak{a}$ is diffeomorphic to the Grassmannian $Gr(n; n_1)$ of n_1-dimensional subspaces in $\mathbb{R}^n$, defined in Chapter 1.6. To see why this is, remark that no generality is lost by taking $\mathfrak{a}$ to be the matrix where the top n_1 elements on the diagonal are equal to 1 and the remaining are equal to 0. This understood, define a map Φ: $M_\mathfrak{a} \to Gr(n; n_1)$ by setting $\Phi(\mathfrak{v})$ for $\mathfrak{v} \in M_\mathfrak{a}$ to be the span of the eigenvectors of $\mathfrak{v}$ with eigenvalue 1. To see about an inverse, suppose that $V \subset \mathbb{R}^n$ is a given n_1-dimensional subspace. Let Π_V denote the orthogonal projection to V. This is a self-adjoint matrix with n_1 eigenvalues equal to 1 and $n - n_1$ equal to zero. This matrix is diagonalizable, and so there exists $\mathfrak{m} \in SO(n)$ such that $\mathfrak{m}^{-1} \Pi_V \mathfrak{m} = \mathfrak{a}$. Thus, $\Pi_V \in M_\mathfrak{a}$. The assignment of Π_V to V inverts the map Φ. To summarize: The proposition identifies $Gr(n; n_1)$ with the manifold $SO(n)/S(O(n_1) \times O(n - n_1))$. The principal bundle defined by the quotient map from SO(n) to $SO(n)/S(O(n_1) \times O(n - n_1))$ can be viewed in the context of $Gr(n; n_1)$ as follows: The total space P is SO(n), and the fiber over a given subspace $V \subset Gr(n; n_1)$ consists of the set of matrices $\mathfrak{m} \in SO(n)$ that rotate $\mathbb{R}^n$ so as to map V to itself.

In the case $n_1 = 1$, the space $Gr(n; 1)$ is the real projective space $\mathbb{RP}^{n-1}$. What is said above thus identifies $\mathbb{RP}^{n-1}$ with $SO(n)/S(O(1) \times O(n - 1))$. Noting that O(1) is the two element group $\{\pm 1\}$, and noting the identification $SO(n)/SO(n - 1) = S^{n-1}$, the proposition in this case says not much more than the fact that $\mathbb{RP}^{n-1} = S^{n-1}/\{\pm 1\}$ where $\{\pm 1\}$ acts on $\mathbb{R}^n$ by sending any given vector x to $-x$.

Example 10.3 A very much analogous example involves U(n). To say more, introduce the Hermitian inner product on $\mathbb{M}(n;\mathbb{C})$ given by $\langle \mathfrak{m}, \mathfrak{m}'\rangle = \mathrm{tr}(\mathfrak{m}^\dagger \mathfrak{m}')$ to view $\mathbb{M}(n;\mathbb{C})$ as a copy of $\mathbb{C}^{n^2}$. This done, the action of U(n) on $\mathbb{M}(n;\mathbb{C})$ that sends a given matrix $\mathfrak{a}$ to $\mathfrak{m}\mathfrak{a}\mathfrak{m}^{-1}$ defines a representation in $U(n^2) \subset Gl(n^2;\mathbb{C})$. If $\mathfrak{a} \in \mathbb{M}(n;\mathbb{C})$ is any given nonzero matrix, then the proposition finds $M_\mathfrak{a} = \{\mathfrak{v} \in \mathbb{M}(n;\mathbb{C}) : \mathfrak{v} = \mathfrak{m}\mathfrak{a}\mathfrak{m}^{-1}\}$ is a smooth manifold. Moreover, the set $H \subset U(n)$ of matrices $\mathfrak{m}$ such that $\mathfrak{m}\mathfrak{a}\mathfrak{m}^{-1} = \mathfrak{a}$ is a subgroup, and the map from U(n) to $M_\mathfrak{a}$ that sends $\mathfrak{m}$ to $\mathfrak{m}\mathfrak{a}\mathfrak{m}^{-1}$ defines a principal H-bundle.

A concrete example takes $\mathfrak{a}$ to be diagonal with some $n_1 \in \{1,\ldots, n-1\}$ entries equal to 1 and the remaining equal to zero. As in the SO(n) example above, the space $M_\mathfrak{a}$ is diffeomorphic to the complex Grassmannian $Gr_\mathbb{C}(n; n_1)$ as defined in Chapter 6.8. In particular, $Gr_\mathbb{C}(n; n_1)$ is diffeomorphic to $U(n)/(U(n_1) \times U(n-n_1))$. When U(n) is viewed as a principal $U(n_1) \times U(n-n_1)$ bundle over $Gr_\mathbb{C}(n; n_1)$, its fiber over a given complex n_1-dimensional subspace consists of the set of matrices in U(n) that act on $\mathbb{C}^n$ so as to map the subspace to itself.

Note by the way, that if $1 \leq n_1 \leq \cdots \leq n_k$ are integers that sum to n, then the quotient $U(n)/(\times_{1\leq m\leq k} U(m))$ is the same as $SU(n)/S(\times_{1\leq m\leq k}U(m))$ where $S(\times_{1\leq m\leq k}U(m))$ is the subgroup of k-tuples in $\times_{1\leq m\leq k}U(k)$ whose corresponding product of determinants is equal to 1.

Example 10.4 This case involves SU(2) acting on $\mathbb{C}^2$ via matrix multiplication. Take the vector $\mathfrak{v} = \begin{pmatrix} 1 \\ 0 \end{pmatrix}$ and set $M_\mathfrak{v}$ to be the set of vectors of the form $\mathfrak{m}\mathfrak{v} \in \mathbb{C}^2$ with $\mathfrak{m} \in SU(2)$. The group H in this case is $U(1) = S^1$ and $M_\mathfrak{v} = S^2$. To be more explicit, view SU(2) as S^3, the unit sphere in $\mathbb{C}^2$. This is done by the usual correspondence that pairs a given $\begin{pmatrix} w \\ z \end{pmatrix} \in S^3$ with the matrix $\begin{pmatrix} w & -\bar{z} \\ z & \bar{w} \end{pmatrix}$. View S^1 as the subgroup U(1) of matrices $\begin{pmatrix} u & 0 \\ 0 & \bar{u} \end{pmatrix}$ with $|u| = 1$, then the quotient SU(2)/U(1) is S^2. Thus, $S^2 = SU(2)/U(1)$.

This generalizes to higher dimensions as follows: View S^{2n+1} as the unit sphere in $\mathbb{C}^{n+1}$. Then $S^{2n+1} = U(n+1)/U(n)$ and $U(n+1)$ can be viewed as the total space of a principal U(n)-bundle over over S^{2n+1}. Note that there is a residual U(1)-action on S^{2n+1}, this the action that sends a given $u \in U(1)$ and $(z_1,\ldots, z_{n+1}) \in S^{2n+1}$ to $(uz_1,\ldots, uz_n)$. This action is free, and the quotient is n-dimensional complex projective space $\mathbb{CP}^n$; this is the space of 1-dimensional vector subspaces in $\mathbb{C}^{n+1}$ from Chapter 6.8. Thus the quotient space projection, $S^{2n+1} \to \mathbb{CP}^n$, defines a principal U(1)-bundle. Another view of this bundle is given in Example 10.9 in the next section.

10.6 Cocycle construction examples

What follows describes some principle bundles that are constructed using the cocycle definition in Chapter 10.2.

Example 10.5 Principle U(1)-bundles over S^2 can be constructed using the cocycle definition as follows: View S^2 as the set of unit length vectors in $\mathbb{R}^3$, and this done, decompose it as S^2 as $U_+ \cup U_-$ where U_+ consists of vectors $x = (x_1, x_2, x_3)$ with $|x| = 1$ and $x_3 > -1$, and where U_- consists of vectors with norm 1 and with $x_3 < 1$. These two sets intersect in the cylindrical region S^2 that consists of the points in $\mathbb{R}^3$ with norm 1 and with $x_3 \in (-1, 1)$. This understood, the complex number $z = x_1 + ix_2$ is nowhere zero on this region, and so the assignment $x \to |z|^{-1} z$ defines a smooth map from $U_+ \cap U_-$ into the unit circle in $\mathbb{C}$, thus the group U(1). Let $g\colon U_+ \cap U_- \to U(1)$ denote this map.

Now, let m denote any given integer. A principal U(1)-bundle is defined by taking for cocycle data the open cover $\mathfrak{U} = \{U_+, U_-\}$ and principal bundle transition function $g^m\colon U_+ \cap U_- \to U(1)$.

There is an analogous construction for any given surface. Let Σ denote the surface in question. Fix a finite set $\Lambda \subset \Sigma$ of distinct points. Let $p \in \Lambda$. Fix an open set $U_p \subset \Sigma$ containing p and with a diffeomorphism $\varphi_p\colon U_p \to \mathbb{R}^2$ that sends p to the origin. Do this for each point in Λ, but make sure to choose the sets in question so that their closures are pair-wise disjoint. Assign an integer, m(p) to each $p \in \Lambda$.

The cover $\mathfrak{U} = \{U_0 = \Sigma - \Lambda\} \cup \{U_p\}_{p \in \Lambda}$ is used for the cocycle data set. A bundle transition function $g_{0p}\colon U_0 \cap U_p \to U(1)$ is defined as follows: If $x \in U_0 \cap U_p$, write $\varphi_p(x)$ as a complex number, z(x). This done, set $g_{0p}(x) = z(x)/|z(x)|^{m(p)}$. As no three distinct sets from the cover intersect, this data defines a principal U(1)-bundle over Σ.

Example 10.6 Principal SU(2)-bundles over S^4 can be constructed as follows: Identify the equatorial S^3 as SU(2). Now define a principal bundle SU(2) over S^4 by declaring that it be isomorphic to the trivial bundle over the complement of the south pole, and also trivial over the complement of the north pole. Use U to denote the former set and V to denote the latter. The intersection of these two sets can be described as follows: View S^4 as the set $x = (x_1, \ldots, x_4, x_5) \in \mathbb{R}^5$ with $|x| = 1$. Then U is the set $x_5 > -1$ and V is the set $x_5 < 1$. Thus, their overlap is the set where $x_5 \neq \pm 1$. Define a map from this set to S^3 by sending y to the point $g(x) = (x_1, x_2, x_3, x_4)/(1 - x_5^2)^{1/2}$. Identify S^3 with SU(2) as done previously. Fix an integer p, and define a principal SU(2)-bundle $P^{(p)} \to S^4$ by declaring its transition function to be $g(x)^p$. As it turns out the bundle $P^{(1)}$ is diffeomorphic to S^7. So is $P^{(-1)}$, but these bundles are not isomorphic as principal SU(2)-bundles.

To see the relation to S^7, it is useful to introduce $\mathbb{H}$ to denote the set of 2×2 complex matrices that can be written as

$$\begin{pmatrix} a & -\bar{b} \\ b & \bar{a} \end{pmatrix}.$$

If I reintroduce the Pauli matrices $\{\tau_1, \tau_2, \tau_3\}$ from Chapter 6.4, then an element in $\mathbb{H}$ can be written as $x_4 + x_1\tau^1 + x_2\tau^2 + x_3\tau^3$. Here, $a = x_4 + ix_3$ and $b = x_2 + ix_1$. Thus, $\mathbb{H}$ can be viewed as $\mathbb{R}^4$, or as $\mathbb{C}^2$. Given the multiplication rule described in Chapter 6.4, this last way of writing $\mathbb{H}$ identifies it with the vector space (and algebra via matrix multiplication) of quaternions.

In any event, the 7-sphere is the set of pairs $(\mathfrak{a}, \mathfrak{b}) \in \mathbb{H} \times \mathbb{H}$ with $|\mathfrak{a}|^2 + |\mathfrak{b}|^2 = 1$. Here, $|\mathfrak{a}|^2 = \mathrm{tr}(\mathfrak{a}^\dagger \mathfrak{a})$. The group SU(2) sits in $\mathbb{H}$ as the set of unit vectors. It acts on S^7 so that $\mathfrak{m} \in SU(2)$ sends $(\mathfrak{a}, \mathfrak{b})$ to $(\mathfrak{a}\mathfrak{m}^{-1}, \mathfrak{b}\mathfrak{m}^{-1})$. It can also act to send $(\mathfrak{a}, \mathfrak{b})$ to $(\mathfrak{m}\mathfrak{a}, \mathfrak{m}\mathfrak{b})$. The quotient of either action is S^4, and one quotient defines $P^{(1)}$, the other defines $P^{(-1)}$.

To see that the quotient is S^4, define a map from S^7 to $\mathbb{R}^5$ by sending a given vector $(\mathfrak{a}, \mathfrak{b})$ to the vector $(2\mathfrak{a}\mathfrak{b}^\dagger, |\mathfrak{a}|^2 - |\mathfrak{b}|^2) \in \mathbb{H} \times \mathbb{R} = \mathbb{R}^4 \times \mathbb{R} = \mathbb{R}^5$. This vector has unit length, so defines a point in S^4. This map is constant along orbits of SU(2) by the action that sends $\mathfrak{m} \in SU(2)$ and $(\mathfrak{a}, \mathfrak{b}) \in S^7$ to $(\mathfrak{a}\mathfrak{m}^{-1}, \mathfrak{b}\mathfrak{m}^{-1})$. As exhibited momentarily, it is the projection to S^4 that defines the principal bundle $P^{(1)}$.

The following description of the cocycle data for $P^{(1)}$ verifies that the map just given from S^7 to S^4 is indeed the principal bundle projection. To start, identify U_+ with $\mathbb{H}$ via the map that sends any given point $\mathfrak{z} \in \mathbb{H}$ to $(2\mathfrak{z}, 1 - |\mathfrak{z}|^2)/(1 + |\mathfrak{z}|^2)$. Define the inverse map from $\mathbb{H} \times SU(2)$ to S^7 so as to send $(\mathfrak{z}, \mathfrak{m})$ to $(\mathfrak{a} = \mathfrak{z}\mathfrak{m}, \mathfrak{b} = \mathfrak{m})/(1 + |\mathfrak{z}|^2)^{-1/2}$. Over the set U_-, define a map from $\mathbb{H}$ to U_- by sending $\mathfrak{r} \in \mathbb{H}$ to $(2\mathfrak{r}^\dagger, |\mathfrak{r}|^2 - 1)/(1 + |\mathfrak{r}|^2)^{1/2}$. Now define a map from $\mathbb{H} \times SU(2)$ to S^7 by sending $(\mathfrak{r}, \mathfrak{n})$ to $(\mathfrak{a} = \mathfrak{n}, \mathfrak{b} = \mathfrak{r}\mathfrak{n})/(1 + |\mathfrak{r}|^2)^{1/2}$. The equator is the set $|\mathfrak{z}| = |\mathfrak{r}| = 1$ and $\mathfrak{r} = \mathfrak{z}^\dagger$. Thus, the points $\mathfrak{m}$ and $\mathfrak{n}$ are sent to the same point if $\mathfrak{n} = \mathfrak{z}\mathfrak{m}$. This is the transition function for $P^{(1)}$.

There is a very much analogous story for $P^{(-1)}$ that identifies the bundle projection with the map from S^7 that sends $(\mathfrak{a}, \mathfrak{b})$ to $(2\mathfrak{a}^\dagger\mathfrak{b}, |\mathfrak{a}|^2 - |\mathfrak{b}|^2)$.

Here is a parenthetical remark: One can define a *fiber* bundle over S^4 with fiber SU(2) (but not a principal bundle) as follows: Choose integers (p, q) and declare the space $X_{p,q}$ to be diffeomorphic over U_+ to $U_+ \times SU(2)$ and also diffeomorphic over U_- to $U_- \times SU(2)$. Over the intersection of these sets, define the transition function so as to send $(x, \mathfrak{m}) \in U_+ \times SU(2)$ to $U_- \times SU(2)$ using the rule $(x, \mathfrak{m}) \to (x, g(x)^p \mathfrak{m} g(x)^q)$. This is a principal bundle if and only if one of p or q is zero. Now restrict to the case where $p + q = 1$, and set $k = p - q$. John Milnor proved that $X_{p,q}$ in this case is homeomorphic to S^7 but not diffeomorphic to S^7 if $k^2 - 1 \neq 0$ mod(7). In fact, each residue class mod(7) gives a different smooth structure on S^7. As it turns out, there are 28 different smooth structures on S^7. All can be realized as follows: Intersect the unit sphere about the origin in $\mathbb{C}^5$ with the submanifold in $\mathbb{C}^5-\{0\}$ where the complex coordinates $(z_1,\ldots,z_5)$ obey the equation

$$z_1^2 + z_2^2 + z_3^2 + z_4^3 + z_5^{6k-1} = 0$$

for $k \in \{1, 2, \ldots, 28\}$. (These are known as Brieskorn spheres.) Note that this submanifold is a complex manifold, and that it intersects S^9 transversely.

To read more about exotic spheres, see John Milnor's original article, "*On manifolds diffeomorphic to the 7-sphere*", Annals of Mathematics 64 (1956) 399–405, and the paper by John Milnor and Michel Kervaire, "*Groups of homotopy spheres I*", Annals of Mathematics 77 (1963) 504–537. See also the entry by J.P. Levine, "*Lectures on groups of homotopy spheres*" on pages 62–95 of the book *Algebraic and Geometric Topology*, Lecture Notes in Mathematics *1126*, Springer, 1985.

Example 10.7 Let M denote any given 4-dimensional manifold. Choose a finite set, Λ, of distinct points in M and associate to each an integer. For each $p \in \Lambda$, fix a coordinate chart U, centered at p, with coordinate map $\varphi: U \to \mathbb{R}^4$. Take U so that charts given by distinct points from Λ are disjoint. Define a principal SU(2)-bundle $P \to M$ as follows: Set $P|_{M-\Lambda}$ to be isomorphic to $(M-\Lambda) \times SU(2)$. For each $p \in \Lambda$, set $P|_U = U \times SU(2)$ also. To define the transition functions, first agree to identify the unit sphere in $\mathbb{R}^4$ with S^3. This done, let q denote the integer associated to the point p, and then the transition function, $g_{U,M-\Lambda}$ to be the map that sends $x \in U \cap (M-p) = U-p$ to $(\varphi(x)/|\varphi(x)|)^q \in SU(2)$.

Example 10.8 Something along the lines of what is done in Example 10.7 can be done any time you have an interesting map from S^{n-1} to SO(k), U(k), or some other group G. Let f denote your map. Pick a set Λ of distinct points in M^n, and associate to each an integer. Choose for each point a coordinate chart as in the previous example. Define the bundle to be trivial over $M-\Lambda$ and over each coordinate chart. If $p \in \Lambda$, the transition function for $U \cap (M-p) = U-p$ sends x to $f(x)^q$. For example, recall from Chapter 6.6 the definition of the matrices $\{\gamma_\alpha\}_{\alpha=1,\ldots,5}$ in $\mathbb{M}(4; \mathbb{C})$. Define a map from $\mathbb{R}^5$ to $\mathbb{M}(4; \mathbb{C})$ by sending a given point $x = (x_1,\ldots,x_5)$ to $f(x) = \sum_j x_j\gamma_j$. Note that $f^\dagger f = |x|^2$, so f maps S^4 to U(4).

Example 10.9 This example returns to the principal U(1)-bundle that is defined by the projection from S^{2n+1} to $\mathbb{CP}^n$ as described at the end of the previous part of this chapter. Recall from Chapter 6.8 that $\mathbb{CP}^n$ has an open cover by $n+1$ sets with each set in the cover, $\mathfrak{U} = \{\mathcal{O}_1,\ldots,\mathcal{O}_{n+1}\}$ diffeomorphic to $\mathbb{C}^n$. By way of reminder, any given point $(z_1,\ldots,z_n) \in \mathcal{O}_k = \mathbb{C}^n$ parametrizes the complex line in $\mathbb{C}^{n+1}$ that is spanned (over $\mathbb{C}$) by the vector with k'th entry is 1, whose i'th entry for $i < k$ is z_i, and whose i'th entry for $i > k$ is z_{i-1}.

This understood, the intersection between $\mathcal{O}_k$ and $\mathcal{O}_{k'}$ consists of the complex lines in $\mathbb{C}^{n+1}$ spanned by vectors of the form $(p_1, \ldots, p_{n+1})$ with both p_k and $p_{k'}$ nonzero. A map from $\mathcal{O}_k \cap \mathcal{O}_{k'}$ to U(1) assigns to any such line the complex number $g_{k,k'} = |p_k|^{-1}|p_{k'}|p_k/p_{k'}$. The collection $\{g_{k,k'}\}_{1\leq k\leq n+1}$ defines principal U(1)-bundle transition functions for the open cover $\mathfrak{U}$. To see that such is the case, it is enough to verify that the cocycle condition is obeyed on any given triple intersection. To do this, suppose that k, k′, and k″ are distinct. Then $g_{k,k'}g_{k',k''}g_{k'',k}$ is observedly 1 as required.

By the way, the identification between the trivial bundle $\mathcal{O}_k \times U(1)$ with a subset of S^{2n+1} is as follows: Use $z = (z_1, \ldots, z_n)$ as before for the coordinates on $\mathcal{O}_k$. Let $u \in U(1)$. The point $(z, u) \in \mathcal{O}_k \times U(1)$ gives the point whose k'th coordinate is $(1 + |z|^2)^{-1/2}u$, whose i'th coordinate for $i < k$ is $(1 + |z|^2)^{-1/2}z_i u$, and whose i'th coordinate for $i > k$ is given by $(1 + |z|^2)^{-1/2}z_{i-1}u$.

Example 10.10 This example constructs what are known as *quaternionic projective spaces*. To start, fix a positive integer, n. Consider $\mathbb{R}^{4(n+1)}$ as $\times^{(n+1)} \mathbb{H}$. Let S^{4n+3} denote the unit sphere in this Euclidean space; and define an action of SU(2) on S^{4n+2} by sending any given $\mathfrak{m} \in SU(2)$ and $(\mathfrak{a}_1, \ldots, \mathfrak{a}_{n+1}) \in S^{4n+3}$ to $(\mathfrak{a}_1\mathfrak{m}^{-1}, \ldots, \mathfrak{a}_{n+1}\mathfrak{m}^{-1})$. The space of orbits, $S^{4n+3}/SU(2)$ is a smooth manifold which is usually denoted by $\mathbb{HP}^n$, the *quaternionic projective space*. It has dimension 4n. The quotient map defines a principal SU(2)-bundle. The proof that such is the case copies in an almost verbatim fashion what is said in the preceding example about the principle U(1)-bundle $S^{2n+1} \to \mathbb{CP}^n$. The details are left to you the reader to work out.

10.7 Pull-backs of principal bundles

Let N and M denote a pair of smooth manifolds, and let $f: M \to N$ denote a smooth map. As explained in Chapter 5.1, the map f can be used to pull-back a vector bundle over N to obtain a vector bundle over M. Principal bundles can also be pulled back using f. To elaborate, suppose that G is a compact Lie group and $\pi: P \to N$ is a principal G-bundle. The pull-back bundle f^*P is a principal bundle over M that is defined as follows: It sits in $M \times P$ as the subset of pairs (x, p) with $f(x) = \pi(p)$. Here is a neat way to see f^*P as a manifold: Let $\Delta_N \subset N \times N$ denote the diagonal, thus the set of pairs (y, y) with $y \in N$. This is a submanifold. Map $M \times P$ to $N \times N$ using the map that sends a pair (x, p) to $(f(x), \pi(p))$. It is a straightforward exercise to verify that this map is transversal to the diagonal in the sense used in Chapter 5.7. This understood, the proposition in Chapter 5.7 guarantees that the inverse image of Δ_N in $M \times P$ is a submanifold. This inverse image is f^*P.

To see that f^*P meets the principal bundle requirements listed in Chapter 10.1, note to start that G acts so that any given $\mathfrak{g} \in G$ sends a point $(x, p) \in f^*P$ to $(x, p\mathfrak{g}^{-1})$. Meanwhile, the map π to M sends (x, p) to x. The last of the three requirements listed in Chapter 10.1 is the following: Given $x \in M$, needed is an open set, $U \subset M$, that contains x and comes with a G-equivariant diffeomorphism from $P|_U$ to $U \times G$ that intertwines π with the projection from $U \times G$ to U. To obtain this data, go to N and select an open set $V \subset N$ that contains f(x) and comes with a G-equivariant diffeomorphism, φ_V, from $P|_V$ to $V \times G$ that intertwines the map from $P|_V$ with the projection to V. Let ψ_V: $P|_V \to G$ denote the composition of first φ_V and then projection to G. Now take $U = f^{-1}(V)$ and set φ_U: $f^*P|_U \to U \times G$ to be the map that sends a pair (x, p) to $(x, \psi_V(p))$.

It is perhaps redundant to say that the bundle f^*P comes with the tautological, G-equivariant map to P, this the map that sends a given pair (x, p) to p.

Example 10.11 Chapter 10.1 declared a pair of principal G-bundles π: $P \to M$ and π': $P' \to M$ to be isomorphic if there exists a G-equivariant diffeomorphism f: $P \to P'$ such that $\pi = \pi' \circ f$. Keeping this in mind, suppose that π: $P \to M$ defines a principal U(1)-bundle. As it turns out, there exists an integer $n \geq 1$, a map f: $M \to \mathbb{CP}^n$, and an isomorphism between P and the pull-back f^*S^{2n+1} with S^{2n+1} viewed here as a principal U(1)-bundle over $\mathbb{CP}^n$; this as described in Example 10.9. What follows gives a construction of this data when M is compact.

To start, fix a finite cover, $\mathfrak{U}$, of M such that each set U from $\mathfrak{U}$ comes with a bundle isomorphism φ_U: $P|_U \to U \times U(1)$. Let ψ_U: $P \to U(1)$ denote the composition of first φ_U and then the projection to the U(1) factor. By refining this cover, I can assume that each open set is also a coordinate chart, and so there is a subordinate partition of unity. Recall from Appendix 1.2 that this is a collection, $\{\chi_U\}_{U\in\mathfrak{U}}$, of nonnegative functions such that any given $U \in \mathfrak{U}$ version of χ_U has support only in U, and such that $\sum_{U\in\mathfrak{U}} \chi_U = 1$ everywhere. Set $h_U(\cdot)$ to denote the function on P that is obtained by composing π with $\chi_U(\sum_{U\in\mathfrak{U}} \chi_U{}^2)^{-1/2}$. Write the number of sets in the cover $\mathfrak{U}$ as $n + 1$, and label the sets in $\mathfrak{U}$ from 1 to $n + 1$.

Define now a map, $\hat{f}$, from P to $S^{2n+1} \subset \mathbb{C}^{n+1}$ by taking the k'th coordinate to be zero on the complement of U_k and equal to $h_{U_k} \psi_{U_k}$ on U_k. This map is U(1)-equivariant, and so the quotient of the image by U(1) defines a map, f, from M to $\mathbb{CP}^n$. The bundle isomorphism between P and f^*S^{2n+1} sends $p \in P$ to $(\pi(p), \hat{f}(p)) \in M \times S^{2n+1}$.

Example 10.12 Example 10.10 describes the quaternionic projective spaces $\{\mathbb{HP}^n\}_{n\geq 1}$; and it explains how to view the map from S^{4n+3} to $\mathbb{HP}^n$ as a principal SU(2)-bundle. Let π: $P \to M$ denote a given principal SU(2)-bundle. A construction that is very much like that in the preceding example will give an integer, $n \geq 1$, a map $f: M \to \mathbb{HP}^n$, and a principal bundle isomorphism between P and the pull-back bundle f^*S^{4n+3}.

Example 10.13 Let N denote a smooth manifold and suppose that $E \to N$ is a vector bundle with fiber dimension n, either real or complex. As explained in Chapter 10.3, one can construct from E the principal bundle of frames for E. This is a principal bundle with group G either $Gl(n; \mathbb{R})$ or $Gl(n; \mathbb{C})$ as the case may be. If E also has a fiber metric, one can use the metric to construct the principal $G = O(n)$-, $SO(n)$-, $U(n)$- or $SU(n)$-bundle depending on the circumstances. I use P_E in what follows to denote any one of these and use G to denote the relevant Lie group. Now let M denote a second smooth manifold and suppose that $f: M \to N$ is a smooth map. As explained in Chapter 5.1, the map f can be used to pull E back so as to give a vector bundle, f^*E, over M. Metrics on bundles pull-back as well, the norm of a vector $(x, v) \in f^*E \subset M \times E$ being $|v|$. If E has a metric, then give f^*E this pull-back fiber metric. In any event, one has the principal G-bundle P_{f^*E} over M. Meanwhile, principal G-bundles pull-back as explained above, so one also has the principal G-bundle f^*P_E. I hope it is no surprise that P_{f^*E} is the same as f^*P_E. Those who doubt can unravel the definitions and so verify the assertion.

10.8 Reducible principal bundles

Suppose that G is a Lie group, $H \subset G$ is a subgroup, and $P \to M$ is a principal H-bundle. A principal G-bundle over M can be constructed from P as follows:

$$P_G = (P \times G)/H = P \times_H G,$$

where the equivalence relation identifies pairs (p, g) with (ph^{-1}, hg). Multiplication by an element $\mathfrak{g} \in G$ sends the equivalence class (p, g) to that of $(p, g\mathfrak{g}^{-1})$. To see that this is a principal bundle (and a manifold), go to any given set $U \subset M$ where $P|_U$ admits an isomorphism, φ, with $U \times H$. This isomorphism identifes P_G with $U \times (H \times G)/H$ which is $U \times G$. If $g_{U,U'}: U \cap U' \to H$ is the transition function for P over an intersection of charts, then this is also the transition function for P_G with the understanding that H is to be viewed as a subgroup in G.

A principal G-bundle $P' \to M$ that is isomorphic to $P \times_H G$ with $P \to M$ some principal H-bundle and with $H \subset G$ a proper subgroup is said to be *reducible*. For example, if $E \to M$ is a complex vector bundle with fiber $\mathbb{C}^n$ with Hermitian metric, then one can construct the U(n) principal bundle $P_{U(E)} \to M$. This bundle is reducible to a principal SU(n)-bundle if the vector bundle $\wedge^n E = \det(E)$ is trivial.

Every bundle has a fiber metric, and this implies that every principal $Gl(n; \mathbb{R})$-bundle is reducible to a principal $O(n)$-bundle. (If the transition functions can be chosen so as to have positive determinant, then the bundle is reducible to an $SO(n)$-bundle.) By the same token, any given principal $Gl(n; \mathbb{C})$-bundle is reducible to a principal $U(n)$-bundle. This is because the metric can be used to find local trivializations for the bundle whose transition functions map to $O(n)$ or $U(n)$ as the case may be.

Here is another example: If a 2n-dimensional manifold M admits an almost complex structure, j, then its principal $Gl(2n; \mathbb{R})$ frame bundle is reducible to a $Gl(n; \mathbb{C})$-bundle. Here, $Gl(n; \mathbb{C}) \subset Gl(2n; \mathbb{R})$ is identified with the subgroup, G_j, of matrices $\mathfrak{m}$ such that $\mathfrak{m} j_0 = j_0 \mathfrak{m}$ where j_0 is the standard almost complex structure on $\mathbb{R}^{2n}$ as defined in Chapter 2.4.

10.9 Associated vector bundles

Chapter 10.3 explains how to construct a principal bundle from a given vector bundle. As explained here, this construction has an inverse of sorts. To set the stage, suppose that G is a Lie group and π: $P \to M$ is a principal G-bundle. The construction of a vector bundle from P requires as additional input a representation of the group G into either $Gl(n; \mathbb{R})$ or $Gl(n; \mathbb{C})$. Use V here to denote either $\mathbb{R}^n$ or $\mathbb{C}^n$ as the case may be; and use $Gl(V)$ to denote the corresponding general linear group, either $Gl(n; \mathbb{R})$ or $Gl(n; \mathbb{C})$. Let ρ denote the representation in question. Recall from Chapter 10.4 that ρ is a smooth map from G to $Gl(V)$ two special properties: First, $\rho(\iota)$ is the identity in the matrix group $Gl(V)$. Second, $\rho(gg') = \rho(g)\rho(g')$ for any given pair of elements g, g′ in G.

The corresponding vector bundle has fiber V and is denoted by $P \times_\rho V$. It is the quotient of $P \times V$ by the equivalence relation that equates a given pair (p, v) to all pairs of the form $(pg^{-1}, \rho(g)v)$ with $g \in G$. Said differently, G acts on $P \times V$ so that any given $g \in G$ sends any given pair (p, v) to $(pg^{-1}, \rho(g)v)$. The space $P \times_\rho V$ is the space of G-orbits via this action.

To verify that this is indeed a vector bundle, note first that the projection to M is that sending the equivalence class of a given (p, v) to $\pi(p)$. The action of $\mathbb{R}$ or $\mathbb{C}$ has a given real or complex number z sending (p, v) to (p, zv). The zero section, $\hat{o}$, is the equivalence class of $(p, 0) \in P \times V$. To see the local structure, let $x \in M$ denote any given point and let $U \subset M$ denote an open set with a principal bundle isomorphism, φ, from $P|_U$ to $U \times G$. Use ψ: $P \to G$ to denote the composition of first φ and then projection from $U \times G$ to G. Now define φ^V: $(P \times_\rho V)|_U \to U \times V$ to be the map that sends the equivalence class of (p, v) to $(\pi(p), \rho(\psi(p))\, v)$. Note that this map is 1–1 and invertible; its inverse sends $(x, v) \in U \times V$ to the equivalence class of $(\varphi^{-1}(x, \iota), v)$.

This same map φ^V defines the smooth structure for $(P \times_\rho V)|_U$. I claim that this smooth structure is compatible with that defined for $(P \times_\rho V)|_{U'}$ when U and U′ are intersecting open sets. To verify this claim, let $\varphi: P|_U \to U \times G$ and $\varphi': P|_{U'} \to U' \times G$ denote the associated principal bundle isomorphisms. The composition $\varphi' \circ \varphi^{-1}$ maps $(U' \cap U) \times G$ as a map, $(x, g) \to (x, \mathfrak{g}_{U',U}(x)\, g)$ where $\mathfrak{g}_{U',U}$ maps $U' \cap U$ to G. The corresponding vector bundle transition function $\varphi'^V \circ \varphi^V$ sends (x, v) to $(x, \rho(\mathfrak{g}_{U',U}(x))\, v)$, and this is, of course, a smooth, invertible map that is linear over $\mathbb{R}$ or $\mathbb{C}$ as the case may be.

What follows are examples of associated vector bundles.

Example 10.14 If $E \to M$ is a given vector bundle, with fiber $V = \mathbb{R}^n$ or $\mathbb{C}^n$, then I defined in Chapter 10.3 the bundle $P_{Gl(E)}$ which is a principal $G = Gl(n; \mathbb{R})$ or $G = Gl(n; \mathbb{C})$-bundle as the case may be. Let ρ now denote the defining representation of G on V. Then $P_{Gl(E)} \times_\rho V$ is canonically isomorphic to E. Of course, they have the same transition functions, so they are isomorphic. Another way to see this is to view $P_{GL(E)} \subset \bigoplus_n E$ as the set of elements $(e_1, \ldots, e_n)$ that span the fiber of E at each point. Define $f: (P_{Gl(E)} \times V) \to E$ by sending $e = (e_1, \ldots, e_n)$ and v to $f(e, v) = \sum_j v_j e_j$. This map is invariant with respect to the G-action on $P_{Gl(E)} \times V$ whose quotient gives $P_{Gl(E)} \times_\rho V$, and so the map descends to the desired bundle isomorphism between $P_{Gl} \times_\rho V$ and E.

If $V = \mathbb{R}^n$ and E has a fiber metric, or if $V = \mathbb{C}^n$ and E has a Hermitian metric, I defined in Chapter 10.3 the respective principal SO(n)- or U(n)-bundles $P_{O(E)}$ and $P_{U(E)}$. Let ρ denote the standard inclusion homomorphism from SO(n) into $Gl(n; \mathbb{R})$ when $V = \mathbb{R}^n$, and let it denote the corresponding homomorphism from U(n) into $Gl(n; \mathbb{C})$ when $V = \mathbb{C}^n$. Reasoning as in the previous paragraph leads to the conclusion that $P_{O(E)} \times_\rho \mathbb{R}^n = E$ in the real case, and $P_{U(E)} \times_\rho \mathbb{C}^n = E$ in the complex case.

Example 10.15 This example describes a sort of converse to what is said in the previous example. To set the stage, let $V = \mathbb{R}^n$ or $\mathbb{C}^n$. In the real case, use G to denote either $Gl(n; \mathbb{R})$ or O(n); and in the complex case, use G to denote either $Gl(n; \mathbb{C})$ or or U(n). In any case, use ρ to denote the standard representation of the relevant group acting on V. Suppose that $P \to M$ is a principal G-bundle, and let $E = P \times_\rho V$. In the real case, P is then isomorphic to $P_{Gl(E)}$ or $P_{SO(E)}$ as the case may be. In the complex case, P is isomorphic to $P_{GL(E)}$ or $P_{U(E)}$.

Now recall from Chapter 5.2 in the real case that there exists $m > n$ and a smooth map from M to the Grassmannian Gr(m; n) with the property that its pull-back of the tautological $\mathbb{R}^n$-bundle is isomorphic to E. Chapter 6.11 tells a similar story in the complex case: There exists $m > n$ and a map from M to $Gr_{\mathbb{C}}(m; n)$ whose pull-back of the tautological $\mathbb{C}^n$-bundle is isomorphic to E. Keeping these facts in mind, let Gr denote Gr(m; n) in the real case and $Gr_{\mathbb{C}}(m; n)$ in the complex case, and let $f: M \to Gr$ denote the relevant map.

Introduce $E_T \to Gr$ to denote the aforementioned tautological G-bundle, and use $P_T \to Gr$ to denote the bundle of frames in E_T. Take these to be orthonormal or unitary in the case that G is SO(n) or U(n) respectively. The map f also pulls back P_T, and $f^*P_T = P_{Gl(f^*E_T)}$ or $P_{O(f^*E_T)}$ or $P_{U(f^*E_T)}$ as the case may be. It follows as a consequence that the original bundle P is isomorphic to the appropriate version of f^*P_T. Here is a formal statement of this last observation:

Proposition 10.16 *Let* G *denote either* Gl(n; $\mathbb{R}$) *or* SO(n) *in the real case, and either* Gl(n; $\mathbb{C}$) *or* U(n) *in the complex case. Let* M *denote a smooth manifold and let* $P \to M$ *denote a principal* G-*bundle. There exists* $m \geq n$ *and a smooth map from* M *to the Grassmannian* Gr *whose pull-back of the principal* G-*bundle* $P_T \to Gr$ *is isomorphic to* P.

Example 10.17 The neat thing about this associated vector bundle construction is that all bundles that are constructed from E via various algebraic operations on the fiber vector space $V = \mathbb{C}^n$ or $\mathbb{R}^n$, such as direct sum, tensor product, dualizing, Hom, etc., can be viewed as arising from $P_{Gl(E)}$ through an associated bundle construction. For example, $\otimes_n E$, $\wedge^p E$, $Sym^p(E)$, $E^* = Hom(E; \mathbb{R}$ or $\mathbb{C})$, etc. correspond to representations of Gl(n; $\mathbb{R}$) or Gl(n; $\mathbb{C}$) into various vector spaces. Thus, all of the latter bundles can be studied at once by focusing on the one principal bundle $P_{Gl(E)}$.

One example is as follows: Let g be a Riemannian metric on a manifold M, and so that there exists the principal O(n)-bundle $P_{O(TM)} \to M$ of oriented, orthonormal frames in TM. Let ρ denote the representation of O(n) on the vector space $\mathbb{A}(n; \mathbb{R})$ of $n \times n$ antisymmetric matrices that has $\mathfrak{g} \in O(n)$ act so as to send $\mathfrak{a} \in \mathbb{A}$ to $\rho(\mathfrak{g})\mathfrak{a} = \mathfrak{g}\mathfrak{a}\mathfrak{g}^{-1}$. This representation is called the *adjoint* representation, and the associated vector bundle is isomorphic to $\wedge^2 T^*M$.

A second example along these same lines concerns the bundle $\wedge^n T^*M$. The latter is associated to $P_{O(TM)}$ via the representation of O(n) in the two group $O(1) = \{\pm 1\}$ that sends a matrix $\mathfrak{g}$ to its determinant.

Vector bundles that are associated to the orthonormal frame bundle of a Riemannian manifold are called *tensor* bundles, and a section of such a bundle is said to be a *tensor* or sometimes a *tensor field*.

Appendix 10.1 Proof of Proposition 10.1

Assume below that G is a subgroup of some $n \geq 1$ version of either Gl(n; $\mathbb{R}$) or Gl(n; $\mathbb{C}$).

To see that H is a subgroup, note that if h and h′ are in H, then $\rho(hh')\mathfrak{v}$ is the same as $\rho(h)\rho(h')\mathfrak{v} = \rho(h)\mathfrak{v} = \mathfrak{v}$. To see that $\mathfrak{h}^{-1} \in H$, note that $\rho(\iota)\mathfrak{v} = \mathfrak{v}$, and because $\iota = \mathfrak{h}^{-1}h$, so $\rho(h^{-1}h)\mathfrak{v} = \mathfrak{v}$. As $\rho(h^{-1}h) = \rho(h^{-1})\rho(h)$, this means that $\rho(h^{-1})\rho(h)\mathfrak{v} = \mathfrak{v}$. If $\rho(h)\mathfrak{v} = \mathfrak{v}$, then this last identity requires that $\rho(h^{-1})\mathfrak{v} = \mathfrak{v}$ also.

To see that H is a submanifold and a Lie group, it is sufficient to give a local coordinate chart near any given element. To start, remark that a chart for a neighborhood of the identity ι supplies one for a neighborhood of any other element in H. Here is why: If $U \subset H$ is an open set containing ι with a diffeomorphism $\varphi: U \to \mathbb{R}^m$ for some m, then the set $U_h = \{h' : h'h^{-1} \in U\}$ is a chart near h. This understood, the diffeomorphism φ_h that sends h' to $\varphi(h'h^{-1})$ gives the desired diffeomorphism for U_h.

Granted all of this, what follows constructs coordinates for some neighborhood of ι. To do this, introduce $\mathfrak{H} \subset \mathfrak{lie}(G)$ to denote the kernel of the differential of ρ at the identity. This is a Euclidean space whose dimension depends, in general, on $\mathfrak{v}$. Let m denote this dimension. As is argued next, the exponential map $\mathfrak{h} \to e^{\mathfrak{h}}$ restricts $\mathfrak{H}$ so as to map the latter into H. For a proof, fix a large integer $N \gg 1$ and write $e^{\mathfrak{h}} = e^{\mathfrak{h}/N} \cdots e^{\mathfrak{h}/N}$ as the product of N factors. This allows $\rho(e^{\mathfrak{h}})\mathfrak{v}$ to be written as $\rho(e^{\mathfrak{h}/N}) \cdots \rho(e^{\mathfrak{h}/N})\mathfrak{v}$. Meanwhile, Taylor's theorem with remainder finds that

$$\rho(e^{\mathfrak{h}/N})\mathfrak{v} = \rho(\iota)\mathfrak{v} + \frac{1}{N}\rho_*(\mathfrak{h})\mathfrak{v} + \mathfrak{e}_1 = \mathfrak{v} + \mathfrak{e}$$

where $|\mathfrak{e}| \le c_0 N^{-2}$. This understood, it follows by iterating this last result N times that

$$\rho(e^{\mathfrak{h}})\mathfrak{v} = \mathfrak{v} + \mathfrak{e}_N$$

where $|\mathfrak{e}_N| \le c_0 N^{-1}$. As N can be as large as desired, this means that $\rho(e^{\mathfrak{h}})\mathfrak{v} = \mathfrak{v}$.

It follows from what was just said that the map $\mathfrak{h} \to e^{\mathfrak{h}}$ embeds a ball about 0 in $\mathfrak{H}$ into H. What follows argues that this embedding restricts to some ball about 0 in $\mathfrak{H}$ so as to map onto a neighborhood of ι. Note in the meantime that this result implies that the map $\mathfrak{h} \to e^{\mathfrak{h}}$ gives local coordinates for a neighborhood of the identity in H. It also implies that H is a totally geodesic submanifold of G when given the metric that is induced from the left invariant metric on G that comes from the latter's identification as a submanifold of $Gl(n; \mathbb{R})$ or $Gl(n; \mathbb{C})$ as the case may be.

To see that this map $\mathfrak{h} \to e^{\mathfrak{h}}$ is onto a neighborhood of the identity, suppose $h \in H$ is near the identity. As noted in Chapter 8.7, this element h can be written in any event as $e^{\mathfrak{m}}$ for some $\mathfrak{m} \in \mathfrak{lie}(G)$. Write $\mathfrak{m} = \mathfrak{h} + \mathfrak{z}$ with $\mathfrak{h} \in \mathfrak{H}$ and with $\mathfrak{z}$ orthogonal to $\mathfrak{H}$ and $|\mathfrak{m}|$ small. Keep in mind that with $|\mathfrak{m}|$ being small, then both $|\mathfrak{h}|$ and $|\mathfrak{z}|$ are small as both are less than $|\mathfrak{m}|$. This understood, note that $e^{\mathfrak{h}+\mathfrak{z}} = e^{\mathfrak{h}}e^{\mathfrak{z}} + \mathfrak{e}$ where $|\mathfrak{e}| \le c_0\,|\mathfrak{h}|\,|\mathfrak{z}|$. It follows as a consequence, that $\rho(e^{\mathfrak{h}+\mathfrak{z}})\mathfrak{v} = \rho(e^{\mathfrak{z}})\mathfrak{v} + \mathfrak{v}'$ with $|\mathfrak{v}'| \le c_0\,|\mathfrak{h}|\,|\mathfrak{z}|$. Meanwhile,

$$\rho(e^{\mathfrak{z}}) = \iota + \rho_*|_\iota \mathfrak{z} + \mathcal{O}(|\mathfrak{z}|^2)$$

and so $\rho(e^{\mathfrak{z}})\mathfrak{v} = \mathfrak{v} + (\rho_*|_\iota \mathfrak{z})\ \mathfrak{v} + \mathfrak{v}''$ where $|\mathfrak{v}''| \leq c_0\ |\mathfrak{m}|\ |\mathfrak{z}|$. Given that $\mathfrak{z}$ is orthogonal to $\mathfrak{H}$, it follows that $|\rho(e^{\mathfrak{z}})\mathfrak{v} - \mathfrak{v}| \geq c_0^{-1}|\mathfrak{z}|\ |\mathfrak{v}|$ when $|\mathfrak{m}|$ is small. Given that h is in H, so it follows that $\mathfrak{z} = 0$ if h is in some neighborhood of ι.

A suitable cover of $M_\mathfrak{v}$ by charts will give it a manifold structure. Keep in mind when defining the charts that any given element in $M_\mathfrak{v}$ is mapped to $\mathfrak{v}$ by the action of G on V. As a consequence, it is sufficient to construct a local chart near $\mathfrak{v}$. Indeed, if $U \subset M_\mathfrak{v}$ is a neighborhood of $\mathfrak{v}$ with a homeomorphism $\varphi: U \to \mathbb{R}^d$ for some d, and if $\mathfrak{v}' = \rho(g)\mathfrak{v}$ for some $g \in G$, then $U_{\mathfrak{v}'} = \{\mathfrak{b} \in M_\mathfrak{a} : \rho(g^{-1})\mathfrak{b} \in U\}$ is a neighborhood of $\mathfrak{v}'$. Moreover, the map $\varphi_{\mathfrak{v}'}$ that sends $\mathfrak{b} \in U_{\mathfrak{v}'}$ to $\varphi(\rho(g^{-1})\mathfrak{b})$ is then a local homeomorphism from $U_{\mathfrak{v}'}$ to $\mathbb{R}^d$.

Let $\mathfrak{H}^\perp \subset \mathfrak{lie}(G)$ denote the orthogonal complement of $\mathfrak{H}$. The chart for a neighborhood of $\mathfrak{v}$ with its homeomorphism is obtained by defining the inverse, this a homeomorphism, ψ, from a ball in $\mathfrak{H}^\perp$ to a neighborhood of $\mathfrak{v}$ that sends the origin to $\mathfrak{v}$. This homeomorphism sends $\mathfrak{z} \in \mathfrak{H}^\perp$ to $\psi(z) = \rho(e^{\mathfrak{z}})\mathfrak{v}$. This map is 1–1 on a small ball about the origin. This because $e^{\mathfrak{z}} - e^{\mathfrak{z}'} = \mathfrak{z} - \mathfrak{z}' + \mathfrak{r}$ with $|\mathfrak{r}| \leq c_0(|\mathfrak{v}| + |\mathfrak{v}'|)\ |\mathfrak{v} - \mathfrak{v}'|$. Thus,

$$\rho(e^{\mathfrak{z}}) - \rho(e^{\mathfrak{z}'}) = \rho_*|_i \mathfrak{z} - \rho_*|_i \mathfrak{z}' + \mathfrak{r}' \quad \textit{where} \quad |\mathfrak{r}'| \leq c_0(|\mathfrak{v}| + |\mathfrak{v}'|)|\mathfrak{v} - \mathfrak{v}'|\,.$$

As a consequence, $\rho(e^{\mathfrak{z}})\mathfrak{v} = \rho(e^{\mathfrak{z}'})\mathfrak{v}$ if and only if $\mathfrak{z} - \mathfrak{z}' \in \mathfrak{H}$ when both $|\mathfrak{z}|$ and $|\mathfrak{z}'|$ are small. This is possible only if $\mathfrak{z} - \mathfrak{z}' = 0$.

To see why this map is onto a neighborhood of $\mathfrak{v}$ in $M_\mathfrak{v}$, suppose for the sake of argument that there is a sequence $\{\mathfrak{v}_k\}_{k\geq 1} \subset M_\mathfrak{v}$ that converges to $\mathfrak{v}$, but is such that no element from this sequence can be written as $\rho(e^{\mathfrak{z}})\mathfrak{v}$ with $\mathfrak{z} \in \mathfrak{H}^\perp$ having small norm. By definition, each $\mathfrak{v}_k$ can be written as $\rho(g_k)\mathfrak{v}$ with $g_k \in G$. Moreover, there exists $\delta > 0$ such that each g_k has distance at least δ from any point in H. Here, the metric is that used in Chapter 8.7. To see why this is, note that if such were not the case, then g_k could be modified by multiplying on the right by the closest element in H so that the result is close to the origin in G. The result could then be written as $\exp(\mathfrak{z})\exp(\mathfrak{h})$ with $\mathfrak{h} \in \mathfrak{H}$ and $\mathfrak{z} \in \mathfrak{H}^\perp$; and this would exhibit $\mathfrak{v}_k$ as $\rho(e^{\mathfrak{z}})\mathfrak{v}$. Meanwhile, the fact that G is compact implies that the sequence $\{g_k\}_{k\geq 1}$ has a convergent subsequence. Let g denote the limit. Given that the elements in the sequence $\{g_k\}$ are uniformly far from H, this matrix g is not in H. Even so $\rho(g)\mathfrak{v} = \mathfrak{v}$. As these last two conclusions contradict each other, there is no sequence $\{\mathfrak{v}_k\}_{k\geq 1}$ as described above. Note that this is the only place where the compactness of G is used.

To finish the proof of Proposition 10.1, consider the claim in the last bullet of the proposition. This claim follows with a proof that the map from G to $M_\mathfrak{v}$ has the property that is described by the third bullet in Chapter 10.1's definition of a principal bundle. This understood, note that it is sufficient to verify the following: There is an open set $U \subset M_{\mathfrak{v}}$ containing $\mathfrak{v}$ with a diffeomorphism $\varphi: G|_U \to U \times H$ that is suitably equivariant. To prove that such is the case, fix a small radius ball $B \subset \mathfrak{H}^{\perp}$ on which the map $\mathfrak{z} \to \rho(e^{\mathfrak{z}})\mathfrak{v}$ is a diffeomorphism onto its image. Let U denote the image of this diffeomorphism, and set $\psi: U \to B$ to be the inverse of this diffeomorphism. Thus, $\rho(e^{\psi(\mathfrak{y})})\mathfrak{v} = \mathfrak{y}$ for all $\mathfrak{y} \in U$. Define φ on $G|_U$ to be the map that sends $g \in G|_U$ to $\varphi(g) = \rho(g)v, e^{-\psi(\rho(g))}g)$. This gives the required H-equivariant diffeomorphism.

Additional reading

- *Topology of Fibre Bundles*, Norman Steenrod, Princeton University Press, 1999.
- *Fibre Bundles*, Dale Husemöller, Springer, 1993.
- *Basic Bundle Theory and K-Cohomology Invariants*, Dale Husemoller, Michael Joachim, Branislaw Jurco and Marten Schottenloher, Springer, 2009.
- *Metric Structures in Differential Geometry*, Gerard Walschap, Springer, 2004.
- *Differential Geometry, Lie Groups and Symmetric Spaces*, Sigurdur Helgason, American Mathematical Society, 2001.

11 Covariant derivatives and connections

Let π: E → M denote a vector bundle, with fiber $V = \mathbb{R}^n$ or $\mathbb{C}^n$. Use $C^\infty(M; E)$ to denote the vector space of smooth sections of E. Recall here that a section, $\mathfrak{s}$, is a map from M to E such that $\pi \circ \mathfrak{s}$ is the identity. This is an infinite dimensional vector space. The question arises as to how to take the derivative of a section of E in a given direction. Of course, one can take the differential of a map, this giving a linear map $\mathfrak{s}_*$: TM → $\mathfrak{s}^*$TE. This notion of derivative turns out to be less than satisfactory. Needed is a derivative that defines a bundle homomorphism from TM to E, not to $\mathfrak{s}^*$TE. At issue here is how to define such a derivative.

As explained in what follows, there are various ways to proceed, though all give the same thing at the end of the day. One such definition involves principal Lie group bundles and the notion of a *connection* on such a bundle. The related notions of covariant derivative and connection are the focus of this chapter.

11.1 Covariant derivatives

The space $C^\infty(M; E)$ is a module for the action of the algebra, $C^\infty(M)$, of smooth functions with values in $\mathbb{R}$ if $V = \mathbb{R}^n$ or values in $\mathbb{C}$ if $V = \mathbb{C}^n$. The action is such that a given function f acts on $\mathfrak{s} \in C^\infty(M; E)$ to give $f\mathfrak{s}$. A *covariant derivative* for $C^\infty(M; E)$ is a map $\nabla: C^\infty(M; E) \to C^\infty(M; E \otimes T^*M)$ with the following two properties: First, it respects the vector space structure. This is to say that $\nabla(c\mathfrak{s}) = c\nabla\mathfrak{s}$ and $\nabla(\mathfrak{s} + \mathfrak{s}') = \nabla\mathfrak{s} + \nabla\mathfrak{s}'$ and when c is in $\mathbb{R}$ or $\mathbb{C}$ (as the case may be), while $\mathfrak{s}$ and $\mathfrak{s}'$ are sections of E. Second, it obeys the analog of Leibnitz's rule:

$$\nabla(f\mathfrak{s}) = f\nabla\mathfrak{s} + \mathfrak{s} \otimes df$$

for all $f \in C^\infty(M)$.

To see that there exist covariant derivatives, it is enough to exhibit at least one. To obtain one, take a locally finite open cover, Λ, of M such that any given open set $U \in \Lambda$ comes with a vector bundle isomorphism $\varphi_U : E|_U \to U \times V$. Let $\{\chi_U\}_{U\in\Lambda}$ denote a subordinate partition of unity. Thus, $\chi_U : M \to [0, 1]$ has support only in U, and $\sum_{U\in\Lambda} \chi_U = 1$ at each point. Save this partition for a moment.

For each $U \in \Lambda$, define the covariant derivative, d, on $C^\infty(U; U\times V)$ as follows: Write a given section, $\mathfrak{v}$, of the trivial bundle $U \times V$ as a map $x \to (x, v(x))$. Now define the covariant derivative $d\mathfrak{v}$ as the map $x \to (x, dv|_x)$.

Now, suppose that $\mathfrak{s}$ is any given section of E. Define

$$\nabla\mathfrak{s} = \sum_{U\in\Lambda} \chi_U \varphi_U^*(d(\varphi_U \circ \mathfrak{s}|_U)).$$

Note that this obeys the Leibnitz rule by virtue of the fact that $\sum_{U\in\Lambda} \chi_U = 1$.

11.2 The space of covariant derivatives

There are lots of covariant derivatives. As is explained next, the space of covariant derivatives is an affine space modeled on $C^\infty(M; \mathrm{Hom}(E; E \otimes T^*M))$. To see this, first note that if $\mathfrak{a} \in C^\infty(M; \mathrm{Hom}(E; E \otimes T^*M))$ and if ∇ is any given covariant derivative, then $\nabla + \mathfrak{a}$ is also a covariant derivative. Meanwhile, if ∇ and ∇' are both covariant derivatives, then their difference, $\nabla - \nabla'$, is a section of $\mathrm{Hom}(E; E \otimes T^*M)$. This is because their difference is linear over the action of $C^\infty(M)$.

The following lemma puts this last remark in a more general context. Lemma 11.1 applies to the situation at hand using for E′ the bundle $E \otimes T^*M$ and taking $\mathcal{L} = \nabla' - \nabla$.

Lemma 11.1 *Suppose that* E *and* E′ *are vector bundles (either real or complex) and that* $\mathcal{L}$ *is an* $\mathbb{R}$*- or* $\mathbb{C}$*-linear map (as the case may be) that takes a section of* E *to one of* E′*. Suppose in addition that* $\mathcal{L}(f\,\mathfrak{s}) = f\,\mathcal{L}(\mathfrak{s})$ *for all functions* f*. Then there exists a unique section,* L*, of* Hom(E; E′) *such that* $\mathcal{L}(\cdot) = L(\cdot)$.

Proof of Lemma 11.1 To find L, fix an open set $U \subset M$ where both E and E′ has a basis of sections. Denote the basis for E and the basis for E′ as $\{e_a\}_{1\le a\le d}$ and $\{e'_b\}_{1\le b\le d'}$ where d and d′ denote here the respective fiber dimensions of E and E′. Since $\{e'_b\}_{1\le b\le d'}$ is a basis of sections of E′, any given $a \in \{1, \ldots, d\}$ version of $\mathcal{L}e_a$ can be written as a linear combination of this basis. This is to say that there are functions $\{L_{ab}\}_{1\le b\le d'}$ on U such that

$$\mathcal{L}e_a = \sum_{1\le b\le d'} L_{ab} e'_b.$$

This understood, the homomorphism L is defined over U as follows: Let $\mathfrak{s}$ denote a section of E over U, and write $\mathfrak{s} = \sum_{1\le a\le d} \mathfrak{s}_a\, e_a$ in terms of the basis $\{e_a\}_{1\le a\le d}$. Then $L\mathfrak{s}$ is defined to be the section of E' given by $L\mathfrak{s} = \sum_{1\le a\le d,\ 1\le b\le d'} L_{ab}\, \mathfrak{s}_a\, e'_b$. The identity $\mathcal{L}\mathfrak{s} = L\mathfrak{s}$ is guaranteed by the fact that $\mathcal{L}(f\ \cdot) = f\ \mathcal{L}(\cdot)$ when f is a function. The same identity guarantees that the homomorphism L does not depend on the choice of the basis of sections. This is to say that any two choices give the same section of Hom(E; E′).

It is traditional to view $\nabla - \nabla'$ as a section of $\mathrm{End}(E) \otimes T^*M$ rather than as a section of $\mathrm{Hom}(E; E \otimes T^*M)$; these bundles being canonically isomorphic.

What was just said about the affine nature of the space of covariant derivatives has the following implication: Let ∇ denote a covariant derivative on $C^\infty(M; E)$, and let $\mathfrak{s}$ denote a section of E. Suppose that U is an open set in M and $\varphi_U\colon E|_U \to U \times V$ is a bundle isomorphism. Write $\varphi_U\mathfrak{s}$ as $x \to (x, \mathfrak{s}_U(x))$ with $\mathfrak{s}_U\colon U \to V$. Then $\varphi_U(\nabla\mathfrak{s})$ appears as the section

$$x \to (x,(\nabla\mathfrak{s})_U) \ \textit{where}\ \nabla\mathfrak{s}_U = d\mathfrak{s}_U + \mathfrak{a}_U\mathfrak{s}_U$$

and $\mathfrak{a}_U$ is some $\mathfrak{s}$-independent section of $(U \times \mathrm{End}V) \otimes T^*M|_U$.

Be forewarned that the assignment $U \to \mathfrak{a}_U$ does *not* define a section over M of $\mathrm{End}(E) \otimes T^*M$. Here is why: Suppose that U′ is another open set with a bundle isomorphism $\varphi_{U'}\colon E|_{U'} \to U' \times V$ that overlaps with U. Let $\mathfrak{g}_{U',U}\colon U \cap U' \to \mathrm{End}(V)$ denote the transition function. Thus, $\mathfrak{s}_{U'} = \mathfrak{g}_{U',U}\, \mathfrak{s}_U$. Meanwhile, $\nabla\mathfrak{s}$ is a bonafide section of $E \otimes T^*M$ and so $(\nabla s)_{U'} = \mathfrak{g}_{U',U}(\nabla\mathfrak{s})_U$. This requires that

$$\mathfrak{a}_{U'} = \mathfrak{g}_{U',U}\mathfrak{a}_U\mathfrak{g}_{U'U}^{-1} - (d\mathfrak{g}_{U',U})\mathfrak{g}_{U'U}^{-1}.$$

Conversely, a covariant derivative on $C^\infty(M; E)$ is defined by the following data: First, a locally finite, open cover Λ of M such that each $U \in \Lambda$ comes with a bundle isomorphism $\varphi_U\colon E|_U \to U \times V$. Second, for each $U \in \Lambda$, a section of $\mathfrak{a}_U\colon (U \times \mathrm{End}V) \otimes T^*M|_U$, this denoted by $\mathfrak{a}_U$. The collection $\{\mathfrak{a}_U\}_{U\in\Lambda}$ define a covariant derivative if and only if the condition above holds for any pair U, $U' \in \Lambda$ that overlap.

11.3 Another construction of covariant derivatives

What follows describes a relatively straightforward construction of a covariant derivative on a given vector bundle $E \to M$ with fiber $\mathbb{V}^n = \mathbb{R}^n$ or $\mathbb{C}^n$. I assume here that M is compact, but with care, you can make this construction work in general. Let $\mathbb{V}$ denote either $\mathbb{R}$ or $\mathbb{C}$ as the case may be.

This construction exploits the fact that the product bundle $M \times \mathbb{V}^N$ has one very simple covariant derivative that is defined as follows:

Let $x \to \mathfrak{f} = (x, (f_1(x), \ldots, f_N(x)))$ denote a section of the bundle $M \times \mathbb{V}^N \to M$. Define the section $d\mathfrak{f}$ of $(M \times \mathbb{V}^N) \otimes T^*M$ by

$$x \to (x, (df_1|_x, \ldots, df_N|_x)).$$

This gives a covariant derivative. Now, suppose that $N \geq 1$ and that E is a subbundle of $M \times \mathbb{V}^N$. Introduce $\Pi \in \mathrm{Hom}(M \times \mathbb{V}^N; E)$ to denote the fiberwise orthogonal projection in $\mathbb{V}^N$ onto E. Let $\mathfrak{s}$: $M \to E$ now denote a section of E. Since E sits in $M \times \mathbb{R}^N$, I can view $\mathfrak{s}$ as a section of $M \times \mathbb{R}^N$ and so make sense of $d\mathfrak{s}$. This understood, then

$$\nabla \mathfrak{s} = \Pi d\mathfrak{s}$$

is a section of $E \otimes T^*M$, and the assignment $\mathfrak{s} \to \nabla\mathfrak{s}$ is a covariant derivative on E.

Now suppose we are given the bundle $E \to M$ in isolation. Here is how to view E as a subbundle of some trivial bundle: Let $E^* \to M$ denote the bundle $\mathrm{Hom}(E; \mathbb{V})$. Recall that one can find some integer $N \geq 1$ and a set $\{\mathfrak{s}_1, \ldots, \mathfrak{s}_N\}$ of sections of E^* with the property that this set spans the fiber of E at each point in M. One way to do this is to use the open cover, Λ, of M that was introduced above, with its associated partition of unity $\{\chi_U\}_{U\in\Lambda}$. Fix a basis $\{v_1, \ldots, v_n\}$ for V. Each $U \in \Lambda$ determines the set of n sections $\{\mathfrak{s}_{1U}, \ldots, \mathfrak{s}_{nU}\}$ where $\mathfrak{s}_{kU} = \varphi_U^{-1}(\chi_U v_k)$. Here φ_U: $E|_U \to U \times V$ is the associated isomorphism. Also, v_k is viewed here as a section of $U \times V$. Take the set $\{\mathfrak{s}_1, \ldots, \mathfrak{s}_N\}$ to be an ordering of the set $\cup_{U\in\Lambda} \{\mathfrak{s}_{1U}, \ldots, \mathfrak{s}_{nU}\}$.

Define a bundle homomorphism ψ: $E \to M \times \mathbb{V}^N$ by the following rule: Let $x \in M$ and $v \in E|_x$. Then $\psi(v) = (x, \mathfrak{s}_1|_x \cdot v, \ldots, \mathfrak{s}_N|_x \cdot v)$. This puts E inside $M \times \mathbb{V}^N$ as a subbundle, and so we can use the definition of ∇ above to define a covariant derivative for sections of E.

Here is a relatively concrete example: View S^n as the unit sphere in $\mathbb{R}^{n+1}$. Now view TS^n as the set of pairs $\{(x, v): x, v \in \mathbb{R}^{n+1}$ *with* $|x| = 1$ *and* $x^T v = 0\}$. This identifies TS^n with a subbundle in $S^n \times \mathbb{R}^{n+1}$. Let $\mathfrak{s}$ now denote a section of TS^n, thus a vector field on S^n. Define the section $\nabla\mathfrak{s}$ of $TS^n \otimes T^*S^n$ by the rule $\nabla\mathfrak{s} = \Pi d\mathfrak{s}$. Note that in this case $\Pi|_x = \iota - xx^T$ where ι is the identity $(n + 1) \times (n + 1)$ matrix. Since $x^T\mathfrak{s}(x) = 0$, another way to write this is $\nabla\mathfrak{s}$ is $\nabla\mathfrak{s} = d\mathfrak{s} + x dx^T \mathfrak{s}$. For instance, consider the following vector field: Let $e \in \mathbb{R}^{n+1}$ denote a constant vector. Then $x \to \mathfrak{s}(x) = e - xx^T e$ is a section of TS^n. Its covariant derivative is $(xx^T - \iota) dx x^T e$.

11.4 Principal bundles and connections

This part the chapter defines the notion of a *connection* on a principal bundle. This notion is of central importance in its own right. In any event, connections are used momentarily to give an alternate and very useful definition of the covariant derivative. The discussion here has seven parts.

11.4.1 Part 1

Suppose that E is associated to some principal bundle $\pi: P \to M$. (We know that E is always associated to $P_{Gl(E)}$, and to perhaps other principal bundles.) In particular, suppose that G is a Lie group, $V = \mathbb{R}^m$ or $\mathbb{C}^m$ as the case may be, and ρ is a representation of G in Gl(V). Suppose further that E is given as $P \times_\rho V$. Recall from Chapter 10.9 that the latter is the quotient of $P \times V$ by the relation $(p, v) \sim (pg^{-1}, \rho(g)v)$ when $g \in G$. A section $\mathfrak{s}$ of E appears in this context as a G-equivariant map from P to V. Said differently, a section $\mathfrak{s}$ defines, and is conversely defined by a smooth map $\mathfrak{s}^P: P \to V$ that obeys $\mathfrak{s}^P(pg^{-1}) = \rho(g)\mathfrak{s}^P(p)$. Indeed, suppose first that one is given such a map. The corresponding section $\mathfrak{s}$ associates to any given point $x \in M$ the equivalence class of $(p, \mathfrak{s}^P(p))$ where p can be any point in $P|_x$. To see that this makes sense, remark that any other point in $P|_x$ can be written as pg^{-1} with $g \in G$. Because $\mathfrak{s}^P(pg^{-1}) = \rho(g)\mathfrak{s}^P(p)$, the pair $(pg^{-1}, \mathfrak{s}^P(pg^{-1}))$ defines the same equivalence class in $P \times_\rho V$ as does the original pair $(p, \mathfrak{s}^P(p))$. To go in reverse, suppose that $\mathfrak{s}$ is a section of $P \times_\rho V$. By definition, $\mathfrak{s}$ associates an equivalence class in $P \times V$ to every point in M. An equivalence class is an assignment to each $p \in P$ of a point $v \in V$ such that pg^{-1} is assigned the point $\rho(g)$ p. But this is just another way to say that such an assignment is a smooth map from P to V that obeys the required condition.

This view from P of a section of $P \times_\rho V$ has the following useful generalization: Suppose that $\mathfrak{p}: E' \to M$ is a second vector bundle either real or complex, and perhaps unrelated to P. The examples that follow take E′ to be either TM, T*M or some exterior power $\wedge^k T^*M$. In any event, a section over M of $(P \times_\rho V) \otimes E'$ appears upstairs on P as a suitably G-equivariant section over P of the tensor product of the product bundle $P \times V$ with the bundle π^*E'. Here, π^*E' is the pull-back of E′ via π; this bundle is defined in Chapter 5.1 as the subspace of pairs $(p, e') \in P \times E'$ that share the same base point in M. Said differently, a section of the tensor product bundle $(P \times_\rho V) \otimes E'$ over M appears up on P as a G-equivariant, fiber-wise linear map from π^*E' to V. The notion of G-equivariance in this context is as in the preceding paragraph: If $\mathfrak{s}$ denotes the section of $(P \times_\rho V) \otimes E'$ in question, then the corresponding map, $\mathfrak{s}^P$, from π^*E' to V is such that $\mathfrak{s}^P(pg^{-1}, e') = \rho(g)\, \mathfrak{s}^P(p, e')$ for each $g \in G$ and pair $(p, e') \in \pi^*E' \subset P \times E'$. Meanwhile, the notion of fiber-wise linear means the following: If r is in either $\mathbb{R}$ or $\mathbb{C}$ as the case may be, then $\mathfrak{s}^P(p, re') = r\, \mathfrak{s}(p, e')$; and if (p, e_1') and (p, e_2') are points in π^*E' over the same point in P, then $\mathfrak{s}^P(p, e_1' + e_2') = \mathfrak{s}^P(p, e_1') + \mathfrak{s}^P(p, e_2')$.

11.4.2 Part 2

We already know how to take the derivative of a map from a manifold to a vector space. This Y is the manifold in question, and $\mathfrak{v}$ the map from Y to the vector space, then the derivative is a vector of differential forms on Y; any given component is the exterior derivative of the corresponding component of $\mathfrak{v}$. In the case when $Y = P$ and $\mathfrak{v}$ is the map $\mathfrak{s}^P$ that corresponds to a section, $\mathfrak{s}$, of $P \times_\rho V$, then this vector of differential forms is denoted by, $(\mathfrak{s}^P)_*$. As a vector of differential forms, it defines a fiber-wise linear map from TP to V. Although this map is suitably G-equivariant, it does not by itself define a covariant derivative. Indeed, a covariant derivative appears upstairs on P as a G-equivariant, fiber-wise linear map from π^*TM to V. Said differently, a covariant derivative appears on P as a suitably G-equivariant section over P of the tensor product of π^*TM with the product bundle $P \times V$.

11.4.3 Part 3

To see what $(\mathfrak{s}^P)_*$ is missing, it is important to keep in mind that TP has some special properties that arise from the fact that P is a principal bundle over M. In particular, there exists over P the sequence of vector bundle homomorphisms

$$(*) \qquad 0 \to \ker(\pi_*) \to TP \to \pi^*TM \to 0,$$

where the notation is as follows: First, $\pi_*: TP \to TM$ is the differential of the projection map π and $\ker(\pi_*)$ designates the subbundle in TP that is sent by π_* to the zero section in TM. This is to say that the vectors in the kernel π_* are those that are tangent to the fibers of π. Thus, $\ker(\pi_*)$ over $P|_x$ is canonically isomorphic to $T(P|_x)$. The arrows in $(*)$ are meant to indicate the following homomorphisms: That from $\ker(\pi_*)$ to TP is the tautological inclusion as a subbundle. That from TP to π^*TM sends a given vector $v \in TP$ to the pair $(\pi(p), \pi_* v)$ in $\pi^*TM \subset P \times TM$.

What follows are three key observations about $(*)$. Here is the first: The sequence in $(*)$ is *exact* in the sense that the image of any one of the homomorphisms is the kernel of the homomorphism to its right. Here is the second: The action of G on P lifts to give an action of G on each of the bundles that appear in $(*)$. To elaborate, suppose that g is a given element in G. Introduce for the moment $m_g : P \to P$ to denote the action of $g \in G$ that sends p to pg^{-1}. This action lifts to TP so as the push-forward map $(m_g)_*$ as defined in Chapter 5.3. If $v \in TP$ is in the kernel of π_*, then so is $(m_g)_* v$ because $\pi \circ m_g = \pi$. Thus, $(m_g)_*$ acts also on $\ker(\pi_*)$. The action lift of m_g to π^*TM is defined by viewing the latter in the manner described above as a subset of $P \times TM$. Viewed in this way, m_g act so as to send a pair $(p, v) \in P \times TM$ to the pair (pg^{-1}, v). Here, is the final observation: The homomorphisms in $(*)$ are equivariant with respect to the lifts just described of the G-action on P. Indeed, this follows automatically for the inclusion map from $\ker(\pi_*)$ to TP, and it follows for the map from TP to π^*TM because $\pi \circ m_g = \pi$.

A very much-related fact is that $\ker(\pi_*)$ is canonically isomorphic to the trivial bundle $P \times \mathfrak{lie}(G)$. This isomorphism is given by the map $\psi\colon P \times \mathfrak{lie}(G) \to \ker(\pi_*)$ that sends a pair $(p, \mathfrak{m}) \in P \times \mathfrak{lie}(G)$ to the tangent vector at $t = 0$ of the path $t \to p\exp(t\mathfrak{m})$. This map ψ is equivariant with respect to the action of G on $\mathfrak{lie}(G)$ that sends $g \in G$ and $\mathfrak{m}$ to $g\mathfrak{m}g^{-1}$. This is to say that the differential of the map $m_g\colon P \to P$ that sends p to pg^{-1} act so that

$$(m_g)_* \psi(p, \mathfrak{v}) = \psi(pg^{-1}, g\mathfrak{v}g^{-1}) .$$

11.4.4 Part 4

A *connection* on the principal bundle P is neither more nor less than a G-equivariant splitting of the exact sequence $(*)$. Thus, a connection, A, is by definition a linear map

$$A\colon TP \to \ker(\pi_*)$$

that equals the identity on the kernel of π_* and is equivariant with respect to the action of G on P. This is to say that if $g \in G$ and $v \in TP$, then $(m_g)_* (Av) = A((m_g)_* v)$.

The isomorphism $\psi\colon P \times \mathfrak{lie}(G) \to \ker(\pi_*)$ is used to view a connection, A, as a map

$$A\colon TP \to \mathfrak{lie}(G)$$

with the following properties:

- $A(\psi(p, \mathfrak{m})) = \mathfrak{m}$.
- *If* $g \in G$ *and* $v \in TP$, *then* $A(m_{g*}v) = gA(v)g^{-1}$.

This last view of A is quite useful. Viewed this way, a connection is a section of the tensor product of P's cotangent bundle with the product bundle $P \times \mathfrak{lie}(G)$ with certain additional properties that concern the G-action on P and the sequence $(*)$.

The kernel of the homomorphism A is a subbundle in TP which is isomorphic via the right most arrow in $(*)$ to π^*TM. This subbundle ker(A) is often called the *horizontal* subbundle of TP and denoted by $H_A \subset TP$. Meanwhile, the bundle $\ker(\pi_*)$ is just as often called the *vertical* subbundle.

With regards to the notation in what follows, it is customary not to distinguish by notation the aforementioned two views of a connection, one as a bundle homomorphism from TP to $\ker(\pi_*)$ and the other as a fiber-wise linear map from TP to $\mathfrak{lie}(G)$. In any event, the latter view is used primarily in what follows. Another notational quirk with regards to the second of these two views uses $T^*P \otimes \mathfrak{lie}(G)$ to denote the tensor product of T^*P with the product vector bundle $P \times \mathfrak{lie}(G)$.

11.4.5 Part 5

If A and A′ are connections on P, then $\mathfrak{a}^P = A - A'$ annihilates $\ker(\pi_*)$. As a consequence, it defines a fiber-wise linear, G-equivariant map from π^*TM to $\mathfrak{lie}(G)$. Given that $\mathfrak{a}^P$ is G-equivariant and linear on the fibers of π^*TM, it corresponds (as described in Chapter 11.4.2 above) to a section over M of $(P \times_{ad} \mathfrak{lie}(G)) \otimes T^*M$. Here, ad denotes the representation of G on its Lie algebra that has $g \in G$ sending any given matrix $\mathfrak{a} \in \mathfrak{lie}(G)$ to $g\mathfrak{a}g^{-1}$. As explained momentarily, the latter matrix is also in $\mathfrak{lie}(G)$. This representation of G on $\mathfrak{lie}(G)$ is called the *adjoint* representation.

To say more about the adjoint representation, recall that the Lie algebra of G was defined to be the tangent space to the identity element ι in G. Given that G is a matrix group, this vector space is a subspace of the relevant $n \in \{1, 2, \ldots\}$ version of either $Gl(n; \mathbb{R})$ or $Gl(n; \mathbb{C})$. What follows explains how to see that $g\mathfrak{a}g^{-1} \in \mathfrak{lie}(G)$ if $\mathfrak{a} \in \mathfrak{lie}(G)$ and $g \in G$. As explained in Chapter 8.7, the exponential map

$$\mathfrak{a} \to e^{\mathfrak{a}} = \iota + \mathfrak{a} + \frac{1}{2}\mathfrak{a}^2 + \cdots$$

maps $\mathfrak{lie}(G)$ to G. Moreover, this map restricts to a ball about the origin in $\mathfrak{lie}(G)$ so as to define a coordinate chart in G for a neighborhood of the origin. This implies in particular the following: Suppose that $\mathfrak{a}$ is any given matrix. If $t \in (0, 1)$ is not too big, the matrix $e^{t\mathfrak{a}}$ is in G if and *only* if $\mathfrak{a} \in \mathfrak{lie}(G)$. With the preceding as background, suppose that $\mathfrak{a} \in \mathfrak{lie}(G)$ and $g \in G$. Fix some small $t > 0$. Use the formula above to see that the matrix $e^{tg\mathfrak{a}g^{-1}} = g\, e^{t\mathfrak{a}}\, g^{-1}$. Since all three matrices in this product are in G, the result is in G. If t is not too large, this implies that $t\, g\, \mathfrak{a}\, g^{-1} \in \mathfrak{lie}(G)$ and thus $g\, \mathfrak{a}\, g^{-1} \in \mathfrak{lie}(G)$.

To continue the discussion about connections, suppose now that $\mathfrak{a}$ is a section over M of the bundle $(P \times_{ad} \mathfrak{lie}(G)) \otimes T^*M$ and that A is a connection on P. Just as a section of $P \times_{ad} \mathfrak{lie}(G)$ can be viewed as a G-equivariant map from P to $\mathfrak{lie}(G)$, so $\mathfrak{a}$ can be viewed on P as a G-equivariant, fiber-wise linear map from π^*TM to $\mathfrak{lie}(G)$. Let $\mathfrak{a}^P$ denote the latter incarnation of $\mathfrak{a}$. Then $A + \mathfrak{a}^P$ defines another connection on P.

These last observations have the following corollary: If P has one connection, then it has infinitely many, and the space of smooth connections on P is an affine space based on $C^\infty(M; (P \times_{ad} \mathfrak{lie}(G)) \otimes T^*M)$.

11.4.6 Part 6

What follows constructs a connection on P. As a preamble to the construction, suppose that $U \subset M$ is an open set with a principal bundle isomorphism $\varphi_U\colon P|_U \to U \times G$. The product principal bundle $U \times G$ has a tautological connection 1-form, this defined as follows when G is a subgroup in $Gl(n; \mathbb{R})$ or $Gl(n; \mathbb{C})$. The connection 1-form at a point $(x, \mathfrak{g}) \in U \times G$ is

$$A_0 = \mathfrak{g}^{-1}d\mathfrak{g},$$

this a matrix of 1-forms as it should be. Here, and in what follows, I view a connection on a principal G-bundle P as a $\mathfrak{lie}(G)$-valued 1-form P with the properties listed in Chapter 11.4.4.

To see that A_0 has the correct properties, fix a matrix $\mathfrak{m} \in \mathfrak{lie}(G)$ with $\mathfrak{lie}(G)$ viewed as a subvector space in $\mathbb{M}(n; \mathbb{R})$ or $\mathbb{M}(n; \mathbb{C})$ as the case may be. Introduce the map $t \to \mathfrak{g}e^{t\mathfrak{m}}$ from $\mathbb{R}$ into G, this as described in Chapter 5.5. The pull-back of A_0 by this map at $t = 0$ is the 1-form on $\mathbb{R}$ given by $(A_0|_{\mathfrak{g}})(\psi(\mathfrak{g}, \mathfrak{m}))\, dt$. This is $\mathfrak{m}\, dt$, as it should be. Note also that A_0 is suitably equivariant under the action of G on G by right translation.

The $\mathfrak{lie}(G)$-valued 1-form A_0 annihilates vectors that are tangent to the U factor of $U \times G$. This understood, the corresponding horizontal subbundle $H_{A_0} \subset TP$ are precisely the tangents to the U factor, this the factor TU in the obvious splitting of $T(U \times G) = TU \oplus TG$.

With this preamble in mind, now fix a locally finite cover $\mathfrak{U}$ of M such that each set $U \in \mathfrak{U}$ comes with a principal G-bundle isomorphism $\varphi_U\colon P|_U \to U \times G$. Let $\{\chi_U\}_{U\in\mathfrak{U}}$ denote a subordinate partition of unity. Such a partition is constructed in Appendix 1.2. Set

$$A = \sum\nolimits_{U\in\mathfrak{U}} \chi_U \varphi_U^* A_0.$$

This $\mathfrak{lie}(G)$-valued 1-form on P is a connection. The verification is left as an exercise.

11.4.7 Part 7

This last part of the story gives another way to view a connection on P. To start, let $U \subset M$ again denote an open set with a principal G-bundle isomorphism $\varphi\colon P|_U \to U \times G$. Now let A denote a given connection on G. Then $(\varphi_U^{-1})^*A$ is a connection on the product bundle $U \times G$. As such, it can be written as

$$(\varphi_U^{-1})^*A = \mathfrak{g}^{-1}d\mathfrak{g} + \mathfrak{g}^{-1}\mathfrak{a}_U\mathfrak{g}$$

where $\mathfrak{a}_U$ is a 1-form on U with values in $\mathfrak{lie}(G)$. The horizontal space H_A can be described as follows: Identify $T(U \times G) = TU \oplus TG$ and identify TG with $\mathfrak{lie}(G)$ as done above. Then

$$(\varphi_U)_*(H_A)|_{(x,\mathfrak{g})} = \{(v, -\mathfrak{g}^{-1}\mathfrak{a}_U(v)\mathfrak{g}) \in TU \otimes \mathfrak{lie}(G): v \in TU|_x\}.$$

11.5 Connections and covariant derivatives

Suppose that $E \to M$ is a vector bundle which is given by $P \times_\rho V$ where ρ is a representation of G on V. Let A denote a connection on P. Then A defines a covariant derivative, ∇_A, on $C^\infty(M; E)$ as follows: View a section $\mathfrak{s}$ of $E = P \times_\rho V$ as a G-equivariant map, $\mathfrak{s}^P$: $P \to V$. The corresponding covariant derivative $\nabla_A\mathfrak{s}$ will be viewed as a G-equivariant section, $(\nabla_A\mathfrak{s})^P$, of $(P \times V) \otimes \pi^*TM$. To define this section, introduce as in Chapter 11.4.2, the differential, $(\mathfrak{s}^P)_*$, of the map $\mathfrak{s}^P$ from P to V. This differential maps TP to V. Now, $(\nabla_A\mathfrak{s})^{(P)}$ is supposed to be a G-equivariant section over P of $(P \times V) \otimes \pi^*T^*M$, which is to say a G-equivariant homomorphism from π^*TM to V. To obtain such a thing, recall first that the connection's horizontal subbundle, $H_A \subset TP$ is canonically isomorphic to π^*TM. This understood, the restriction to H_A of the homomorphism $(\mathfrak{s}^P)_*$: $TP \to V$ defines a covariant derivative of $\mathfrak{s}$. The latter is, by definition, the covariant derivative $\nabla_A\mathfrak{s}$.

Here is a reinterpretation of this definition. Let $x \in M$ and let $v \in TM|_x$. The covariant derivative $\nabla_A\mathfrak{s}$ is supposed to assign to v a section of E. To obtain this section, pick any point $p \in P|_x$. There is a unique *horizontal* vector $v_A \in H_A|_p$ such that $\pi_* v_A = v$. The covariant derivative $\nabla_A\mathfrak{s}$ sends v to the equivalence class in $E|_x = (P|_x \times_\rho V)$ of the pair $(p, (\mathfrak{s}^P)_* v_A)$. To see that this is well defined, I need to check that a different choice for $p \in P|_x$ gives the same equivalence class. That such is the case is a consequence of the fact that A and $\mathfrak{s}^P$ are suitably equivariant with respect to the G-action on $P|_x$. This is proved by unwinding all of the definitions—a task that I will leave to you.

To see that this defines a covariant derivative, one need only check Leibnitz's rule. To do so, suppose that f is a smooth function on M. Then $(f\mathfrak{s})^P = \pi^*(f)\, \mathfrak{s}^P$ and as a consequence, $(f\mathfrak{s}^P)_* = \pi^*(f)\, (\mathfrak{s}^P)_* + \mathfrak{s}^P \otimes \pi^*(df)$. This implies that

$$(\nabla_A(f\mathfrak{s}))^P = (f\nabla_A\mathfrak{s}^A)^P + \mathfrak{s}^P \otimes \pi^*(df).$$

The latter asserts Leibnitz's rule.

If A and A′ are connections on P, then the difference $\nabla_{A'} - \nabla_A$ is supposed to be a section over M of $End(E) \otimes T^*M$. What follows identifies this section: Recall from Chapter 11.4.5 that the difference $A' - A = \mathfrak{a}$ can be viewed as a section over M of $(P \times_{ad} \mathfrak{lie}(G)) \otimes T^*M$. Meanwhile, the representation ρ induces a bundle homomorphism from $P \times_{ad} \mathfrak{lie}(G)$ to End(E) as follows: The bundle End(E) is the associated vector bundle $P \times_{ad(\rho)} Gl(V)$ where ad(ρ) here is the representation of G on GL(V) that has any given $g \in G$ act on $\mathfrak{m} \in Gl(V)$ as $\rho(g)\mathfrak{m}\rho(g^{-1})$. This understood, the differential of ρ at the identity in G defines a G-equivariant homomorphism $\rho_*: \mathfrak{lie}(G) \to Gl(V)$; this intertwines the representation ad on $\mathfrak{lie}(G)$ with the representation ad(ρ). The latter homomorphism gives a corresponding homomorphism from $(P \times_{ad} \mathfrak{lie}(G)) \otimes T^*M$ to $End(E) \otimes T^*M$. This corresponding homomorphism is also denoted by ρ_* because it is defined by ignoring the T^*M factor. To elaborate, it is defined so as to send a given decomposable element $\mathfrak{m} \otimes \upsilon$ to $\rho_*(\mathfrak{m}) \otimes \upsilon$; and as any given element is a linear combination of decomposable elements, the latter rule is all that is needed. Granted all of the above, it then follows by unwinding the definitions that

$$\nabla_{A'} - \nabla_A = \rho_*(\mathfrak{a}) .$$

Note that if P is the bundle $P_{Gl(E)}$ of orthonormal frames in E, then $P \times_{ad} \mathfrak{lie}(G)$ is the bundle End(E). This and the preceding equation gives a 1–1 correspondence between covariant derivatives on E and connections on $P_{Gl(E)}$.

11.6 Horizontal lifts

Let $\pi: P \to M$ denote a principal G-bundle. Fix a connection, A on P. The connection gives a way to lift any given smooth map $\gamma: \mathbb{R} \to M$ to a map $\gamma_A: \mathbb{R} \to P$ such that $\pi \circ \gamma_A = \gamma$. This is done as follows: Fix a point p_0 in the fiber of P over the point $\gamma(0)$. The lift γ_A is defined by the following two conditions:

- $\gamma_A(0) = p_0$
- *Let* ∂_t *denote the tangent vector to* $\mathbb{R}$. *Then* $(\gamma_A)_*\partial_t \in H_A$.

Thus, the lift starts at p_0 and its tangent vectors are everywhere horizontal as defined by the connection A. This is to say that $A((\gamma_A)_*\partial_t) = 0$.

To see what this looks like, suppose that γ crosses a given open set $U \subset M$ where there is a bundle isomorphism $\varphi_U: P|_U \to U \times G$. Then $\varphi_U(\gamma_A)$ is the map $t \to (\gamma(t), \mathfrak{g}(t))$ where $t \to \mathfrak{g}(t)$ makes the tangent vector to the path $t \to (\gamma(t), \mathfrak{g}(t))$ horizontal as defined using the connection $(\varphi_U^{-1})^*A = \mathfrak{g}^{-1}d\mathfrak{g} + \mathfrak{g}^{-1}\mathfrak{a}_U\mathfrak{g}$ on $U \times G$. With $T(U \times G)$ identified as before with $TU \oplus \mathfrak{lie}(G)$, this tangent vector is

$$\left(\gamma_*\partial_t, \mathfrak{g}^{-1}\frac{d}{dt}\mathfrak{g}\right).$$

The latter is horizontal if and only if it has the form $(v, -g^{-1}\mathfrak{a}_U(v)g)$ with $v \in TU$. This understood, the map $t \to g(t)$ must obey

$$g^{-1}\frac{d}{dt}g = -g^{-1}\mathfrak{a}_U(\gamma_*\partial_t)|_{\gamma(t)}g$$

which is to say that

$$\frac{d}{dt}g + \mathfrak{a}_U(\gamma_*\partial_t)|_\gamma g = 0.$$

If you recall the vector field theorem from Chapter 8.3 about integrating vector fields, you will remember that an equation of this sort has a unique solution given its starting value at some $t = t_0$. This being the case, horizontal lifts always exist.

11.7 An application to the classification of principal G-bundles up to isomorphism

The notion of a horizontal lift can be used to say something about the set of isomorphism classes of principal G-bundles over any given manifold. To this end, suppose that X is a smooth manifold and $\pi: P_X \to X$ is a principal G-bundle. Let M denote another smooth manifold.

Theorem 11.2 *Let* f_0 *and* f_1 *denote a pair of smooth maps from* M *to* X. *The principal G-bundles* $f_0^*P_X$ *and* $f_1^*P_X$ *are isomorphic if* f_0 *is homotopic to* f_1.

Proof of Theorem 11.2 There is an interval $I \subset \mathbb{R}$ containing [0, 1] and a smooth map, F, from $I \times M$ to X such that $F(0, \cdot) = f_0$ and $F(1, \cdot) = f_1$. Keeping this in mind, fix a connection, A, on the pull-back bundle F^*P. Define a map $\psi_A: f_0^*P \to f_1^*P$ as follows: Let $p \in f_0^*P$ and let $t \to \gamma_{A,p}(t) \in F^*P$ denote the horizontal lift of the path $t \to (t, \pi(p)) \in I \times M$ that starts at p when $t = 0$. Set $\psi_A(p)$ to be $\gamma_{A,p}(1)$. This is an invertible, fiber preserving map. It also obeys $\psi_A(pg^{-1}) = \psi_A(p)g^{-1}$ because of the equivariance properties of the connection. In particular, these guarantee that $\gamma_{A,pg^{-1}} = \gamma_{A,p}\,g^{-1}$ at all times $t \in I$.

The classification problem is completed with a second theorem which asserts the existence, for any given lie group G, of a *universal classifying space*, this a smooth manifold X such that any given principal G-bundle over any given manifold is the pull-back of a fixed principal bundle $P_X \to X$. One can take dim(X) to be finite if a bound has been given a priori on the dimension of the manifolds under consideration. In any event, the result is that there is a 1–1 correspondence {*principal* G-*bundles on* M *up to isomorphism*} ⇔ {*homotopy classes of maps from* M *to* X}.

For example, the proposition in Example 10.14 asserts that the large m versions of real Grassmannian Gr(m; n) and the complex Grassmannian GrC (m; n) can serve as respective classifying spaces for principle O(n)- and U(n)-bundles over manifolds of a given dimension.

Theorem 11.2 has the following as a corollary.

Corollary 11.3 *Let* $P \to M$ *denote a principal* G-*bundle and let* U *denote a contractible, open set in* M. *Then there is an isomorphism* $\varphi\colon P|_U \to U \times G$.

What follows is another corollary with regards to the case when M has dimesion 2. To set the stage, recall that Chapters 6.4, 6.5 and 6.11 described various complex, rank 1 vector bundles over surfaces in $\mathbb{R}^3$ that were pull-backs of a fixed bundle over S^2 that was defined using the Pauli matrices.

Corollary 11.4 *Let* M *denote a compact, 2-dimensional manifold and let* f_0 *and* f_1 *denote homotopic maps from* M *to* S^2. *Let* $\pi\colon E \to S^2$ *denote a complex, rank 1 bundle. Then* f_0^*E *is isomorphic to* f_1^*E *if* f_0 *is homotopic to* f_1.

11.8 Connections, covariant derivatives and pull-back bundles

Suppose that M and N are manifolds, that $\phi\colon M \to N$ is a smooth map and that $\pi\colon P \to N$ is a principal bundle with fiber a given group G. As explained in Chapter 10.2, the map ϕ can be used to define pull-back principal bundle, $\pi_M\colon \phi^*P \to M$, this the subspace in $M \times P$ of points (x, p) with $\phi(x) = \pi(p)$. The projection, π_M is induced by the obvious projection from $M \times P$ to its M factor. Meanwhile, there exists a canonical G-equivariant map $\hat{\phi}\colon \phi^*P \to P$ that covers ϕ, this induced by the projection from $M \times P$ to its P factor.

By the way, another way to view ϕ^*P is to consider $\mathcal{P} = M \times P$ as a principal G-bundle over the product $M \times N$. Then M embeds in $M \times N$ as the graph of ϕ, this the subset of pairs $(x, \phi(x))$. The bundle ϕ^*P is the restriction of the principal bundle $\mathcal{P}$ to the graph of ϕ.

Suppose that A is a connection on P; thus a $\mathfrak{lie}(G)$-valued 1-form on P that behaves in a specified manner when pulled back by the action of any given element in G. Maps pull-back 1-forms, and the pull-back $\hat{\phi}^*A$ is a $\mathfrak{lie}(G)$-valued 1-form on ϕ^*P. This 1-form satisfies all of the required conditions for a connection because $\hat{\phi}$ is G-equivariant. You can check directly using the definition of pull-back from Chapter 5.3; or you can see this by viewing $\mathcal{P} = M \times P$ as a principal G-bundle over $M \times N$. Then A defines a connection on $\mathcal{P}$, since it is a $\mathfrak{lie}(G)$-valued 1-form on $\mathcal{P}$ with all of the required properties. It just happens to be a 1-form with no dependence on the M factor in M × P. Granted all of this, then $\hat{\phi}^*A$ is the restriction of A (viewed as a connection on the bundle $\mathcal{P} \to M \times N$) to the graph of the map ϕ.

Now suppose that V is either $\mathbb{R}^n$ or $\mathbb{C}^n$ and $\rho: G \to Gl(V)$ is a representation. The connection A on P gives the bundle $E = (P \times_\rho V)$ a covariant derivative, ∇. The pull-back bundle, $\phi^*E = (\phi^*P \times_\rho V)$, then has the covariant derivative, $\phi^*\nabla$, this defined by the connection $\hat{\phi}^*A$. Here is another way to see this: View A again as a connection on the bundle $\mathcal{P} = M \times P$ over M × N. Let E denote the vector bundle $M \times E \to M \times N$. Of course, this is $\mathcal{P} \times_\rho V$. The restriction of E to the graph of ϕ is the bundle ϕ^* E. The connection A, now viewed as living on $\mathcal{P}$, defines a covariant derivative for E, and hence for the restriction of E to the graph of ϕ. This covariant derivative is $\phi^*\nabla$.

What was just said about vector bundles can be said with no reference to connections. To elaborate, suppose that $\pi: E \to N$ is a vector bundle, with no particular principal bundle in view. View M × E as a vector bundle, E, over M × N. As before, the pull-back bundle ϕ^*E is the restriction of E to the graph of ϕ. If ∇ is a covariant derivative for sections of E, then its restriction to sections of E over the graph of ϕ defines the covariant derivative $\phi^*\nabla$ for sections of ϕ^*E.

Additional reading

- *Foundations of Differential Geometry, Volume 1*, Shoshichi Kobayshi and Katsumi Nomizu, Wiley Interscience, 1996.
- *Differential Forms and Connections*, R. W. R. Darling, Cambridge University Press, 1984.
- *Geometry, Topology and Physics*, Mikio Nakahara, Taylor and Francis, 2003.
- *Lectures on Seiberg–Witten Invariants*, John D. Moore, Springer, 2001.
- *Geometry of Four-Manifolds*, S. K. Donaldson and P. B. Kronheimer, Oxford University Press, 1997.
- *Instantons and Four-Manifolds*, Daniel S. Freed and Karen K. Uhlenbeck, Springer, 1990.

12 Covariant derivatives, connections and curvature

This chapter continues the discussion of connections and covariant derivatives with the introduction of the notion of the *curvature* of a covariant derivative or connection. This notion of curvature is defined in Chapter 12.4. The first three sections of the chapter constitute a digression of sorts to introduce two notions involving differentiation of differential forms and vector fields that require *no* auxiliary choices. Both are very important in their own right. In any event, they are used to define curvature when covariant derivatives reappear in the story.

12.1 Exterior derivative

This first part of the digression introduces the *exterior derivative*. As explained below, the exterior derivative is a natural extension to differential forms of the notion of the differential of a function.

To start, suppose that f is a function on M. Chapter 3.9 describes how its differential, df, can be viewed as a section of T^*M. The association of any given function to its differential defines a linear map from $C^\infty(M)$ to $C^\infty(M; T^*M)$. This understood, the exterior derivative extends this linear map so as to be a linear map from $C^\infty(M; \wedge^p T^*M)$ to $C^\infty(M; \wedge^{p+1} T^*M)$ for any given $p \in \{0, \ldots, n = \dim(M)\}$. Note in this regard that $\wedge^p T^*M$ is the 0-dimensional bundle $M \times \{0\}$ for $p > n$. This extension is also denoted by d.

What follows defined d by induction on the degree, p, of the differential form. If $f \in C^\infty(M) = C^\infty(M; \wedge^0 T^*M)$, define df as in Chapter 3.9. (Recall that in a coordinate patch with coordinates $(x_1, \ldots, x_n)$, this 1-form is $df = \sum_j \frac{\partial}{\partial x_j} f\, dx_j$.) Granted the preceding, suppose next that $d: C^\infty(M; \wedge^p T^*M) \to C^\infty(M; \wedge^{p+1} T^*M)$ has been defined for all integers $p = 0, \ldots, q-1$ for $1 \le q \le n$. The following rules specify how d is to act on $C^\infty(M; \wedge^q T^*M)$:

- *If* $\omega \in C^\infty(M;\ \wedge^q T^*M)$ *and* $\omega = d\alpha$ *for some* $\alpha \in C^\infty(M;\ \wedge^{q-1} T^*M)$, *then* $d\omega = 0$.
- *If* $\omega = f\ d\alpha$ *for some* α *as before, and with* f *a smooth function, then* $d\omega = df \wedge d\alpha$.
- *If* ω_1 *and* ω_2 *are two q-forms, then* $d(\omega_1 + \omega_2) = d\omega_1 + d\omega_2$.

Note that this induction covers all cases. This is because any given point in M has a neighborhood on which $\wedge^p T^*M$ is spanned by a set of sections of the form $dg_1 \wedge \cdots \wedge dg_p$ for $g_1, \ldots, g_p$ chosen from a set of dim(M) functions. (To see this, take a coordinate chart centered on the point. Then the p-fold wedge products of the differentials of the coordinate functions span $\wedge^p T^*M$ on this coordinate neighborhood.)

This definition must be checked for consistency. The point being that a given p-form ω can be written as a sum of terms, each of the form $g_{p+1}\, dg_1 \wedge \cdots \wedge dg_p$, in many different ways. The consistency is equivalent to the following hallmark of partial derivatives: They commute: If f is a smooth function on $\mathbb{R}^n$, then $\frac{\partial}{\partial x_j}\frac{\partial}{\partial x_k} f = \frac{\partial}{\partial x_k}\frac{\partial}{\partial x_j} f$.

To see why this is key, fix a coordinate chart so as to identify a neighborhood of given point in M with a neighborhood of the origin in $\mathbb{R}^n$. Let f denote a function on M. Then df appears in this coordinate chart as

$$df = \Sigma_j \frac{\partial}{\partial x_j} f dx_j.$$

Now, if $d(df)$ is to be zero, and also $d(dx_j)$ is to be zero, then the consistency of the definition requires that

$$\Sigma_{k,j} \frac{\partial}{\partial x_k}\frac{\partial}{\partial x_j} f dx_k \wedge dx_j = 0.$$

The left-hand side above is indeed zero. The reason is that $dx_k \wedge dx_j = -dx_j \wedge dx_k$ and so what is written here is the same as

$$\frac{1}{2}\Sigma_{k,j}\left(\frac{\partial}{\partial x_k}\frac{\partial}{\partial x_j} f - \frac{\partial}{\partial x_j}\frac{\partial}{\partial x_k} f\right) dx_k \wedge dx_j,$$

as can be seen by changing the order of summation. Meanwhile, what is written above is zero because partial derivatives in different directions commute.

By the same token, suppose that α is a (p−1)-form for $p > 1$. In this coordinate patch,

$$\alpha = \Sigma_{1 \le i_1 < \cdots < i_{p-1} \le n}\, \alpha_{i_1 i_2 \ldots i_{p-1}}\, dx_{i_1} \wedge \cdots \wedge dx_{i_{p-1}}$$

where each $\alpha_{.}$ is a function. This understood, then

- $d\alpha = \Sigma_{1\leq i_1<\cdots<i_{p-1}\leq n\ j}\frac{\partial}{\partial x_j}\alpha_{i_1 i_2\ldots i_{p-1}}dx_j \wedge dx_{i_1} \wedge \ldots \wedge dx_{i_{p-1}}$,
- $d(d\alpha)=\Sigma_{1\leq i_1<\cdots<i_{p-1}\leq n}\frac{\partial}{\partial x_k}\frac{\partial}{\partial x_j}\alpha_{i_1 i_2\ldots i_{p-1}}dx_k \wedge dx_j\wedge dx_{i_1} \wedge \ldots \wedge dx_{i_{p-1}}$;

and so the fact that $d(d\alpha) = 0$ follows as before from the fact that the matrix of second derivatives is symmetric.

Here is a related fact: The assertion that the matrix of second derivatives of a given function vanishes is not a coordinate independent statement. Indeed, suppose that $\frac{\partial}{\partial x_k}\frac{\partial}{\partial x_j}f = 0$ at a given point. Suppose now that $x_j = \phi_j(y)$ where $\phi: \mathbb{R}^n \to \mathbb{R}^n$ is a diffeomorphism on a neighborhood of the origin. Let $f' = \phi^* f$. This is to say that $f'(y) = f(\phi(y))$. The Chain rule tells us that

$$\frac{\partial}{\partial y_j}f' = \Sigma_m \left(\frac{\partial \phi_m}{\partial y_j}\right)\Big|_y \left(\frac{\partial}{\partial x_m}f\right)\Big|_{\phi(y)},$$

and then that

$$\frac{\partial}{\partial y_k}\frac{\partial}{\partial y_j}f' = \Sigma_{p,m}\frac{\partial \phi_p}{\partial y_k}\frac{\partial \phi_m}{\partial y_j}\frac{\partial^2}{\partial x_p \partial x_m}f + \Sigma_m \frac{\partial^2 \phi_m}{\partial y_k \partial y_j}\frac{\partial}{\partial x_m}f.$$

Thus, the vanishing of the matrix of second derivatives of f in all coordinate charts requires both the first and second derivatives of f to vanish in all coordinate charts.

By the way, the preceding identity implies that the matrix of second derivatives of a function does not define by itself a section over M of a vector bundle. This because of the right most contribution to its transformation law.

Here is one additional observation about the exterior derivative: Suppose that α is a p-form and β is a q-form. Then

$$d(\alpha\wedge\beta) = d\alpha\wedge\beta + (-1)^p\, \alpha\wedge d\beta.$$

This is consistent with the fact that $\alpha \wedge \beta = (-1)^{pq}\, \beta \wedge \alpha$.

12.2 Closed forms, exact forms, diffeomorphisms and De Rham cohomology

A p-form ω on M is said to be *closed* if $d\omega = 0$. Such a form is said to be *exact* of $\omega = d\alpha$. The *p'th De Rham cohomology* of M is, by definition, the vector space quotient of the linear space of closed forms by the linear subspace of exact forms. This vector space is denoted by $H^p_{\text{De Rham}}(M)$. Thus,

$$H^p_{\text{De Rham}}(M) = \text{kernel}(d)/\text{image}(d).$$

As is explained momentarily, these spaces are, a priori, invariants of the differentiable structure on M. (Even so, they are actually dependent only on the topological structure. If M and M′ are homeomorphic, smooth manifolds, then they have isomorphic De Rham cohomologies.) For example, $H^p_{\text{De Rham}}(S^n)$ is zero if $p \neq 0$, n and $\mathbb{R}$ if $p = 0$ or n.

To see why the De Rham cohomology is a differentiable structure invariant, it is enough to prove that the exterior derivative, d, commutes with pull-back by a map. This is to say the following: If M and N are smooth manifolds, if $f: M \to N$ is a smooth map, and if α is a $(p-1)$-form on N, then

$$f^*(d\alpha) = d(f^*\alpha).$$

The proof is an exercise with the Chain rule. To get you going on this task, suppose that $p \in M$ and that $(y_1, \ldots, y_m)$ are coordinates centered at p. Let $(x_1, \ldots, x_n)$ denote coordinates for a neighborhood of f(p) centered at f(p). Near p, the map f appears in these coordinates as a map $x \to (f_1(x), \ldots, f_n(x))$ of a neighborhood of $0 \in \mathbb{R}^m$ to a neighborhood of $0 \in \mathbb{R}^n$. Use the formula given in the just-completed Chapter 12.1 for α and $d\alpha$. Then use the Chain rule to compute their pull-backs to the via f.

On a related topic, suppose that f_0 and f_1 are smooth maps from M to a manifold N which are homotopic in the usual sense: There is a smooth map $\psi: [0, 1] \times M \to N$ with $\psi(0, \cdot) = f_0$ and $\psi(1, \cdot) = f_1$. Let ω denote a closed form on N. Then both $f_0^*\omega$ and $f_1^*\omega$ are closed forms. However, they define the same De Rham cohomology class as they differ by the image of d. To see why this is the case, write $T^*(\mathbb{R} \times M) = \mathbb{R} \oplus T^*M$. This is to say that a cotangent vector at (t, x) can be written as $\alpha = \alpha_0\, dt + \alpha_M$ where $\alpha_M \in T^*M|_x$ and where $\alpha_0 \in \mathbb{R}$. By the same token

$$\wedge^p T^*(\mathbb{R} \times M) = (\wedge^{p-1} T^*M) \oplus (\wedge^p T^*M)$$

this is to say that a p-form α on $\mathbb{R} \times M$ at any given point (t, x) can be written as

$$dt \wedge \alpha^{p-1} + \alpha_p$$

where $\alpha_{p-1} \in \wedge^{p-1} T^*M|_x$ and $\alpha_p \in \wedge^p T^*M|_x$. The form α_{p-1} can be viewed as a time $t \in \mathbb{R}$ dependent section of $\wedge^{p-1} T^*M|_x$ and the form α_p is a time $t \in \mathbb{R}$ dependent section of $\wedge^p T^*M$. The exterior derivative $d\alpha$ is a $(p + 1)$-form on $\mathbb{R} \times M$ and so it has corresponding components $dt \wedge (d\alpha)_p + (d\alpha)_{p+1}$. These are given by

$$(d\alpha)_p = -d^\perp \alpha_{p-1} + \frac{\partial}{\partial t}\alpha_p \ \textit{and}\ (d\alpha)_{p+1} = -d^\perp \alpha_p$$

where $d^\perp$ is used here to denote the exterior derivative at any fixed $t \in \mathbb{R}$ of a differential form on $\{t\} \times M$.

Now suppose that $\alpha = \psi^*\omega$ where ψ: $[0, 1] \times M \to N$ is the homotopy from f_0 to f_1. If ω is closed, then $d\psi^*\omega = \psi^* d\omega = 0$, so $\psi^*\omega$ is a closed form. This understood, the preceding formula for $(d\psi^*\omega)_{p+1}$ says that the pull-back of ω by $\psi(t, \cdot)$ is closed for any given $t \in [0, 1]$, and the formula for $(d\psi^*\omega)_p$ can be integrated to see that

$$f_1^*\omega = f_0^*\omega + d\int_0^1 (\psi^*\omega)_{p-1}.$$

This proves that $f_1^*\omega$ differs from $f_0^*\omega$ by an exact form.

This fact just stated has the following corollary:

Proposition 12.1 *Let* $U \subset M$ *denote a contractible open set. This is to say that there exists a smooth map* ψ: $[0, 1] \times U \to M$ *such that* $\psi(1, \cdot)$ *is the identity map on* U *and* $\psi(0, \cdot)$ *maps* U *to a point* $p \in M$. *Let* $p \geq 1$ *and let* ω *denote a closed* p*-form on* M. *Then there exists a* $(p-1)$*-form* α *on* U *such that* $d\alpha = \omega$.

Proof of Proposition 12.1 It follows that $\omega|_U = \psi(1, \cdot)^*\omega$ and this differs from $\psi(0, \cdot)^*\omega$ by an exact form. Meanwhile, $\psi(0, \cdot)$ maps U to a point, and since $\wedge^p T^*\{\text{point}\} = 0$ for $p > 0$, so $\psi(0, \cdot)^*\omega$ is equal to zero.

The proposition has as its corollary what is usually called the Poincaré lemma:

Corollary 12.2 (the Poincaré lemma) *Let* U *denote a contractible manifold, for instance* $\mathbb{R}^n$. *Then any closed form* p*-form on* U *for* $p \geq 1$ *is the exterior derivative of a* $(p-1)$*-form.*

This corollary can be used to prove the De Rham cohomology of a compact manifold is a finite dimensional vector space over $\mathbb{R}$. See Chapter 9 in the book *Differential Forms and Algebraic Topology* by R. Bott and L. Tu (1982).

12.3 Lie derivative

The preceding discussion about the pull-back of a differential form via a homotopy gives a useful formula when the homotopy is a map ψ: $\mathbb{R} \times M \to M$ that is obtained by moving the points in M some given time along the integral curves of a vector field. To elaborate, suppose that M is compact and that v is a vector field on M. The vector field theorem in Chapter 8.3 supplies a smooth map ψ: $\mathbb{R} \times M \to M$ with the following properties:

- $\psi(0, x) = x$.
- $\psi_*\partial_t = v|_\psi$.

Let ω now denote a p-form on M. Then $\psi^*\omega$ is a p-form on $\mathbb{R} \times M$. What can be said about the components $(\psi^*\omega)_{p-1}$ and $(\psi^*\omega)_p$? The time derivative of $(\psi^*\omega)_p$ at $t = 0$ is said to be *the Lie derivative of* ω *with respect to* v. It is denoted $\mathfrak{L}_v\,\omega$ and is such that

$$\frac{\partial}{\partial t}(\psi * \omega)_p|_{t=0} = \mathfrak{L}_v\omega = (d\omega)(v,\cdot) + d(\omega(v,\cdot))$$

where the notation is as follows: If υ is a p-form on M and v is a vector field, then $\upsilon(v, \cdot)$ is the $(p-1)$-form that is obtained by the homomorphism $TM \otimes (\bigotimes^p T^*M) \to \bigotimes^{p-1} T^*M$ that comes by writing $TM \otimes (\bigotimes^p T^*M) = (TM \otimes T^*M) \otimes (\bigotimes^{p-1} T^*M)$ and then using the defining homomorphism from $(TM \otimes T^*M)$ to $\mathbb{R}$. The form $(\psi^*\omega)_{p-1}$ at $t = 0$ is equal to $\omega(v, \cdot)$. The values of $(\psi^*\omega)_p$ and $(\psi^*\omega)_{p-1}$ at times $t \neq 0$ are obtained by using these $t = 0$ formulae with ω replaced its pull-back via the diffeomorphism $\psi(t, \cdot)$: $M \to M$.

The Lie derivative acting on p-forms can be viewed as a derivation on an extension of the algebra of smooth functions. Recall that $C^\infty(M; \mathbb{R})$ is an algebra with addition and multiplication being point-wise addition and multiplication of functions. The derivations of this algebra are the vector fields, thus the sections of TM. Addition of forms and wedge product give $\Omega_M = \bigoplus_{0 \le p \le n} C^\infty(M; \wedge^p T^*M)$ the structure of an algebra, albeit one in which multiplication is not commutative. This is an example of what is called a *super algebra*, which is a vector space with a $\mathbb{Z}/2\mathbb{Z} = \{0, 1\}$ grading and a multiplication whereby $\alpha\beta = (-1)^{\deg(\alpha)\deg(\beta)}\beta\alpha$. The grading for Ω_M takes the p = even forms to have grading degree zero and the p = odd forms to have grading degree one. The point is that $C^\infty(M)$ is a subalgebra of Ω_M and the Lie derivative, $v \to \mathfrak{L}_v$ is an extension of the derivation that v defines on $C^\infty(M)$ to give a derivation of Ω_M. This is to say that $\mathfrak{L}_v(\alpha \wedge \beta) = \mathfrak{L}_v\alpha \wedge \beta + \alpha \wedge \mathfrak{L}_v\beta$. You can check that this formula holds using the definition given above for $\mathfrak{L}_v$.

12.4 Curvature and covariant derivatives

Suppose that π: $E \to M$ is a vector bundle with fiber $V = \mathbb{R}^n$ or $\mathbb{C}^n$. Also suppose that ∇ is a covariant derivative on $C^\infty(M; E)$. As explained below, there is a nice extension of ∇ that maps $C^\infty(M; E \otimes (\wedge^p T^*M))$ to $C^\infty(M; E \otimes (\wedge^{p+1} T^*M))$ for each $p \in \{0, 1, \ldots, \dim(M)\}$. Here, as before, $\wedge^0 T^*M$ is the vector bundle $M \times \mathbb{R}$, and $\wedge^{\dim(M)+1} T^*M$ is the bundle with 0-dimensional fiber $M \times \{0\}$. This extension is called the *exterior covariant derivative*, it is denoted as d_∇, and it is defined by the following rules:

- *If* ω *is a p-form and* $\mathfrak{s}$ *is a section of* E, *then* $d_\nabla(\mathfrak{s}\,\omega) = \nabla\mathfrak{s} \wedge \omega + \mathfrak{s}\, d\omega$.
- *If* $\mathfrak{w}_1$ *and* $\mathfrak{w}_2$ *are sections of* $E \otimes \wedge^p T^*M$, *then* $d_\nabla(\mathfrak{w}_1 + \mathfrak{w}_2) = d_\nabla\mathfrak{w}_1 + d_\nabla\mathfrak{w}_2$.

These rules are sufficient to define d_∇ on all sections of $E \otimes \wedge^p T^*M$.

Although $d^2 = 0$, this is generally not the case for d_∇. It is the case, however, that d_∇^2 defines a section of $C^\infty(M; \mathrm{End}(E) \otimes \wedge^2 T^*M)$. This section is denoted by F_∇ and it is characterized by the fact that $d_\nabla^2\mathfrak{w} = F_\nabla \wedge \mathfrak{w}$. The section F_∇ is called the *curvature* of the covariant derivative. It measures the extent to which the covariant derivatives in different directions fail to commute. More is said about this momentarily.

What follows directly explains why $d_\nabla^2\mathfrak{w}$ can be written as $F_\nabla \wedge \mathfrak{w}$ with F_∇ an End(E)-valued 2-form. Here is a first justification: Suppose that ω is a smooth p-form and $\mathfrak{s}$ is a section of E. As $d_\nabla(\mathfrak{s}\,\omega) = d_\nabla\mathfrak{s} \wedge \omega + \mathfrak{s} \wedge d\omega$, so

$$d_\nabla^2(\mathfrak{s}\omega) = d_\nabla(d_\nabla\mathfrak{s} \wedge \omega) + d_\nabla(\mathfrak{s}\, d\omega).$$

Now, remember that if α is a 1-form, then $d(\alpha \wedge \beta) = d\alpha \wedge \beta - \alpha \wedge d\beta$. This being the case, the left-most term on the right-hand side above is $d_\nabla^2\mathfrak{s} \wedge \omega - d_\nabla\mathfrak{s} \wedge d\omega$. Meanwhile, the right-most term on the right-hand side is $d_\nabla\mathfrak{s} \wedge d\omega + s \wedge d^2\omega$. Since $d^2\omega$ is zero, the two terms on the right side above add to $d_\nabla^2\mathfrak{s} \wedge \omega$. In particular, if $\omega = f$ and thus a function, this says that $d_\nabla^2(f\mathfrak{s}) = f{d_\nabla}^2\mathfrak{s}$ and so d_∇^2 commutes with multiplication by a function. This understood, apply Lemma 11.1 in Chapter 11.2 to see that d_∇^2 is given by the action of a section of $\mathrm{End}(E) \otimes \wedge^2 T^*M$.

This claim can be seen explicitly on a set $U \subset M$ where there is a bundle isomorphism $\varphi\colon E|_U \to U \times V$. Let d denote the covariant derivative that acts on a section $x \to \mathfrak{v}(x) = (x, v(x))$ of $U \times V$ so as to send it to the section $x \to d\mathfrak{v}|_x = (x, dv|_x)$ of the bundle $(U \times V) \otimes T^*M|_U$. Then $\varphi(d_\nabla(\varphi^{-1}\mathfrak{v}))$ acts to send the section $\mathfrak{v}$ to the section given by $x \to (x, dv + \mathfrak{a}\, v)$ where a is a section of $\mathrm{End}(E) \otimes T^*M|_U$. This understood, $\varphi(d_\nabla^2(\varphi^{-1}\mathfrak{v}))$ is the section

$$x \to \Big(x,\ (d\mathfrak{a} + \mathfrak{a} \wedge \mathfrak{a})v\Big).$$

This identifies the curvature 2-form F_∇ on U:

$$\varphi(F_\nabla) = d\mathfrak{a} + \mathfrak{a} \wedge \mathfrak{a}.$$

What follows says more about the meaning of $\mathfrak{a} \wedge \mathfrak{a}$ and, more generally, $\mathfrak{a} \wedge \mathfrak{b}$ for End(E)-valued p and q forms $\mathfrak{a}$ and $\mathfrak{b}$. To start, fix any given point in M and fix a basis $\{\eta_\alpha\}$ for $\wedge^p T^*M$ and a basis $\{\sigma_\beta\}$ for $\wedge^q T^*M$ at this point. Here, the index α runs from 1 to $d!/p!(d-p)!$ where $d = \dim(M)$. Likewise, the index β runs from 1 to $d!/q!(d-q)!$. Write $\mathfrak{a} = \sum_\alpha \mathfrak{a}_\alpha\, \eta_\alpha$ and $\mathfrak{b} = \sum_\beta \mathfrak{b}_\beta\, \sigma_\beta$ where $\{\mathfrak{a}_\alpha\}$ and $\{\mathfrak{b}_\beta\}$ are in End(E). This done, then

$$\mathfrak{a} \wedge \mathfrak{b} = \Sigma_{\alpha,\beta}\, \mathfrak{a}_\alpha \mathfrak{b}_\beta \eta_\alpha \wedge \sigma_\beta,$$

where $\mathfrak{a}_\alpha, \mathfrak{b}_\beta \in \mathrm{End}(E)$ denotes the composition of the indicated endomorphisms. (This is just matrix multiplication.) In the case where $\mathfrak{a} = \mathfrak{b}$ and both are 1-forms, then what is written above is equivalent to the formula

$$\mathfrak{a} \wedge \mathfrak{a} = \sum_{1\leq\alpha<\beta\leq d}[\mathfrak{a}_\alpha, \mathfrak{a}_\beta]\eta_\alpha \wedge \eta_\beta.$$

In particular, $\mathfrak{a} \wedge \mathfrak{a} = 0$ if and only if the various components of $\mathfrak{a}$ pair-wise commute as endomorphisms of E.

If U' is a second open set with an isomorphism $\varphi'\colon E|_{U'} \to U' \times V$, then $\varphi' d_\nabla(\varphi')^{-1}$ can be written as $d + \mathfrak{a}'$. On $U' \cap U$, the End(E)-valued 1-form $\mathfrak{a}'$ can be obtained from $\mathfrak{a}$ by the rule

$$\mathfrak{a}' = g_{U',U}\mathfrak{a}g_{U',U}^{-1} + dg_{U',U}\, g_{U',U}^{-1}$$

where $g_{U',U} : U' \cap U \to Gl(V)$ is the transition function. This transformation rule is exactly what is needed so as to guarantee that

$$d\mathfrak{a}' + \mathfrak{a}'\wedge\mathfrak{a}' = g_{U',U}(d\mathfrak{a} + \mathfrak{a}\wedge\mathfrak{a})\, g_{U',U}^{-1}.$$

This last formula says that the identification of $\varphi(F_\nabla)$ with the section

$$x \to (x,\ d\mathfrak{a} + \mathfrak{a} \wedge \mathfrak{a})$$

of $(U \times Gl(V) \otimes \wedge^2 T^*M|_U$ does indeed specify consistently over the whole of M an End(E)-valued 2-form.

12.5 An example

Let $V = \mathbb{R}^N$ or $\mathbb{C}^N$ and suppose that $E \to M$ is a subbundle of the product vector bundle $M \times V$. Let $\Pi\colon M \to \mathrm{End}(V)$ denote the map that sends any given point $x \in M$ to the orthogonal projection of V onto $E|_x$. Chapter 11.3 uses this setting to define a covariant derivative on sections of E. By way of a reminder, this is done as follows: View a section $\mathfrak{s}\colon M \to E$ as a map to V whose value at each $x \in M$ happens to land in $E|_x$. Then define

$$d_\nabla\mathfrak{s} = \nabla\mathfrak{s} = \Pi d\mathfrak{s}$$

where d is the usual exterior derivative on maps from M to the fixed vector space V. Thus, $d\mathfrak{s}$ is an N-dimensional vector at each point whose entries are 1-forms on M; any given entry is the differential of the function that defines the corresponding entry of $\mathfrak{s}$. To compute d_∇^2, use the fact that $\Pi\mathfrak{s} = \mathfrak{s}$ to write $d_\nabla\mathfrak{s}$ as $d\mathfrak{s} - (d\Pi)\mathfrak{s}$. Here and in what follows, $d\Pi$ is the $N \times N$ matrix of 1-forms that is obtained by differentiating the entries of Π. Use this rewriting of $d_\nabla\mathfrak{s}$ with the fact that $d^2 = 0$ to see that

$$d_\nabla^2 \mathfrak{s} = \Pi d(\Pi) \wedge d\mathfrak{s} = \Pi d\Pi \wedge d(\Pi^2 \mathfrak{s}) = (\Pi d\Pi \wedge d\Pi\Pi)\mathfrak{s}.$$

Here I have used two facts about Π. First, $\Pi d\Pi\Pi = 0$ which can be seen by first differentiating the identity $\Pi^2 = \Pi$ to get $d\Pi\Pi + \Pi d\Pi = d\Pi$ and then multiplying on the right by Π. Second, $\Pi s = s$. The preceding equation identifies the curvature 2-form:

$$F_\nabla = \Pi d\Pi \wedge d\Pi\Pi.$$

Note that what is written on the right-hand side here is bracketed fore and aft by Π and so it defines a bonafide section of $\mathrm{End}(E) \otimes \wedge^2 T^*M$.

By way of a concrete example, suppose that M is an $(n-1)$-dimensional submanifold in $\mathbb{R}^n$. As explained in Chapter 3.4, the tangent bundle TM can be viewed as the set of pairs $(x, v) \in M \times \mathbb{R}^n$ with $n(x)^T v = 0$ where $n(x)$ is the unit length normal vector to M at x. The corresponding covariant derivative is defined by the projection $\Pi = \iota - n\, n^T$ where ι is the $n \times n$ identity matrix. Then $\Pi d\Pi = dn\, n^T$ and $d\Pi\Pi = n\, dn^T$. This understood, it follows that

$$F_\nabla = dn \wedge dn^T.$$

On the face of it, this defines at each point of M a skew-symmetric, $n \times n$ matrix of 1-forms whose (i, j) component is $dn_i \wedge dn_j$ where n_i is the i'th component of the normal vector with the latter viewed as a map from M to $\mathbb{R}^n$. The fact that $|n| = 1$ at each point of M implies that this $n \times n$ matrix of 1-forms annihilates the vector n. Indeed, multiplying this matrix against n gives the $\mathbb{R}^n$-valued 1-form $dn \wedge (\sum_i dn^i n^i) = dn \wedge \frac{1}{2} d(|n|^2) = 0$. This being the case, what is written above does indeed define a 2-form on M with values in End(TM).

For example, if M is the sphere S^{n-1} in $\mathbb{R}^n$, then $n(x) = x$. Near the north pole where $x_n = 1$, the collection of functions $y = (x_1, \ldots, x_{n-1})$ can serve as coordinates for S^{n-1}. This understood, then $n(x(y))$ has i'th coordinate y_i if $i < n$ and n'th coordinate $(1 - |y|^2)^{1/2}$. Thus, dn is the vector of 1-forms whose i'th coordinate is dy_i if $i < n$ and whose n'th coordinate is $-(1 - |y|^2)^{-1/2}\, y_j dy_j$; where I am summing over repeated indices. Note that this vanishes at the north pole, as it should since n at the north pole has i'th coordinate zero for $i < n$ and n'th coordinate 1. Meanwhile, $dn \wedge dn^T$ is a priori the $n \times n$ matrix of 2-forms whose i, j entry for i and j both less than n is $dy_i \wedge dy_j$, whose (i, n) entry for $i < n$ is $-(1 - |y|^2)^{-1/2}\, dy_i \wedge y_j dy_j$, whose (n, i) entry is minus what was just written, and whose (n, n) entry is zero. Note in particular that this matrix of 1-forms at the north pole has zero in its (i, n) and (n, i) entries for any i, and so defines (as it should) a 2-form with values in the endomorphisms of TS^{n-1} at the north pole. This 2-form has (i, j) entry $dy^i \wedge dy^j$.

12.6 Curvature and commutators

Let v and u denote vector fields on M. Viewed as derivations on $C^\infty(M)$, they have a commutator, denoted [v, u], which is also a derivation, and thus also a vector field on M. The derivation [v, u] sends a given function f to v(uf) − u(vf). Here, vf is shorthand for the action of v on the function f. It is also equal to the pairing between the 1-form df, a section of T^*M, and the vector field v, a section of TM. This expression for [v, u] = v(uf) − u(vf) obeys the required rule of a derivation: D(fg) = Df g + f Dg. In a local coordinate chart where $v = v^i(x)\frac{\partial}{\partial x^i}$ and $u = u^i\frac{\partial}{\partial x^i}$, then

$$[v, u] = \left(v^j \frac{\partial}{\partial x^j}(u^i) - u^j\left(\frac{\partial}{\partial x^j}v^i\right)\right)\frac{\partial}{\partial x^i}.$$

Here, I am summing over repeated indices.

The commutator and exterior derivative are related in the following way: Let α denote a 1-form on M. Then

$$(d\alpha)(v, u) = -\alpha([v, u]) + v(\alpha(u)) - u(\alpha(v)).$$

This formula views a section of $\wedge^p T^*M$ as a homomorphism from $\otimes_p TM$ to $\mathbb{R}$ which happens to be completely antisymmetric with respect to permutations of the factors in the p-fold tensor product. For example, α(v) denotes the function on M that is obtained by evaluation of α on the vector field v and (dα)(v, u) that obtained by evaluating dα on $v \otimes u$.

The preceding formula has the following application: Let E → M denote a vector bundle and let ∇ denote a covariant derivative for $C^\infty(M; E)$. Let v and u denote vector fields on M, and let ∇_v and ∇_U denote the respective directional derivations on $C^\infty(M; E)$ that are obtained from v and u using ∇. This is to say that if $\mathfrak{s}$ is a section of E, then $\nabla_v\mathfrak{s}$ is the section of E that is obtained by pairing the section $\nabla\mathfrak{s}$ of $C^\infty(M; E \otimes T^*M)$ with v. The commutator of these directional covariant derivatives is given by

$$[\nabla_v, \nabla_u] + \nabla_{[v, u]} = F_\nabla(v, u).$$

This gives another way to define F_∇. Here, F_∇ is viewed as an endomorphism from $\otimes_2 T^*M$ to $P \times_{ad} \mathfrak{lie}(G)$.

12.7 Connections and curvature

Suppose that P → M is a principal G-bundle and A is a connection on P. The connection defines a covariant derivative on sections of any vector bundle that is associated to P via a representation of G into a general linear group. Each such covariant derivative has an associated curvature 2-form. All of these 2-forms are derived from a single section, F_A, of the bundle $(P \times_{ad} \mathfrak{lie}(G)) \otimes \wedge^2 T^*M$. The latter section is called the *curvature 2-form* or just *curvature* of the connection. What follows defines F_A and explains how it gives the curvatures of all of the various covariant derivatives.

A first way to proceed invokes what has been said about P over open sets in M where P is isomorphic to the product principal G-bundle. To start, suppose that $U \subset M$ is an open set and that there is a given bundle isomorphism $\varphi: P|_U \to U \times G$. As explained in Chapter 11.4.7, the $\mathfrak{lie}(G)$-valued 1-form $(\varphi^{-1})^*A$ on $U \times G$ at a given point $(x, \mathfrak{g}) \in U \times G$ can be written as

$$(\varphi^{-1})^*A = \mathfrak{g}^{-1}d\mathfrak{g} + \mathfrak{g}^{-1}\mathfrak{a}_U\mathfrak{g},$$

where $\mathfrak{a}_U$ is a 1-form on U with values in $\mathfrak{lie}(G)$. Here, as before, I am assuming that G is a subgroup of $Gl(n; \mathbb{V})$ with $\mathbb{V} = \mathbb{R}$ or $\mathbb{C}$ as the case may be. Thus, $\mathfrak{lie}(G)$ is a subvector space of $n \times n$ matrices and $d\mathfrak{g}$ is an $n \times n$ matrix whose entries are the coordinate 1-forms on $Gl(n; \mathbb{V})$.

Now, let V denote a vector space, and let $\rho: G \to Gl(V)$ denote a representation of G. As explained in Chapter 11.5, the connection A induces a covariant derivative, ∇_A, on the sections of $P \times_\rho V$. Meanwhile, the isomorphism φ_U supplies a bundle isomorphism (also denoted by φ) from $(P \times_\rho V)|_U \to U \times V$. Let $\mathfrak{s}$ denote a section of $P \times_\rho V$. Then $\varphi(\mathfrak{s})$ sends $x \in U$ to a section of the form $(x, \mathfrak{s}_U(x))$, where $\mathfrak{s}_U$ is a smooth map to the vector space V. In addition, $\varphi(\nabla_A\mathfrak{s})$ is the section over U of $(U \times V) \otimes T^*M|_U$ given by

$$x \to (x, d\mathfrak{s}_U + \rho_*(\mathfrak{a}_U)\mathfrak{s}_U),$$

where $\rho_*: \mathfrak{lie}(G) \to End(V)$ is the differential of ρ at the identity, ι. It follows from this equation that the section $\varphi(F_{\nabla_A})$ of $(U \times End(V)) \otimes \wedge^2 T^*M$ has the form

$$x \to (x, \rho_*(d\mathfrak{a}_U + \mathfrak{a}_U \wedge \mathfrak{a}_U)).$$

Note in particular how this expression for the curvature of ∇_A is using the $\mathfrak{lie}(G)$-valued 1-form $\mathfrak{a}_U$ that defines the connection A.

With the preceding in mind, introduce the $\mathfrak{lie}(G)$-valued 2-form

$$(F_A)_U = d\mathfrak{a}_U + \mathfrak{a}_U \wedge \mathfrak{a}_U.$$

The assignment of $x \in U$ to $(x, (F_A)_U|_x)$ is a section over U of $(U \times \mathfrak{lie}(G)) \otimes \wedge^2 T^*M$. The curvature 2-form F_A can be defined by declaring that this section be its image under φ. Note in this regard that if $U' \subset M$ is another open set with an isomorphism $\varphi': P|_U \to U' \times G$ and such that $U \cap U' \neq \emptyset$, then $\mathfrak{a}_{U'}$ and $\mathfrak{a}_U$ are related by the rule

$$\mathfrak{a}_{U'} = g_{U',U}\, \mathfrak{a}_U (g_{U',U})^{-1} - dg_{U',U}\, (g_{U',U})^{-1}$$

where $g_{U',U} = \varphi' \circ \varphi^{-1}$: $U \cap U' \to G$ is the transition function. This transformation rule guarantees that $d\mathfrak{a}_{U'} + \mathfrak{a}_{U'} \wedge \mathfrak{a}_{U'} = g_{U',U}(d\mathfrak{a}_U + \mathfrak{a}_U \wedge \mathfrak{a}_U)g_{U',U}^{-1}$. This transformation rule asserts the following: The assignment of $(F_A)_U = d\mathfrak{a}_U + \mathfrak{a}_U \wedge \mathfrak{a}_U$ to any given open set $U \to M$ with isomorphism φ: $P|_U \to U \times G$ defines the image via φ of a bonafide section of $(P \times_{ad} \mathfrak{lie}(G)) \otimes \wedge^2 T^*M$. This section is the curvature 2-form F_A.

Granted this definition of F_A, what was said two paragraphs back tells us that the curvature 2-form F_{∇_A} of the covariant derivative ∇_A as defined on sections of $P \times_\rho V$ is equal to $\rho_*(F_A)$.

12.8 The horizontal subbundle revisited

What follows descibes a definition of F_A that does not introduce isomorphisms of P with the product bundle over open sets in M. To start, go up to P and reintroduce the horizontal subbundle $H_A \subset TP$. Suppose that v_A and u_A are a pair of sections of H_A. These are, of course, sections of TP also. This understood, one can consider their commutator $[v_A, u_A]$. This may or may not be a section of H_A. If it lies in H_A, then its image via A in $\mathfrak{lie}(G)$ is zero.

Let v, u denote vector fields on M and let v_A, u_A denote horizontal lifts as sections over P of H_A as defined in Chapter 11.6. This is to say that $v_A|_{p \in P}$ is the unique lift of $v|_{\pi(p)}$ to $TP|_p$ that lies in the horizontal subspace $H_A|_p$. Then the assignment of the ordered pair $(v, u) \to A([v_A, u_A])$: $P \to \mathfrak{lie}(G)$ has the following properties:

- $A([v_A, u_A])|_{pg^{-1}} = g\, A([u_A, v_A])\, g^{-1}$.
- *If* $f, f' \in C^\infty(M)$, *then* $A([(fv)_A, (f'u)_A]) = f f' A([v_A, u_A])$.
- $A([v_A, u_A]) = -A([u_A, v_A])$.

The first properties assert the fact that the assignment of the ordered pair (v, u) to $A([v_A, u_A])$ defines a section over M of the associated vector bundle $P \times_{ad} \mathfrak{lie}(G)$. The second property implies that the assignment of the ordered pair (v, u) to $A([v_A, u_A])$ does not involve the derivatives of v and u. This the case, it can be viewed as the result of evaluating a section over M of $(P \times_{ad} \mathfrak{lie}(G)) \otimes (\otimes^2 T^*M)$ on the pair (v, u). The third property asserts that this section of $(P \times_{ad} \mathfrak{lie}(G)) \otimes (\otimes^2 T^*M)$ changes sign when v and u are interchanged and so defines a section of $(P \times_{ad} \mathfrak{lie}(G)) \otimes \wedge^2 T^*M$. This section is $-F_A$.

To see that this section of $(P \times_{ad} \mathfrak{lie}(G)) \otimes \wedge^2 T^*M$ is the same as that defined previously, one need only consider this just completed definition over an open set $U \subset M$ where there is an isomorphism φ: $P|_U \to U \times G$. As explained in

Chapter 11.6, the push-forward of the horizontal lift, v_A, of a vector field v is the vector field on $U \times G$ whose components at any given $(x, \mathfrak{g})$ with respect to the identification $T(U \times G) = TU \oplus \mathfrak{lie}(G)$ are $(v, -\mathfrak{g}\mathfrak{a}_U(v)\mathfrak{g}^{-1})$. This understood, it is an exercise with the definition of the Lie bracket to verify that the two definitions of F_A agree on any given pair of vectors $(v, u) \in TU$.

Additional reading

- *Introduction to Smooth Manifolds*, John M. Lee, Springer, 2002.
- *An Introduction to Manifolds*, Loring W. Tu, Springer, 2007.
- *Geometry of Differential Forms*, Shigeyuki Morita and Teruko Nagase, American Mathematical Society, 2001.
- *The Geometry of Physics*, Theodore Frankel, Cambridge University Press, 2003.
- *Geometry, Topology and Physics*, Mikio Nakahara, Taylor and Francis, 2003.
- *Manifolds and Differential Geometry*, Jeffrey M. Lee, American Mathematical Society, 2009.

13 Flat connections and holonomy

Let G be a Lie group and π: P $\to$ M a principal G-bundle. As explained in Chapter 11.4.5, there are infinitely many connections on P. This being the case, it has proved profitable to look for connections that satisfy certain auxiliary conditions. These have been, for the most part, constraints on the curvature 2-form. In particular, connections obeying such constraints have played a central role in many of the recent and most exciting research in differential geometry and in applications of differential geometry to other fields. The simplest of the curvature constraints asks for connections with vanishing curvature 2-form. This sort of connection is said to be *flat*. This chapter studies the flat connections. Various chapters to come discuss some of the other useful curvature constraints.

The story told here introduces some auxiliary notions that are important in their own right; and many are used in subsequent chapters. What follows is a list of these auxiliary notions in their order of appearance:

- *foliations*
- *the group of automorphisms of a principal bundle—the gauge group*
- *the fundamental group of a manifold*
- *covering spaces.*

The last two of these notions are used at the end of this chapter to describe the set of all flat connections on all of the principal G-bundles over a given manifold M.

13.1 Flat connections

A connection on a principal bundle is said to be *flat* when its curvature 2-form is identically zero. The simplest example is the connection on the product principal bundle P = M$\times$G that is defined by the identification

$T(M \times G)|_{(x,\mathfrak{g})} = TM|_x \oplus TG|_{\mathfrak{g}}$. This connection is the $\mathfrak{lie}(G)$-valued 1-form given by the formula $A|_{(x,\mathfrak{g})} = \mathfrak{g}^{-1}d\mathfrak{g}$. The latter 1-form is pulled back from the projection $M \times G \to G$.

There are other flat connections. For example, let $\mathfrak{h}$: $M \to G$ denote any given smooth map. This defines a principal bundle isomorphism φ: $M \times G \to M \times G$ that sends any given pair $(x, \mathfrak{g})$ to $(x, \mathfrak{h}(x)\mathfrak{g})$. The pull-back, φ^*A, is the connection given by

$$(\varphi^*A)|_{(x,\mathfrak{g})} = \mathfrak{g}^{-1}d\mathfrak{g} + (\mathfrak{h}^{-1}\, d\mathfrak{h})|_x;$$

here $\mathfrak{h}^{-1}d\mathfrak{h}$ is shorthand for the $\mathfrak{lie}(G)$-valued section of T^*M that sends a vector $v \in TM|_x$ to the vector in $\mathfrak{lie}(G)$ that corresponds via the isomorphism $TG = G \times \mathfrak{lie}(G)$ to the push-forward $\mathfrak{h}_*v$. To be even more explicit, assume that G is a subgroup of $Gl(n; \mathbb{V})$ for some n and for $\mathbb{V} = \mathbb{R}$ or $\mathbb{C}$. Then $\mathfrak{h}$ is just a map from M to the vector space of $n \times n$ matrices; and as such, any given component is a function on M. Granted that such is the case, $d\mathfrak{h}$ is the $n \times n$ matrix valued 1-form whose i, j component is the differential of the function that defines the i, j component of $\mathfrak{h}$. A calculation finds that the curvature of φ^*A is also zero. Indeed, the latter is the $\mathfrak{lie}(G)$-valued 1-form

$$d(\mathfrak{h}^{-1}\, d\mathfrak{h}) + \mathfrak{h}^{-1}\, d\mathfrak{h} \wedge \mathfrak{h}^{-1}d\mathfrak{h}$$

which is zero because $d\mathfrak{h}^{-1} = -\mathfrak{h}^{-1}d\mathfrak{h}\ \mathfrak{h}^{-1}$. There may be other sorts of flat connections on $M \times G$. The case where $M = S^1$ supplies an instructive example.

13.2 Flat connections on bundles over the circle

The manifold M is the circle, S^1. Assume that G is connected. Let π: $P \to S^1$ denote a principal G-bundle over S^1. Every connection on P is flat as all 2-forms on S^1 are zero. This being the case, the following is the truly interesting question:

> *Suppose that* A *is any connection on* P. *Is there an isomorphism* φ: $P \to S^1 \times G$ *such that* $(\varphi_*A)_{(x,\mathfrak{g})} = \mathfrak{g}^{-1}d\mathfrak{g}$?

To answer this question, note first that there is, in all cases, a principal G-bundle isomorphism φ: $P \to S^1 \times G$. To find one, it proves convenient to write S^1 as $\mathbb{R}/2\pi\mathbb{Z}$. Fix an open interval $U \subset S^1$ containing 0 with a principal G-bundle isomorphism, φ, from $P|_U$ to $U \times G$. Also, fix a connection, A, on P and a point $p \in P|_0$. Let $\gamma_{A,p}$: $[0, 2\pi] \to P$ denote the horizontal lift of the path given by the projection map from $[0, 2\pi] \in \mathbb{R}$ to S^1. This lifted path is defined in Chapter 11.6. As $P|_{2\pi} = P|_0$, there exists an element $\mathfrak{h}_{A,p} \in G$ such that $\gamma_{A,p}(2\pi)\ \mathfrak{h}_{A,p} = p$. As G is path connected, there is a map, h: $[0, 2\pi] \to G$ such that h near 0 is equal to ι, and $h(2\pi) = \mathfrak{h}_{A,p}$. This understood, the map σ: $[0, 2\pi] \to P$ given by $t \to \gamma_{A,p}(t)h(t)$ is such that $\sigma(0) = \sigma(2\pi) = p$. Thus, σ defines a continuous section of P. As it turns out, the section σ can be chosen so that the resulting section is smooth. This follows from what is said in Appendix 13.1 of this chapter on the topic of smoothing continuous maps. Note in this regard that the smoothing discussion from Appendix 13.1 can be applied directly after first fixing an interval $I \subset S^1$ that contains 0 with a principal bundle isomorphism φ: $P|_I \to I \times G$. Having done this, write the φ-image of the original section of P as a map $t \to (t, g(t))$ with g a continuous map from I to G that is smooth except at 0. It can be approximated by a smooth map to obtain the desired smooth section of P.

Granted that σ is smooth, then an isomorphism φ: $S^1 \times G \to P$ is given by the rule that sends $(t, g) \in S^1 \times G$ to $\sigma(t)\, g \in P|_t$. Note that $\varphi(t, g\, g') = \varphi(t, g)g'$ for any given $g' \in G$, so this does indeed define a principal G-bundle isomorphism.

To return to the question posed in italics above, choose a principal G-bundle isomorphism $P \to S^1 \times G$ so as to view the given connection A on P as a connection on the product bundle $S^1 \times G$. This done, then the connection can be written as the $\mathfrak{lie}(G)$-valued 1-form $\mathfrak{g}^{-1}d\mathfrak{g} + \mathfrak{g}^{-1}\mathfrak{n}\, \mathfrak{g}\, dt$ where $\mathfrak{n}$ is a smooth map from S^1 to $\mathfrak{lie}(G)$. Thus, $\mathfrak{a} = \mathfrak{n}\, dt$ is a $\mathfrak{lie}(G)$-valued 1-form on S^1. With P written as $S^1 \times G$, the question in italics above is equivalent to the following:

Let $\mathfrak{a}$ denote a given $\mathfrak{lie}(G)$-valued 1-form on S^1. Is there a principal G-bundle isomorphism φ: $S^1 \times G \to S^1 \times G$ such that

$$\varphi^*(\mathfrak{g}^{-1}d\mathfrak{g} + \mathfrak{g}^{-1}\mathfrak{n}\, \mathfrak{g}\, dt) = \mathfrak{g}^{-1}d\mathfrak{g}.$$

To answer this question, introduce again the horizontal lift, $\gamma_{A,(0,\iota)}$, of the tautological path $[0, 2\pi] \subset S^1$ that starts at the point $(0, \iota) \in S^1 \times G$. Write $\gamma_{A,(0,\iota)}(2\pi)$ as $(0, \mathfrak{h}_{A,(0,\iota)})$ with $\mathfrak{h}_{A,(0,\iota)} \in G$.

The desired isomorphism exists if and only if the element $\mathfrak{h}_{A,(0,\iota)} \in G$ as defined above is the identity element ι.

To see why this is, note first that $\mathfrak{h}_{\mathfrak{g}^{-1}d\mathfrak{g},(0,\iota)} = \iota$. Note second that any given bundle automorphism, φ, of $S^1 \times G$ can be written as $(t, g) \to (t, u(t)\, g)$ where u: $S^1 \to G$. Indeed, u(t) is such that $\varphi(t, \iota)$ is $(t, u(t))$. This understood, it follows that $\mathfrak{h}_{\varphi^*A,(1,\iota)} = u(0)\, \mathfrak{h}_{A,(0,\iota)}\, u(0)^{-1}$. Thus, $\varphi^*A = \mathfrak{g}^{-1}d\mathfrak{g}$ if and only if ι is conjugate in G to $\mathfrak{h}_{A,(0,\iota)}$; and as ι is the only element conjugate to ι, so $\mathfrak{h}_{A,(0,\iota)} = \iota$.

What follows describes how to find a connection with any given value of $\mathfrak{h}_{A,(0,\iota)}$: Suppose that the desired value is $h \in G$. Fix a map $t \to \mathfrak{k}(t)$ from $[0, 2\pi]$ that sends a neighborhood of 0 to ι and a neighborhood of 2π to h. Let $\mathfrak{m}(t) = (\frac{d}{dt}\mathfrak{k})\mathfrak{k}^{-1}$. Then the connection $A = \mathfrak{g}^{-1}d\mathfrak{g} - \mathfrak{g}^{-1}\mathfrak{m}\mathfrak{g}\, dt$ has $\mathfrak{h}_{A,(0,\iota)} = h$.

To see why, recall from Chapter 11.6 that the horizontal lift of the path $t \to [0, t) \in \mathbb{R}/2\pi\mathbb{Z}$ is the path in $S^1 \times G$ that sends any given $t \in \mathbb{R}/(2\pi\mathbb{Z})$ to $(t, \mathfrak{g}(t))$ where $t \to \mathfrak{g}(t)$ obeys the equation

$$\frac{d}{dt}\mathfrak{g} = -\mathfrak{a}\mathfrak{g}.$$

In the case at hand, this equation asserts that $\frac{d}{dt}\mathfrak{g} = (\frac{d}{dt}\mathfrak{k})\mathfrak{k}^{-1}\mathfrak{g}$ and the solution that starts at ι when $t = 0$ is $\mathfrak{g}(t) = \mathfrak{k}(t)$.

13.3 Foliations

Let M denote a given manifold and $P \to M$ a given principal G-bundle. Suppose that A is a flat connection on P. Let v_A and u_A denote any two sections of the horizontal bundle $H_A \subset TP$ as defined in Chapter 11.4.4. Since $F_A = 0$, it follows that their commutator is also a section of H_A. This is to say that H_A is *involutive*.

The notion of an involutive subbundle occurs in a wider context of great import in its own right. To elaborate, let X denote any given manifold, and let H denote a subbundle of TX. The subbundle H is said to be involutive when the commutator of any two sections of H is also a section of H. The following theorem gives some indication of the significance of this notion. The proof of this theorem is given in Appendix 13.2 to this chapter.

Theorem 13.1 (the Frobenius theorem) *Suppose that* X *has dimension* n *and that* $H \subset TX$ *is an involutive subbundle of dimension* $m \le n$. *Then any point* $p \in X$ *has a neighborhood*, U, *with a coordinate chart* $\varphi\colon U \to \mathbb{R}^n = \mathbb{R}^m \times \mathbb{R}^{n-m}$ *such that* $\varphi_* H$ *consists of the tangent vectors to the* $\mathbb{R}^m$ *factor in* $\mathbb{R}^m \times \mathbb{R}^n$.

One says in this case that H is tangent to an m-dimensional *foliation* of X. The image via φ^{-1} of $\mathbb{R}^m \times \{x\}$ for any $x \in \mathbb{R}^{n-m}$ is part of what is called a *leaf* of the foliation. Each leaf is a manifold in its own right, and any given point in this leaf has a neighborhood that is a submanifold, this being the image via φ^{-1} of $\mathbb{R}^m \times \{x\}$ for some $x \in \mathbb{R}^{n-m}$. However, the leaf itself need not be a submanifold unless it is compact. An example to ponder is the 1-dimensional foliation of the torus $S^1 \times S^1$ that is defined by the subbundle H spanned by the vector field $\frac{\partial}{\partial t_1} + \alpha \frac{\partial}{\partial t_2}$ where α is an irrational number. Here, I write S^1 as $\mathbb{R}/2\pi\mathbb{Z}$ and use $\frac{\partial}{\partial t}$ to denote the corresponding Euclidean vector field.

As noted at the outset, the horizontal subbundle $H_A \subset TP$ is involutive if $F_A = 0$. It follows from the definition of F_A that the converse is also true: If H_A is involutive, then $F_A = 0$. Thus, H_A is tangent to the leaves of a foliation of P. This is called the *horizontal* foliation. The action of G on P maps leaves to leaves since H_A is a G-equivariant subbundle. Furthermore, the cotangent space to each leaf is isomorphic to the pull-back via π of T^*M.

What follows is a converse of sorts to what was just said. Let $\mathcal{F}$ denote a foliation of P with two properties: First, the G-action maps leaves to leaves. Second, the cotangent bundle of each leaf is isomorphic to the pull-back via π of T^*M. Then there is a flat connection A on P whose horizontal space is the tangent space to the leaves of the foliation. To elaborate, let $H \subset TP$ denote the subbundle of vectors that are tangent to the leaves of $\mathcal{F}$. This will be the horizontal subbundle of the desired connection. To obtain the connection, note that the equation labeled (*) in Chapter 11.4.3 implies that there is a unique homomorphism from TP to $\ker(\pi_*)$ which has the following two properties: It annihilates vectors in the subbundle H and it acts as the identity on vectors in $\ker(\pi_*)$. This map from TP to $\ker(\pi_*)$ is the connection A, here viewed as in Chapter 11.4.4.

13.4 Automorphisms of a principal bundle

Suppose that P is a principal bundle over a given manifold. An automorphism of P is an isomorphism ϕ: P → P. The set of automorphisms is denoted by Aut(P). This is a group. It is also called the *gauge group*, which is a term that comes from our physics brethren. This group is, in fact, the group of sections of the *associated fiber bundle* $P \times_{Ad} G$; this being $(P \times G)/\sim$ where the equivalence relation identifies any given pair $(p, \mathfrak{g}) \in P \times G$ with $(pg^{-1}, g\,\mathfrak{g}\,g^{-1})$ when $g \in G$. A section of this bundle is neither more nor less than a G-equivariant map from P to G. This is to say that a section of $P \times_{Ad} G$ can be viewed as a map $\mathfrak{u}: P \to G$ with the property that $\mathfrak{u}(pg^{-1}) = g\,\mathfrak{u}(p)g^{-1}$. Such a map defines the automorphism $\phi_{\mathfrak{u}}$ that sends any given $p \in P$ to $p\,\mathfrak{u}(p)$. Note that $\phi_{\mathfrak{u}}(pg^{-1}) = \phi_{\mathfrak{u}}(p)g^{-1}$ is guaranteed by the equivariance condition that is imposed on the map $\mathfrak{u}$. To see that all automorphisms arise in this way, suppose that ϕ is any given automorphism. As ϕ preserves the base point of p, the point $\phi(p)$ can be written as $p\,\mathfrak{u}(p)$ for some map $\mathfrak{u}: P \to G$. The fact that $\phi(pg^{-1}) = \phi(p)g^{-1}$ now demands that $\mathfrak{u}(pg^{-1}) = g\,\mathfrak{u}(p)\,g^{-1}$.

The group Aut(P) is, in some well-defined sense, infinite dimensional. To see why it is so big, remark that its 'Lie algebra' is the space of sections of the vector bundle $C^\infty(M; P \times_{ad} \mathfrak{lie}(G))$. The point being that if $\mathfrak{w}$ is a section of $C^\infty(M; P \times_{ad} \mathfrak{lie}(G))$, then $\exp(\mathfrak{w})$ is a section of $P \times_{Ad} G$. Here, I view $\mathfrak{w}$ as a map from P to $\mathfrak{lie}(G)$ so as to view $\exp(\mathfrak{w})$ as the map that sends any given $p \in P$ to $\exp(\mathfrak{w}(p)) \in G$. It is a direct consequence of Chapter 5.4's definition of the exponential map as

$$\exp(\mathfrak{w}) = 1 + \mathfrak{w} + \frac{1}{2!}\mathfrak{w}^2 + \frac{1}{3!}\mathfrak{w}^3 + \cdots$$

that its application in this context defines a map from $P \times_{ad} \mathfrak{lie}(G)$ to $P \times_{Ad} G$ that fixes the base point in M.

Connections A and A′ on P are said to be *gauge equivalent* when there is an element $\phi \in \mathrm{Aut}(P)$ such that $A' = \phi^*A$. An important point here is that the respective curvature 2-forms of A and ϕ^*A are related as follows: Let $\mathfrak{h} \in C^\infty(M; P \times_{Ad} G)$ denote the equivariant map that defines ϕ. Then

$$(F_{\varphi^*A})|_x = \phi^*(F_A|_x) = (\mathfrak{h}^{-1} F_A h)|_x.$$

To elaborate, remember that F_A is a section of $(P \times_{ad} \mathfrak{lie}(G)) \otimes \wedge^2 T^*M$. Meanwhile, an automorphism, ϕ, of P, induces an automorphism of all bundles that are associated to P, in particular, the bundle $P \times_{ad} \mathfrak{lie}(G)$. The automorphism here is just pointwise conjugation by the element $\mathfrak{h} \in C^\infty(P \times_{Ad} G)$ that defines ϕ. Indeed, if V is a vector space and $\rho\colon G \to Gl(V)$ is a group homomorphism, the automorphism ϕ acts on the bundle $P \times_\rho V = (P \times V)/\sim$ as follows: Let $\mathfrak{h}\colon P \to G$ denote the equivariant map that defines ϕ. Then ϕ sends the equivalence class of any given pair $(p, v) \in P \times V$ to that of $(p\, \mathfrak{h}(p)^{-1}, v)$; and this is the same equivalence class as $(p, \rho(\mathfrak{h}(p)^{-1})\, v)$.

Let $U \subset M$ denote an open set where there is an isomorphism $\varphi_U\colon P|_U \to U \times G$. Suppose that ϕ is an automorphism of P. Then $\varphi_U \circ \phi \circ \varphi_U^{-1}$ is an automorphism of $U \times G$. As such, it acts so as to send any given pair $(x, \mathfrak{g})$ to $(x, \mathfrak{z}(x, \mathfrak{g}))$. Here, the map $\mathfrak{z}$ cannot be just any map from $U \times G$ to G. It must be equivariant with respect to the G-action and this implies that $\mathfrak{z}(x, \mathfrak{g}) = \mathfrak{h}_U(x)\, \mathfrak{g}$ where $\mathfrak{h}_U(x) = \mathfrak{z}(x, \iota)$ is a map from U to G. Note that if U′ is another open set with an isomorphism $\varphi_{U'}\colon P|_{U'} \to U' \times G$, then there is a corresponding map $\mathfrak{h}_{U'}\colon U' \to G$ and the pair $\mathfrak{h}_U$ and $\mathfrak{h}_{U'}$ are determined one from the other on $U \cap U'$ by the transition function $g_{U',U}$ via the rule $\mathfrak{h}_{U'} = g_{U',U}\, \mathfrak{h}_U\, g_{U',U}^{-1}$.

Let A denote a connection on P. The action of ϕ on a connection A can be seen on $P|_U$ as follows: Write $(\varphi_U^{-1})^*A = \mathfrak{g}^{-1}d\mathfrak{g} + \mathfrak{g}^{-1}\mathfrak{a}_U\mathfrak{g}$ where $\mathfrak{a}_U$ is a $\mathfrak{lie}(G)$-valued 1-form on U. Then

$$(\varphi_U^{-1})^*(\varphi^*A) = \mathfrak{g}^{-1}d\mathfrak{g} + \mathfrak{g}^{-1}(\mathfrak{h}_U^{-1}d\mathfrak{h}_U + \mathfrak{h}_U^{-1}\mathfrak{a}_U\, \mathfrak{h}_U)\mathfrak{g}.$$

As can be seen from the formula for $F_{\phi^*A} = \phi^*(F_A)$, the group Aut(P) preserves the set of flat connections on P.

13.5 The fundamental group

The list of all flat connections on a given manifold M refers to the *fundamental group* of the manifold M. This is a discrete group that is determined by the underlying topological manifold. It is usually denoted by $\pi_1(M)$. To define this group, fix a point $x \in M$. Let $\mathcal{L}_x$ denote the space of continuous maps from [0, 1] into M such that $\gamma(0) = \gamma(1) = x$. Define an equivalence relation, $\sim$, on $\mathcal{L}_x$ by declaring that $\gamma_0 \sim \gamma_1$ if there is a smooth homotopy, f, from $[0, 1] \times [0, 1]$ into M such that $f(0, \cdot) = \gamma_0$, $f(1, \cdot) = \gamma_1$, and such that $f(\cdot, 0) = f(\cdot, 1) = x$. This can be viewed as a path in the space $\mathcal{L}_x$ from γ_0 to γ_1. The set of equivalence classes is the set of points in $\pi_1(M)$. This set has the structure of a group, with constant map to x playing the role of the identity element. The inverse of an element represented by a loop γ is the loop that traverses γ in the reverse direction; this is the loop $t \to \gamma(1 - t)$. Group multiplication is defined as follows: Let $\mathfrak{z}_0$ and $\mathfrak{z}_1$ denote elements in $\pi_1(M)$. Choose respective loops γ_0 and γ_1 to represent these elements. Then $\mathfrak{z}_1\mathfrak{z}_0$ is represented by the loop that traverses first γ_0 at twice normal speed so as to arrive back at x at time $t = \frac{1}{2}$, and it then traverses γ_1 at twice normal speed to arrive back at x at $t = 1$. Note that $\pi_1(M)$ does depend on the choice of the point x. Even so, a different choice produces an isomorphic group.

The manifold M is said to be *simply connected* when $\pi_1(M) = \{1\}$. For example, $\mathbb{R}^n$ is simply connected. So is S^n for $n > 1$. However $\pi_1(S^1) = \mathbb{Z}$. Let $q > 0$ and let M denote a compact surface of genus q. If $q = 1$, then M is the torus and $\pi_1(M) = \mathbb{Z} \oplus \mathbb{Z}$. If $q > 1$, then $\pi_1(M)$ is the quotient of the free group on 2q generators $\{z_1, \ldots, z_{2q}\}$ by the relation

$$(z_1 z_2 z_1^{-1} z_2^{-1}) \cdots (z_{2q-1} z_{2q} z_{2q-1}^{-1}\, z_{2q}^{-1}) = \iota.$$

By way of reminder, a free group on generators $\{z_1, \ldots, z_m\}$ consists of the set of finite, ordered strings of the form $y_1 y_2 \ldots$ where each y_i is from the set $\cup_{k \in \mathbb{Z}-\{0\}} \{z_1^k, \ldots, z_m^k\}$ where it is understood that if y_i is a power of z_k then neither y_{i-1} nor y_{i+1} is a power of z_k.

A homomorphism from $\pi_1(M)$ into a Lie group G is a map $\rho\colon \pi_1(M) \to G$ such that $\rho(\mathfrak{z}_1\mathfrak{z}_0) = \rho(\mathfrak{z}_1)\rho(\mathfrak{z}_0)$. For example, a homomorphism from $\mathbb{Z}$ $(= \pi_1(S^1))$ into G consists of the specification of a single element $g \in G$. With g specified, ρ sends any given integer k to g^k. A homomorphism from $\pi_1(S^1 \times S^1) = \mathbb{Z} \oplus \mathbb{Z}$ into G requires the specification of elements $g_1, g_2 \in G$ that *commute* in the sense that $g_1 g_2 g_1^{-1} g_2^{-1} = \iota$. Given such elements, ρ sends the ordered pair of integers (n, m) to $g_1^n g_2^m$. The condition $g_1 g_2 g_1^{-1} g_2^{-1} = \iota$ is what is needed to guarantee that ρ is a homomorphism. Let M denote a compact surface of genus $q > 1$. Then a homomorphism of $\pi_1(M)$ into G is given by the specification of 2q elements $\{g_1, \ldots, g_{2q}\} \in G$ subject to the condition that $(g_1 g_2 g_1^{-1} g_2^{-1}) \ldots (g_{2q-1} g_{2q} g_{2q-1}^{-1} g_{2q}^{-1}) = \iota$.

The set of homomorphisms from $\pi_1(M)$ to G is denoted by $\mathrm{Hom}(\pi_1(M); G)$. Homomorphism ρ and ρ' from $\pi_1(M)$ to G are said to be conjugate when there exists a fixed element $g \in G$ such that $\rho'(\cdot) = g\, \rho(\cdot) g^{-1}$. The notion of conjugacy defines an equivalence relation on the set $\mathrm{Hom}(\pi_1(M); G)$. The set of equivalence classes is often written as $\mathrm{Hom}(\pi_1(M); G)/G$.

13.6 The flat connections on bundles over M

The theorem below describes all of the flat connections on principal G-bundles over a given manifold M. The theorem refers to the set $\mathfrak{F}_{M,G}$ defined as follows: This is the set of equivalence classes of pairs (P, A) where $P \to M$ is a principal G-bundle and A is a flat connection on P; and where the equivalence relation $(P, A) \sim (P', A')$ when there exists an isomorphism $\varphi: P \to P'$ such that $\varphi^* A' = A$.

Theorem 13.2 (classification theorem for flat connections) *The set* $\mathfrak{F}_{M,G}$ *is in 1–1 correspondence with the set* $\mathrm{Hom}(\pi_1(M); G)/G$.

This theorem is proved shortly.

The next two sections of this chapter introduce two important notions that enter the proof of this classification theorem.

13.7 The universal covering space

Let M denote a given smooth manifold. A *covering space* is a smooth manifold, $\tilde{M}$, with a map $\pi: \tilde{M} \to M$ with the following property: Any given point $p \in M$ has a neighborhood, U, such that π's restriction to each component of $\pi^{-1}(U)$ is a diffeomorphism onto U. The basic facts about covering spaces can be found in most any book on algebraic topology, for example the book *Algebraic Topology: A First Course*, by Greenberg and Harper (1981). What follows summarizes the facts that are used in the proof of the classification theorem (Theorem 13.2).

The *universal covering space* of M is a simpy connected covering space. The choice of a base point $x \in M$ and a base point $\tilde{x} \in \pi^{-1}(x)$ gives a canonical identification between $\pi^{-1}(x)$ and the fundamental group of M. Note that any two universal covering spaces of the same manifold M are diffeomorphic via a diffeomorphism that intertwines the respective maps to M. Moreover, such a diffeomorphism is canonical given that it maps the base point in one to the base point in the other.

Every manifold has a universal covering space. What follows explains how to construct one: Let $\mathcal{P}_x$ denote the set of maps from [0, 1] to M that send 0 to x. Then $\tilde{M}$ is the quotient of $\mathcal{P}_x$ by the equivalence relation that identifies γ_0 with γ_1 when $\gamma_0(1) = \gamma_1(1)$ and the two maps are homotopic via a map $f: [0, 1] \to \mathcal{P}_x$ such that $f(s, 1) = \gamma_0(1) = \gamma_1(1)$ for all s.

The group $\pi_1(M)$ acts as a group of diffeomorphisms of $\tilde{M}$ and $M = \tilde{M}/\pi_1(M)$. The action of a class $\mathfrak{z} \in \pi_1(M)$ can be defined as follows: Fix a representative loop $\nu \in \mathcal{P}_x$ for the class $\mathfrak{z}$. Let ν^{-1} denote the loop that is obtained from ν by traversing the latter in the reverse direction. Given $\gamma \in \mathcal{P}_x$, construct the concatenation $\gamma \cdot \nu^{-1} \in \mathcal{P}_x$. Then $\mathfrak{z}$ sends the equivalence class of γ to that of $\gamma \cdot \nu^{-1}$. This diffeomorphism is denoted by $\psi_{\mathfrak{z}}$. (Note that ν^{-1} must be used instead of ν so as to have $\psi_{\mathfrak{z}\mathfrak{z}'} = \psi_{\mathfrak{z}} \circ \psi_{\mathfrak{z}'}$.)

13.8 Holonomy and curvature

Fix an orientation on S^1 and a base point $0 \in S^1$; then fix an orientation preserving identification between S^1 with $\mathbb{R}/2\pi\mathbb{Z}$ so that $0 \in \mathbb{R}$ is mapped to the base point.

Let $\pi: P \to M$ denote a principal G-bundle with connection A. Meanwhile, suppose that $\gamma: S^1 \to M$ is a smooth map. Fix a point $p \in P|_{\gamma(0)}$. As explained in Chapter 11.6, the loop γ has canonical lift to a path $\gamma_{A,p}: [0, 2\pi] \to P$ which is characterized by the fact that $\gamma_{A,p}(0) = p$ and $(\gamma_{A,p})_* \frac{\partial}{\partial t}$ is horizontal. In general, there is no need for $\gamma_{A,p}(2\pi)$ to equal p. However, as $\gamma_{A,p}(2\pi) \in P|_{\gamma(0)}$, one can write $\gamma_A(2\pi) = p\, \mathfrak{h}_{A,\gamma}(p)$ where $\mathfrak{h}_{A,\gamma}$ is a map from $P|_{\gamma(0)} \to G$ that obeys $\mathfrak{h}_{A,\gamma}(pg^{-1}) = g\, \mathfrak{h}_{A,\gamma}(p)\, g^{-1}$. This map is called the *holonomy* of the connection A around the based, oriented loop γ. The holonomy $\mathfrak{h}_{A,\gamma}$ depends on the chosen base point $0 \in \gamma$, and it depends on the orientation of γ; but it is otherwise independent of the chosen identification of S^1 as $\mathbb{R}/2\pi\mathbb{Z}$. The assignment of a based, oriented loop in M to the holonomy has four key properties. What follows are the first three.

Property 13.3 *Let* $*: S^1 \to M$ *map all of* S^1 *to a point in* M. *Then* $\mathfrak{h}_{A,*}(p) = \iota$.

Property 13.4 *Let* $\gamma^{-1}: S^1 \to M$ *denote the traverse of* γ *in the direction opposite to that defined by its given orientation. Then* $\mathfrak{h}_{A,\gamma^{-1}}(p) = (\mathfrak{h}_{A,\gamma}(p))^{-1}$.

Property 13.5 *Let* $\mu: S^1 \to M$ *and* $\nu: S^1 \to M$ *denote smooth, oriented maps such that* $\mu(0) = \nu(0)$. *Define the concatenation* $\nu \cdot \mu$ *by first traversing* μ *and then traversing* ν. *The holonomy of the concatentated loop is given by* $\mathfrak{h}_{A,\nu\cdot\mu}(p) = \mathfrak{h}_{A,\nu}(p)\, \mathfrak{h}_{A,\mu}(p)$.

All of these properties follow directly from what was said in Chapter 11.6 about horizontal lifts of paths in M. The details are left for you to work out.

The fourth property is a formula for how $\mathfrak{h}_{A,\gamma}$ changes when γ is changed. What follows is a digression to set the stage for this formula. To start this digression, fix a smooth map $\gamma: S^1 \to M$. Suppose that $f: (-1, 1) \times S^1 \to M$ is a smooth map with the property that $f(0, \cdot) = \gamma$ and $f(\cdot, 0) = \gamma(0)$. Each $f(s, \cdot)$ is a map from S^1 to M, so the assignment $s \to f(s, \cdot)$ can be viewed as a 1-parameter family of deformations of the based loop γ. Granted this view, then each

$s \in (-1, 1)$ version of $f(s, \cdot)$ has its corresponding holonomy; and these together give a smooth, G-equivariant map

$$\mathfrak{h}_{A,f(\cdot)}: (-1, 1) \times P|_{\gamma(0)} \to G.$$

This map is G-equivariant in the sense that $\mathfrak{h}_{A,f(\cdot)}(pg^{-1}) = g\, \mathfrak{h}_{A,f(\cdot)}(p)\, g^{-1}$.

For $s \in (-1, 1)$ near 0, one can write

$$\mathfrak{h}_{A,f(s)}(p) = \mathfrak{h}_{A,\gamma}(p)\, e^{\mathfrak{m}(s,p)}$$

where $\mathfrak{m}(s, p) \in \mathfrak{lie}(G)$ is such that $\mathfrak{m}(s, pg^{-1}) = g\, \mathfrak{m}(s, p)\, g^{-1}$. Moreover, $\mathfrak{m}$ is smooth on some neighborhood of $0 \times P|_{\gamma(0)}$. Note that this can be done because of what is said in Chapter 5.5 about the exponential map giving a diffeomorphism from a neighborhood of the origin in $\mathfrak{lie}(G)$ to a neighborhood of the identity element in G. Thus, if $h \in G$ is any given element, then the map $\mathfrak{n} \to h\, e^{\mathfrak{n}}$ defines a diffeomorphism from a neighorhood of the origin in $\mathfrak{lie}(G)$ onto a neighborhood of $\mathfrak{h}$ in G.

Of interest is the derivative with respect to the variable $s \in (-1, 1)$ at $\mathfrak{s} = 0$ of the map $\mathfrak{m}(\cdot, \cdot)$. The fourth property of $\mathfrak{h}$ is an expression for this derivative in terms of the curvature of the connection A. This expression for this derivative requires some additional notation. For the latter, introduce $f_s = (f_* \frac{\partial}{\partial s})|_{s=0}$ and $f_t = (f_* \frac{\partial}{\partial t})|_{s=0} = \gamma_* \frac{\partial}{\partial t}$ to denote the respective sections of $\gamma^*TM \to S^1$ that are obtained using the differential of f. Meanwhile, introduce $\gamma^*P \to S^1$ to denote the pull-back bundle. The pull-back of F_A via γ, this γ^*F_A, is a section over S^1 of the vector bundle $(\gamma^*P \times_{ad} \mathfrak{lie}(G)) \otimes \wedge^2\gamma^*T^*M)$ over S^1; thus $\gamma^*F_A(f_s, f_t)$ is a section over S^1 of the bundle $\gamma^*P \times_{ad} \mathfrak{lie}(G)$. (Here, $\wedge^2 T^*M$ should be thought of as the subbundle of antisymmetric homomorphisms in the bundle $\mathrm{Hom}(T^*M \otimes T^*M; \mathbb{R})$.)

As a section of $\gamma^*P \times_{ad} \mathfrak{lie}(G)$, the element $\gamma^*F_A(f_s, f_t)$ can be viewed as an equivariant map from γ^*P to $\mathfrak{lie}(G)$. Meanwhile, the lift $\gamma_{A,(\cdot)}$ defines an isomorphism $\gamma^*P|_{[0,2\pi)} \to [0, 2\pi) \times P_{\gamma(0)}$, and this isomorphism identifies $\gamma^*F_A(f_s, f_t)$ as a G-equivariant map from $[0, 2\pi] \times P|_{\gamma(0)}$ to $\mathfrak{lie}(G)$. I use $\gamma_A^* F_A(f_s, f_t)$ to denote the latter map. This map is G-equivariant in the sense that

$$(\gamma_A^* F_A(f_s, f_t))|_{(t,\, pg^{-1})} = g(\gamma_A^* F_A(f_s, f_t))|_{(t,\, p)} g^{-1}.$$

With the digression now over, what follows is the fourth property of the holonomy:

Property 13.6 *The derivative at* $s = 0$ *of* $\mathfrak{m}$ *is*

$$\left(\frac{\partial}{\partial s}\mathfrak{m}\right)\Big|_{s=0} = -\int_0^{2\pi} (\gamma_A^*(F_A(f_s, f_t))|_{(t,p)}\, dt.$$

Note for the future that this formula tells us that $\mathfrak{m}$ is independent of s if A is a flat connection. This implies, in particular, that it has the same value for homotopic loops.

The derivation of this formula is a nice exercise unwinding all of the definitions. As such, it is also left to the reader save for the following remarks.

Remark 13.7 Let $U \subset M$ denote an open set with an isomorphism $\varphi\colon P|_U \to U \times G$. Write φ^*A as $-\mathfrak{g}^{-1}d\mathfrak{g} + \mathfrak{g}^{-1}\mathfrak{a}\,\mathfrak{g}$. As in Chapter 12.7, write φ^*F_A as the G-equivariant section over $U \times G$ of $\mathfrak{lie}(G) \otimes \wedge^2 T^*M|_U$ whose value at $(x, \mathfrak{g})$ is $(x, \mathfrak{g}^{-1}(F_A)_U\, \mathfrak{g})$ where $(F_A)_U$ is the $\mathfrak{lie}(G)$-valued 2-form given by $(d\mathfrak{a} + \mathfrak{a} \wedge \mathfrak{a})$.

Remark 13.8 Suppose that $\gamma(t_0) \in U$. It follows from what is said in Chapter 11.6 that the horizontal lift of $f(s, \cdot)$ for t near t_0 and s near 0 is mapped by φ to the path in the product bundle $U \times G$ that send t to $(f(s, t)), \mathfrak{g}(s, t))$ where $\mathfrak{g}(s, \cdot)$ obeys the equation

$$\mathfrak{g}^{-1}\frac{\partial}{\partial t}\mathfrak{g} = -\mathfrak{a}(f_t)|_{f(s,\cdot)}.$$

Here, $\mathfrak{a}(f_t)$ is obtained by evaluating the $\mathfrak{lie}(G)$-valued 1-form $\mathfrak{a}$ on the vector $f_t = f_* \frac{\partial}{\partial s}$.

Remark 13.9 Take the covariant derivative of this last equation with respect to the variable s, and use the formula in Chapter 12.6 that expresses F_A in terms of commutators of covariant derivatives to φ^*F_A to obtain an equation for the covariant derivative in the t-direction of $(\mathfrak{g}^{-1}\frac{\partial}{\partial s}\mathfrak{g})$. Integrate this last equation to get the desired equation for $(\frac{\partial}{\partial s}\mathfrak{m})|_{s=0}$.

13.9 Proof of the classification theorem for flat connections

The proof is given below in four parts.

13.9.1 Part 1

The correspondence is given by a map of the following sort: Fix an orientation and base point for S^1. Also, fix a base point $x \in M$. Let $P \to M$ denote a principal G-bundle, and let A denote a flat connection on P. Fix a point $p \in P|_x$. As explained momentarily, the assignment of $\mathfrak{h}_{A,\gamma}$ to a loop $\gamma \in \mathcal{L}_x$ defines a homomorphism, ρ_A, from $\pi_1(M)$ to G. The image of ρ_A in $\mathrm{Hom}(\pi_1(M); G)/G$ gives the desired correspondence.

To start the explanation, remark that it is a consequence of the four properties of the holonomy listed in the preceding section of this chapter that $\mathfrak{h}_{A,\gamma}$ defines a homomorphism from $\pi_1(M)$ to G. The image in $\mathrm{Hom}(\pi_1(M); G)/G$ of this homomorphism does not depend on the chosen point $p \in P|_x$; this is a consequence of the fact that $\mathfrak{h}_{A,\gamma}(pg^{-1}) = g^{-1}\mathfrak{h}_{A,\gamma}(p)g$.

To go further, suppose that $\varphi: P' \to P$ is a bundle isomorphism. With the point $p \in P|_x$ chosen, set $p' = \varphi^{-1}(p)$. It is then the case that $\mathfrak{h}_{\varphi^*A,\gamma}(p') = \mathfrak{h}_{A,\gamma}(p)$. This is because $\varphi^{-1}(\gamma_{A,p})$ has horizontal tangent vectors, and so must equal $\gamma_{\varphi^*A,p'}$. As a consequence, the image of $\mathfrak{h}_{A,\gamma}$ in $\mathrm{Hom}(\pi_1(M); G)/G$ depends only on the equivalence class of the pair (P, A) in $\mathfrak{F}_{M,G}$.

13.9.2 Part 2

Suppose next that pairs (P, A) and (P′, A′) define equivalence classes in $\mathfrak{F}_{M,G}$ with the same image in $\mathrm{Hom}(\pi_1(M); G))/G$. What follows is a construction of an isomorphism $\varphi: P' \to P$ with the property that $\varphi^*A = A'$. To start, fix points $p \in P|_x$ and $p' \in P'|_x$. Let $y \in M$, and choose a path $\gamma: [0, 1] \to M$ such that $\gamma(0) = x$ and $\gamma(1) = y$. Define $\varphi_\gamma: P|_y \to P'|_y$ as follows: Recall that the lift $\gamma_{A,(\cdot)}$ defines an isomorphism from $P|_x$ to $P|_y$ and that $\gamma_{A',(\cdot)}$ defines one from $P'|_x$ to $P'|_y$. This understood, a G-equivariant map from $P|_y$ to G is defined by writing $\gamma_{A,(\cdot)}^{-1}(p_y) = p\,\mathfrak{h}(p_y)$. The map $p_y \to \mathfrak{h}(p_y)$ is G-equivariant in the sense that $\mathfrak{h}(p_y g^{-1}) = g^{-1}\mathfrak{h}(p_y)g$. This understood, set

$$\varphi_\gamma(p_y) = \gamma_{A',\,p'}(1)\mathfrak{h}(p_y).$$

If γ' is homotopic to γ, then $\varphi_{\gamma'} = \varphi_\gamma$ as can be seen using Property 13.6. To see about other choices for γ', suppose that v is a loop based at x, and let $\gamma' = \gamma \cdot v$ denote the concatenation of v and γ. All homotopy classes of paths from x to y can be realized by some loop of this sort. Given that (P, A) and (P′, A′) have the same image in $\mathrm{Hom}(\pi_1(M);G)/G$, it follows that $\varphi_{\gamma'} = \varphi_\gamma$ for such γ'. This understood, set $\varphi: P|_y \to P'|_y$ to equal φ_γ for any choice of γ. To see that φ is smooth, consider a point y′ very near y, and use a path that is obtained by concatenating first γ and then a short arc (say a geodesic with respect to some fixed metric) from y to y′. You can use local trivializations of P and P′ near y and the differential equation from Chapter 11.6 for the horizontal lift to see that φ is smooth. The fact that $\varphi^*A = A'$ follows from the fact that the construction implies that a horizontal lift to P as defined by A of any given path γ is mapped by φ to a horizontal lift of γ to P′ as defined by A′. This implies that the differential of φ maps $H_A \subset TP$ to $H_{A'} \subset TP'$. This is to say that $A' = \varphi^*(A)$.

13.9.3 Part 3

It remains now to prove that all elements in $\mathrm{Hom}(\pi_1(M); G)/G$ can be obtained from elements in $\mathfrak{F}_{M,G}$. To do this, reintroduce the universal covering space for

M, denoted by $\tilde{M}$. Suppose that ρ is a given homomorphism from $\pi_1(M)$ to G. Use ρ to construct the principal G-bundle $P_\rho = \tilde{M} \times_\rho G = (\tilde{M} \times G)/\sim$ where the equivalence relation identifies pair $(\tilde{y}, \mathfrak{g})$ with $(\psi_\mathfrak{z}(\tilde{y}), \rho(z)\, g)$ for $\mathfrak{z} \in \pi_1(M)$. This bundle P_ρ comes with a canonical flat connection, A_ρ, which is defined as follows: The $\mathfrak{lie}(G)$-valued 1-form $A_0 = \mathfrak{g}^{-1}d\mathfrak{g}$ on the product bundle $\tilde{M} \times G$ is pulled back to itself by any diffeomorphism that sends $(\tilde{y}, \mathfrak{g})$ to $(\psi_\mathfrak{z}(\tilde{y}), \rho(\mathfrak{z})\, \mathfrak{g})$ for $\mathfrak{z} \in \pi_1(M)$. This understood, this A_0 is the pull-back of a $\mathfrak{lie}(G)$-valued 1-form on TP_ρ. The latter is the connection A_ρ. This connection is flat because the pull-back to $\tilde{M}$ of its curvature is that of A_0 and A_0 is flat.

13.9.4 Part 4

To see what the holonomy is for the connection A_ρ around loops in M based at x, let γ: [0, 1] $\to$ M denote a loop based at x and let $\mathfrak{z}$ denote its image in $\pi_1(M)$. Fix a point $\tilde{x} \in \tilde{M}$ that projects to x. Then γ has a unique lift as a path $\tilde{y}$: [0, 1] $\to \tilde{M}$ that maps 0 to $\tilde{x}$ and maps 1 to $\psi_\mathfrak{z}^{-1}(\tilde{x})$ where $\mathfrak{z}^{-1}$ is the inverse of $\mathfrak{z}$ in $\pi_1(M)$. The parallel transport of A_ρ along γ starting from the image in P_ρ of a point $(\tilde{x}, \mathfrak{g})$ is, by definition, the parallel transport of A_0 along the path $\tilde{y}$. This gives the point $(\psi_\mathfrak{z}^{-1}(\tilde{x}), \mathfrak{g})$. The equivalence class of this point in P_ρ is not that of $(\tilde{x}, \mathfrak{g})$, but rather that of $(\tilde{x}, \mathfrak{g}(\mathfrak{g}^{-1}\rho(\mathfrak{z})\mathfrak{g}))$. This implies that the image of the representation ρ in $\mathrm{Hom}(\pi_1(M); G)/G$ is the same as that defined by the pair (P_ρ, A_ρ) via the map from $\mathfrak{F}_{M,G}$ to $\mathrm{Hom}(\pi_1(M); G)/G$.

Appendix 13.1 Smoothing maps

Suppose that M and N are smooth manifolds, and that f: M $\to$ N is a continuous map. As it turns out, the map f can be approximated by a smooth map. The discussion that follows considers only the case when M is compact.

As a warm-up, consider first the case where $N = \mathbb{R}^n$. Fix a finite, open cover of M by coordinate charts. Let $\mathfrak{U}$ denote this cover; when $U \in \mathfrak{U}$, let $\varphi_U: U \to \mathbb{R}^m$ denote the corresponding coordinate chart map. Here, m denotes the dimension of M. Fix also a subordinate partition of unity for this cover. These are described in Appendix 1.2. Denote the partition of unity by $\{\chi_U\}_{U\in\mathfrak{U}}$. Finally, fix a small positive number r. This number r will determine the accuracy of the approximation. One more item is needed, this the function χ: $[0, \infty) \to [0, 1]$ that is described in Appendix 1.2. Let $c = \int_{\mathbb{R}^m} \chi(|x|)d^m x$.

Granted all of the preceding, set f_r: M $\to \mathbb{R}^n$ equal to the function

$$p \to f_r(p) = \sum\nolimits_{U \in \mathfrak{U}} \chi_U(p) r^{-n} c^{-1} \int_{\mathbb{R}^m} f(\varphi_U^{-1}(x)) \chi(|x - \varphi_U(p)|/r) d^m x.$$

You can think of the value of f_r at any given $p \in M$ as an average of the values of f over the part of M with distance on the order of r or less from p as defined by a given Riemannian metric. In any event, f_r is smooth because the various $U \in \mathfrak{U}$ versions of φ_U and χ_U are smooth.

To see how close f_r comes to f, note that a change of integration variable allows $f_r(p) - f(p)$ to be written as

$$f_r(p) - f(p) = \sum\nolimits_{U \in \mathfrak{U}} \chi_U(p) r^{-n} c^{-1} \int_{R^m} (f(\varphi_U^{-1}(x + \varphi_U(p))) - f(p)) \chi(|x|/r) d^m x.$$

Since f is continuous and M is compact, the following can be said: Given $\varepsilon > 0$, there exists $r_\varepsilon > 0$ such that $\chi_U(p)\ |f(\varphi_U{}^{-1}(x + \varphi_U(p))) - f(p)| \le \varepsilon\ \chi_U(p)$ when $|x| < r_\varepsilon$. This being the case, if $r < r_\varepsilon$, then the equation above for $f_r(p) - f(p)$ implies that $|f_r - f| < \varepsilon$ at all points in M if $r < r_\varepsilon$.

The construction just given has the property that f_r will in general differ from f at all points in M. However it may be the case that f fails to be smooth only on a subset of M; and in this case, it can be useful to have f_r differ from f only near this subspace. A simple modification of the construction will have this property. Let $Z \subset M$ denote a compact subspace. Suppose the goal is to modify f only on some given open neighborhood of Z. To do so, let $U_Z \subset M$ denote the given neighborhood and let $U_Z' \subset U_Z$ denote an open set with compact closure in U_Z that also contains Z. Choose the open cover $\mathfrak{U}$ so that it has the form $\mathfrak{U} = \{U_0\} \cup \mathfrak{U}_Z$, where $U_0 = M - U_Z'$, and where $\mathfrak{U}_Z$ consists of open sets with compact closure in U_Z. It is not necessary that U_0 come with a coordinate chart map. Such a map won't exist in general. However, each $U \in \mathfrak{U}_Z$ should come with such a map. Now define f_r by the rule:

$$p \to f_r(p) = \chi_{U_0} f(p) + \sum\nolimits_{U \in \mathfrak{U}_z} \chi_U(p) r^{-n} c^{-1} \int_{\mathbb{R}^m} f(\varphi_U^{-1}(x)) \chi(|x - \varphi_U(p)|/r) d^m x.$$

This smoothing agrees with f on $M - U_Z$.

Turn now to the case where N is a smooth manifold. Let n denote the dimension of N. Because M is compact, so $f(M) \subset N$ is a compact set. This understood, the constructions that are described in Chapter 1.3 can be used to obtain the following: An integer $d \gg n$, an open set $Y \subset N$ that contains f(M), and an embedding $\psi: Y \to \mathbb{R}^d$ whose image is a submanifold of $\mathbb{R}^d$. Granted this data, then $\psi(f)$ defines a map from M to $\mathbb{R}^d$. I can smooth this map in the manner just described so as to get a smooth map, $\psi(f)_r$, from M to $\mathbb{R}^d$ that comes as close as desired to $\psi(f)$. However, in general, $\psi(f)_r$ will not map into

$\psi(Y)$, but just near to $\psi(Y)$ in $\mathbb{R}^d$ when r is small. So, more must be done to obtain an approximation to f as a map into Y.

To start this task, introduce the (d−n)-dimensional vector bundle $\pi: E \to Y$ whose fiber at any given point y is the subspace of vectors $v \in \mathbb{R}^d$ that are orthogonal to all vectors tangent at $\psi(p)$ to $\psi(Y)$. This is to say that $E \subset Y \times \mathbb{R}^d$ consists of the pairs (y, v) such that v is orthogonal to the image of $TY|_y$ via Chapter 5.3's push-forward map, ψ_*. Note that E can also be viewed as the pull-back via ψ of the normal bundle to the submanifold $\psi(Y)$. (The normal bundle to a submanifold is defined in Chapter 4.2.)

Define next a map, $\Psi: E \to \mathbb{R}^d$, by the rule $\Psi(p, v) = \psi(p) + v$. By construction, the differential of this map along the zero section is an isomorphism from $TE|_0$ to $T\mathbb{R}^d|_{\psi(Y)}$. This fact can be used with what is said in Chapter 1.4 about immersions and submersions to find $r_0 > 0$ and an open neighborhood $Y_0 \subset Y$ of f(M) with the following properties: Let $E_0 \subset E$ denote the open subset consisting of the points (y, v) with $y \in Y_0$ and $|v| < r_0$. Then Ψ restricts to E_0 as a diffeomorphism of E_0 onto an open neighborhood of $\psi(Y_0)$.

With the preceding understood, remark next that if r is sufficiently small, then the smoothing $\psi(f)_r$ of $\psi(f)$ will map M into $\Psi(E_0)$. Granted such a choice for r, then the assignment $p \to \Psi^{-1}(\psi(f)_r(p))$ defines a smooth map from M to E. This the case, then a smoothing of f is given by the map $p \to f_r(p) = \pi(\Psi^{-1}(\psi(f)_r(p)))$.

Appendix 13.2 The proof of the Frobenius theorem

This appendix gives a proof of the Frobenius theorem (Theorem 13.1). The proof is by induction on the dimension n. To set up the induction, remark first that the Frobenius theorem is automatically true for n = 1. This said, suppose that n > 1 and that the Frobenius theorem is true for all dimensions from 1 through n−1. The five steps that follow explain how to prove the Frobenius theorem for dimension n given this assumption.

Step 1

Fix a point $p \in X$. Since H is a vector bundle in its own right, there is a neighborhood, U, of p with a vector bundle isomorphism $\varphi: H|_U \to U \times \mathbb{R}^m$. Introduce $\{e_1, \ldots, e_n\}$ to denote a basis for $\mathbb{R}^m$. The vectors fields

$$\{\varphi^{-1}e_1, \varphi^{-1} e_2, \ldots, \varphi^{-1}e_m\}$$

span H at each point of U. By taking a somewhat smaller neighborhood of p if necessary, I can require that U be a coordinate chart for X, and so U comes with a diffeomorphism $\psi: U \to \mathbb{R}^n$ that sends p to the origin. Compose this diffeomorphism with the action on $\mathbb{R}^n$ of a matrix in $Gl(n; \mathbb{R})$ if necessary so that $(v_1 = \psi_*\varphi^{-1}e_1, v_2 = \psi_*\varphi^{-1}e_2, \ldots, v_m = \psi_*\varphi^{-1}e_m)$ at the origin $0 \in \mathbb{R}^n$ are the first m elements of the coordinate basis, thus $(\frac{\partial}{\partial x_1}, \ldots, \frac{\partial}{\partial x_m})$. This is assumed in what follows. As $v_1|_0 = \frac{\partial}{\partial x_1}$, there is a neighborhood of 0 on which the vector fields $(v_1, \frac{\partial}{\partial x_2}, \ldots, \frac{\partial}{\partial x_n})$ are linearly independent.

Step 2

This step constructs a diffeomorphism from a neighborhood of the origin in $\mathbb{R}^n$ to $\mathbb{R}^n$ that fixes 0 and whose differential maps the first coordinate vector, $\frac{\partial}{\partial x_i}$, to v_1. The desired diffeomorphism is denoted by σ. The map σ is defined by the following rule:

- $\sigma(0, x_2, \ldots, x_n) = (0, x_2, \ldots, x_n)$
- $\sigma(x_1, x_2, \ldots, x_n)$ *is the point at time* x_1 *on the integral curve of* v_1 *that starts when* $x_1 = 0$ *at the point* $(0, x_2, \ldots, x_n)$.

Invoke the vector field theorem (Theorem 8.2) to construct the map σ. To see that σ is a diffeomorphism near 0, use the fact that $v_1|_0 = \frac{\partial}{\partial x_1}$ to conclude that the differential of σ at the origin is the identity. Granted that such is the case, invoke the inverse function theorem (Theorem 1.1) to conclude that σ defines a diffeomorphism on some neighborhood of the origin.

With σ as just described, the inverse map is a diffeomorphism whose differential sends v_1 to $\frac{\partial}{\partial x_1}$ and maps 0 to 0 with differential equal the identity at 0. Compose the diffeomorphism from Step 1 with this inverse and so obtain a new diffeomorphism of a neighborhood of p in X with a neighborhood of 0 in $\mathbb{R}^n$ that takes p to 0, whose differential *on a neighborhood* of p sends $\varphi^{-1}e_1$ to $\frac{\partial}{\partial x_1}$, and whose differential *at* p sends $(\varphi^{-1}e_2, \ldots, \varphi^{-1}e_m)$ to $(\frac{\partial}{\partial x_2}, \ldots, \frac{\partial}{\partial x_m})$.

Use ψ now to denote this new coordinate chart map; and for each $k \in \{2, \ldots, m\}$, use v_k now to denote the vector field given by the push-forward via this new version of ψ of $\varphi^{-1}e_k$.

Step 3

Since H is involutive, any $k \in \{2, \ldots, m\}$ version of the vector field defined by the commutator $\left[\frac{\partial}{\partial x_1}, v_k\right]$ can be written as $c_k \frac{\partial}{\partial x_1} + \sum_{2\le j\le m} c_{kj} v_j$ where c_k and the various $j \in \{2, \ldots, m\}$ versions of c_{jk} are functions defined on a neighborhood of 0 in $\mathbb{R}^n$. I use c in what follows to denote the map from this neighborhood of the origin to the vector space $\mathbb{M}(m-1, \mathbb{R})$ of $(m-1) \times (m-1)$ matrices whose (k, j) component is $c_{k+1,j+1}$.

Use the vector field theorem (Theorem 8.2) to construct a map, $\mathfrak{g}$, from a somewhat smaller neighborhood of $0 \in \mathbb{R}^n$ to $Gl(m-1; \mathbb{R})$ that obeys

$$\frac{\partial}{\partial x_1}\mathfrak{g} = \mathfrak{g}c \textit{ and } \mathfrak{g}|_{x_1=0} = \iota.$$

Let $\vec{c}$ denote the map from a neighborhood of 0 to $\mathbb{R}^{m-1}$ whose k'th component is c_{k+1}. Use $\mathfrak{g}$ and this map to define a second such map, b, by the rule

$$\vec{b}(x_1, x_2, \ldots, x_n) = \int_0^{x_1} (\mathfrak{g}\vec{c})(s, x_2, \ldots, x_n)ds.$$

Step 4

For $k \in \{2, \ldots, m\}$, use w_k to denote the vector field

$$w_k = \sum\nolimits_{2\leq j\leq m} \mathfrak{g}_{k-1,j-1} v_j - b_{k-1}\frac{\partial}{\partial x_1}.$$

Since $b_{k-1} = 0$ where $x_1 = 0$ and $\mathfrak{g} = \iota$ where $x_1 = 0$, the set $\{\frac{\partial}{\partial x_1}, w_2, \ldots, w_m\}$ of vector fields span $\psi_* H$ at each point on some neighborhood of 0 in $\mathbb{R}^n$. This new spanning set for $\psi_* H$ has been introduced for the following reason: For each $k \in \{2, \ldots, m\}$, the following is true:

- $\left[\frac{\partial}{x_1}, w_k\right] = 0$,
- w_k *is orthogonal to* $\frac{\partial}{\partial x_1}$,
- $w_k|_0 = \frac{\partial}{\partial x_k}$.

Indeed, the maps $\mathfrak{g}$ and b are defined precisely so as to guarantee this.

Step 5

The first two bullets above are fancy ways of saying the following: When written in terms of the coordinate basis $\{\frac{\partial}{\partial x_i}\}_{1\leq j\leq n}$, the vector field w_k has the form

$$w_k = A_{k2}\frac{\partial}{\partial x_2} + \cdots + A_{kn}\frac{\partial}{\partial x_n}$$

where each coefficient function A_{ki} is a function only of the coordinates $(x_2, \ldots, x_n)$. As a consequence, the x_1 coordinate is superfluous. In particular, the collection $\{w_2, \ldots, w_m\}$ can be viewed as a set of linearly independent vector fields defined on a neighborhood of $0 \in \mathbb{R}^{n-1}$. Moreover, they span an involutive subbundle, $H' \subset T\mathbb{R}^{n-1}$, of dimension $m - 1$. This understood, I now invoke the induction hypothesis to complete the proof of the theorem.

Additional reading

- *Geometry, Topology and Physics*, Mikio Nakahara, Taylor and Francis, 2003.
- *Topology and Geometry*, Glen E. Bredo, Springer, 2009.
- *Differential Manifolds*, Lawrence Conlon, Birkhäuser, 2008.
- *Geometry of Differential Forms*, Shigeyuki Morita, American Mathematical Society, 2001.
- *Geometry of Four-Manifolds*, S. K. Donaldson and P. B. Kronheimer, Oxford University Press, 1997.
- *Algebraic Topology: A First Course*, M. J. Greenberg and J. R. Harper, Benjamin/Cummings, 1981.

14 Curvature polynomials and characteristic classes

This chapter explains how the curvatures of connections can be used to construct De Rham cohomology classes that distinguish isomorphism classes of vector bundles and principal bundles. These classes are known as *characteristic* classes. They are of great import to much of modern topology, differential topology, differential geometry, algebraic geometry and theoretical physics.

14.1 The Bianchi Identity

The definition of the characteristic classes involves new machinery, a part of which is introduced here. To start, let $P \to M$ denote a principal G-bundle and let A denote a connection on P. The curvature 2-form of A is again denoted by F_A, this, as I hope you recall, is a section of $(P \times_{ad} \mathfrak{lie}(G)) \otimes \wedge^2 T^*M$. As explained below, this 2-form always obeys the following differential equation:

$$d_A F_A = 0,$$

where d_A: $C^\infty(M; (P \times_{ad} \mathfrak{lie}(G)) \otimes \wedge^2 T^*M) \to C^\infty(M; (P \times_{ad} \mathfrak{lie}(G)) \otimes \wedge^3 T^*M)$ is the exterior covariant derivative (from Chapter 12.4) that is defined by the connection A. This identity is known as the *Bianchi identity.* There are various proofs of this identity; these depending on your point of view about connections.

Here is a quick proof: Let ρ denote a representation of G into some $n \geq 1$ version of either $Gl(n; \mathbb{R})$ or $Gl(n; \mathbb{C})$. Use V in what follows to denote either $\mathbb{R}^n$ or $\mathbb{C}^n$ as the case may be and use Gl(V) to denote the corresponding space of invertible, real or complex $n \times n$ matrices. Recall from Chapter 12.4 that the connection A induces an exterior covariant derivative, d_A, for the sections of $(P \times_\rho V) \otimes \wedge^p T^*M$, and that F_A enters in the following way: Let $\mathfrak{s}$ denote a section of $P \times_\rho V$. Then $d_A^2 \mathfrak{s}$ is equal to $\rho_*(F_A)\, \mathfrak{s}$ where ρ_*: $\mathfrak{lie}(G) \to \mathfrak{lie}(Gl(V))$ is the differential at the identity of ρ. This understood, one can write $d_A^3\mathfrak{s}$ in two ways. Here is the first:

$$d_A^3\mathfrak{s} = d_A(d_A^2\mathfrak{s}) = \rho_*(d_A F_A)\mathfrak{s} + \rho_*(F_A) \wedge d_A\mathfrak{s}.$$

On the other hand,

$$d_A^3\mathfrak{s} = d_A^2(d_A\mathfrak{s}) = \rho_*(F_A) \wedge d_A\mathfrak{s}.$$

These two equations are compatible if and only if $\rho_*(d_A F_A) = 0$. Take V such that ρ_* is injective to see that $d_A F_A = 0$.

What follows is a second proof of the Bianchi identity. Recall the definition of F_A that uses local isomorphisms of P with a product bundle. By way of a reminder, let $U \subset M$ denote an open set with an isomorphism $\varphi: P|_U \to U \times G$. Write the pull-back $(\varphi^{-1})^*A$ at (x, g) as $g^{-1}dg + g^{-1}\mathfrak{a}_U g$ where $\mathfrak{a}_U$ is a $\mathfrak{lie}(G)$-valued 1-form on U. Meanwhile, φ supplies an isomorphism (also denoted by φ) from $(P \times_{ad} \mathfrak{lie}(G))|_U$ to $U \times \mathfrak{lie}(G)$, and with respect to this isomorphism, the curvature F_A appears as the section $x \to (x, (F_A)_U)$ of the bundle $(U \times \mathfrak{lie}(G)) \otimes T^*M|_U$ where

$$(F_A)_U = d\mathfrak{a}_U + \mathfrak{a}_U \wedge \mathfrak{a}_U.$$

Keeping this in mind, recall now the exterior covariant derivative for sections of the bundle $P \times_{ad} \mathfrak{lie}(G)$: If $\mathfrak{s}$ is a section of this bundle, then $\varphi(\mathfrak{s})$ can be written as the section of $U \times \mathfrak{lie}(G)$ that sends x to the pair $(x, \mathfrak{s}_U(x))$ where $\mathfrak{s}_U$ maps U to $\mathfrak{lie}(G)$. Then $\varphi(d_A\mathfrak{s})$ is the section of $(U \times \mathfrak{lie}(G)) \otimes T^*M$ that sends $x \to (x, (d_A\mathfrak{s})_U)$ with

$$(d_A\mathfrak{s})_U = d\mathfrak{s}_U + \mathfrak{a}_U\mathfrak{s}_U - \mathfrak{s}_U\mathfrak{a}_U = d\mathfrak{s}_U + [\mathfrak{a}_U, \mathfrak{s}_U]$$

where [,] indicates the commutator of matrices. Here, $\mathfrak{s}_U\mathfrak{a}_U$ and $\mathfrak{a}_U\mathfrak{s}_U$ are matrices of 1-forms, these obtained by multiplying the matrix $\mathfrak{s}_U$ and matrix of 1-forms $\mathfrak{a}_U$ in the indicated order. Granted this, it then follows that $\varphi(d_A F_A)$ is the section $x \to (x, (d_A F_A)_U)$ of $(U \times \mathfrak{lie}(G)) \otimes \wedge^3 T^*M$ where

$$(d_A F_A)_U = d(d\mathfrak{a}_U + \mathfrak{a}_U \wedge \mathfrak{a}_U) + \mathfrak{a}_U \wedge (d\mathfrak{a}_U + \mathfrak{a}_U \wedge \mathfrak{a}_U) - (d\mathfrak{a}_U + \mathfrak{a}_U \wedge \mathfrak{a}_U) \wedge \mathfrak{a}_U.$$

This is zero. (Keep in mind that $d(\mathfrak{a}_U \wedge \mathfrak{a}_U) = d\mathfrak{a}_U \wedge \mathfrak{a}_U - \mathfrak{a}_U \wedge d\mathfrak{a}_U$.)

14.2 Characteristic forms

Suppose that $m > 0$ and that $\rho: G \to Gl(n; \mathbb{C})$ is a representation of G on $V = \mathbb{C}^m$. This induces the representation $ad(\rho)$ of G on the vector space $\mathbb{M}(n; \mathbb{C})$ whereby a given $g \in G$ acts on a matrix $\mathfrak{m} \in \mathbb{M}(n; \mathbb{C})$ to give the matrix $\rho(g)\mathfrak{m}\rho(g)^{-1}$. Let tr: $\mathbb{M}(n; \mathbb{C}) \to \mathbb{C}$ denote the trace. This induces a bundle homomorphism

$$\mathrm{tr}_\rho\colon\ P\times_{ad(\rho)} \mathbb{M}(n;\ \mathbb{C}) \to M \times \mathbb{C}.$$

To see that this is well defined, note that a given point $\mathfrak{m} \in (P \times_{ad(\rho)} \mathbb{M}(n; \mathbb{C}))$ is an equivalence class of pairs $(p, \mathfrak{m})$ with $p \in P$ and $\mathfrak{m} \in \mathbb{M}(n; \mathbb{C})$ where the equivalence relation identifies such a pair with $(pg^{-1}, \rho(g)\mathfrak{m}\, \rho(g)^{-1})$ when $g \in G$. Given that the trace obeys $\mathrm{tr}(\mathfrak{g}\mathfrak{m}\mathfrak{g}^{-1}) = \mathrm{tr}(\mathfrak{m})$ for any pair $\mathfrak{m} \in \mathbb{M}(n; \mathbb{C})$ and $\mathfrak{g} \in Gl(n; \mathbb{C})$, it follows that tr_ρ gives the same value on any two representatives of a given equivalence class.

The homomorphism tr_ρ as just defined extends to give

$$\mathrm{tr}_\rho\colon\ \left(P\times_{ad}\, \mathfrak{lie}\,(G)\right) \otimes\ \wedge^p\, T^*M) \to\ \wedge^p\, T^*M_{\mathbb{C}}$$

for any given integer p. Here, $T^*M_{\mathbb{C}} = T^*M \otimes C$. This extension is defined by first specifying how it acts on the decomposable elements: This action sends $\mathfrak{m} \otimes \omega$ with $\mathfrak{m} \in P \times_{ad(\rho)} \mathbb{M}(n; \mathbb{C})$ and $\omega \in \wedge^p T^*M$ having the same base point as $\mathfrak{m}$ to $\mathrm{tr}_\rho(\mathfrak{m})\, \omega$. The action of tr_ρ on a sum of decomposable elements is uniquely determined by the linearity requirement.

Granted all of this, suppose now that A is a connection on P. Fix an integer p. The p-fold wedge product of $\rho_*(F_A)$ is a section of $(P \times_{ad(\rho)} \mathbb{M}(n; \mathbb{C})) \otimes \wedge^{2p} T^*M$. This understood, define the form

$$s_p(\rho,\ F_A) = \left(\tfrac{i}{2\pi}\right)^p\ \mathrm{tr}_\rho(\rho_*(F_A) \wedge \cdots \wedge \rho_*(F_A)).$$

The following two comments elaborate on the meaning of the expression on the right hand side of this last equation.

Comment 14.1 Any given element $\mathfrak{z} \in (P \times_{ad(\rho)} \mathbb{M}(n; \mathbb{C})) \otimes \wedge^{2p} T^*M$ can be written as a finite sum $\mathfrak{z} = \Sigma_k\, m_k \otimes \omega_k$ where each $m_k \in P \times_{ad(\rho)} \mathbb{M}(n; \mathbb{C})$, where each $\omega_k \in \wedge^{2p} T^*M$, and where all sit over the same base point in M. Then $\mathrm{tr}_\rho(\mathfrak{z}) = \Sigma_k\, \mathrm{tr}_\rho(m_k)\, \omega_k$.

Comment 14.2 With regards to the meaning of $\rho_*(F_A) \wedge \cdots \wedge \rho_*(F_A)$, suppose that $\mathfrak{a}$ and $\mathfrak{b}$ are respectively matrix valued q and r forms on M. This is to say that $\mathfrak{a}$ is a section of $(P \times_{ad(\rho)} \mathbb{M}(n; \mathbb{C})) \otimes \wedge^q T^*M$ and $\mathfrak{b}$ is a section of $(P \times_{ad(\rho)} \mathbb{M}(n; \mathbb{C})) \otimes \wedge^r T^*M$. Then their product $\mathfrak{a} \wedge \mathfrak{b}$ is the section of $(P \times_{ad(\rho)} \mathbb{M}(n; \mathbb{C})) \otimes \wedge^{q+r} T^*M$ that is defined as follows: Write $\mathfrak{a}$ as a sum of decomposable elements, thus as $\mathfrak{a} = \Sigma_k\, \mathfrak{a}_k \otimes \mu_k$; and likewise write $\mathfrak{b} = \Sigma_i\, \mathfrak{b}_i \otimes \nu_i$. Here, each $\mathfrak{a}_k$ and $\mathfrak{b}_i$ is a section of $P \times_{ad(\rho)} \mathbb{M}(n; \mathbb{C})$ and each μ_k and ν_i is a q-form and r-form, respectively. Then

$$\mathfrak{a} \wedge \mathfrak{b} = \Sigma_{k,i}\, (\mathfrak{a}_k \mathfrak{b}_i) \otimes (\mu_k \wedge \nu_i).$$

This last formula is used in an iterative fashion to define $\rho_*(F_A) \wedge \cdots \wedge \rho_*(F_A)$ by taking $\mathfrak{a}$ above to be $\rho_*(F_A)$ and $\mathfrak{b}$ to be the $p-1$ fold version of $\rho_*(F_A) \wedge \cdots \wedge \rho_*(F_A)$.

As defined, $s_p(\rho, F_A)$ is a $\mathbb{C}$-valued 2p-form on M. However, if ρ maps G into U(n), then $s_p(\rho, F_A)$ is in fact an $\mathbb{R}$-valued 2p-form. This is so by virtue of the fact that $(\rho_*(F_A))^\dagger = -\rho_*(F_A)$. Indeed, it follows from this last identity that the complex conjugate of the right-hand side of the $\mathrm{tr}_\rho(\rho_*(F_A) \wedge \cdots \wedge \rho_*(F_A))$ differs from the original by p factors of -1, these coming from the replacement of $(\rho_*(F_A))^\dagger$ with $-\rho_*(F_A)$. Keep in mind here that the trace on $\mathbb{M}(n; \mathbb{C})$ obeys $\overline{\mathrm{tr}(\mathfrak{m})} = \mathrm{tr}(\mathfrak{m}^\dagger)$. Meanwhile, there are another p factors of -1 from the complex conjugation of $(\frac{i}{2\pi})^p$.

The form s_p has the following two important features:

- *The form* s_p *is a closed 2p-form. This is to say that* $ds_p(\rho, F_A) = 0$.
- *The cohomology class of* $s_p(\rho, F_A)$ *does not depend on* A*: If* A *and* A′ *are two connections on* P*, then* $s_p(\rho, F_{A'}) - s_p(\rho, F_A) = d\alpha$ *where* α *is a* $(2p-1)$*-form.*

To see about the top bullet, note first that

$$d\,\mathrm{tr}_\rho(\mathfrak{m}) = \mathrm{tr}_\rho(d_A \mathfrak{m})$$

for any section $\mathfrak{m}$ of $P \times_{ad(\rho)} \mathbb{M}(n; \mathbb{C})$ and any connection A on P. You can check that this is true on an open set $U \subset M$ with an isomorphism to $U \times G$. Note in addition that

$$d_A(\mathfrak{m} \wedge \mathfrak{m}') = d_A\mathfrak{m} \wedge \mathfrak{m}' + (-1)^q \mathfrak{m} \wedge d_A\mathfrak{m}'$$

when A is any connection on P, $\mathfrak{m}$ is any section of $(P \times_{ad(\rho)} \mathbb{M}(n; \mathbb{C})) \otimes \wedge^q T^*M$, and $\mathfrak{m}'$ one of $(P \times_{ad(\rho)} \mathbb{M}(n; \mathbb{C})) \otimes \wedge^{q'} T^*M$. Again, this can be verified by using the formula for d_A on an open set U as above. Granted these points, it follows directly that

$$d\,\mathrm{tr}_\rho(F_A \wedge \cdots \wedge F_A) = \mathrm{tr}_\rho(d_A F_A \wedge F_A \wedge \cdots \wedge F_A) + \mathrm{tr}_\rho(F_A \wedge d_A F_A \wedge \cdots \wedge F_A) + \cdots = 0$$

as each term in this sum has a factor of $d_A F_A$ which is zero because of the Bianchi identity.

To see about the lower bullet, introduce $[0, 1] \times M$ and the map $\mathfrak{p}\colon [0, 1] \times M \to M$ given by the projection. In the meantime, recall that A′ can be written as $A' = A + \pi^*\mathfrak{a}$ where $\pi\colon P \to M$ is the projection and $\mathfrak{a}$ is a section of $(P \times_{ad} \mathfrak{lie}(G)) \otimes T^*M$. Granted this formula for A′, then the formula $\hat{A} = A + t\mathfrak{a}$ for $t \in (0, 1)$ defines a connection on the pull-back bundle $\mathfrak{p}^*P \to (0, 1) \times M$. The curvature 2-form of $\hat{A}$ at any given $t \in [0, 1]$ is

$$F_{\hat{A}}|_t = dt \wedge \mathfrak{a} + F_{A_t}$$

where F_{A_t} is the curvature on M of the connection $A + t\mathfrak{a}$. Note in particular that the latter has no dt component. The 2p-form $s_p(\rho, F_{\hat{A}})$ is closed on $(0, 1) \times M$. Write this form in the manner of Chapter 12.2 as

$$s_p(\rho, F_{\hat{A}}) = dt \wedge \alpha_{p-1} + \alpha_p$$

where α_{p-1} is a t-dependent $(2p-1)$-form on M and α_p is a t-dependent 2p-form on M. The latter is $s_p(\rho, F_{A_t})$. The fact that s_ρ is closed requires that

$$\frac{\partial}{\partial t}\alpha_p = d\alpha_{p-1}.$$

Integrate this with respect to $t \in [0, 1]$ to see that

$$s_p(\rho, F_{A'}) - s_p(\rho, F_A) = d\left(\int_0^1 \alpha_{p-1}|_t dt\right).$$

This last equation tells us that $s_p(\rho, F_{A'}) - s_p(\rho, F_A)$ can be written as $d\alpha$.

14.3 Characteristic classes: Part 1

Since the form $s_p(\rho, F_A)$ is closed, it defines a class in $H^{2p}_{\text{De Rham}}(M)$. This class is independent of the chosen connection, and so depends only on the chosen principal bundle and the representation ρ. The s_p version is denoted here by $s_p(P_\rho)$.

As explained next, if P and P′ are isomorphic principal bundles, then the corresponding s_p cohomology classes are identical. Indeed, this is a consequence of the following observations. Suppose that $\varphi: P \to P'$ is a principal G-bundle isomorphism and that A is a connection on P′. Then φ^*A is a connection on P. This understood, it is enough to prove that

$$s_p(\rho, F_A) = s_p(\rho, F_{\varphi^*A}).$$

This last identity can be seen by viewing P and P′ over an open set $U \subset M$ where there are isomorphisms $\varphi_U : P|_U \to U \times G$ and also $\varphi'_U : P'|_U \to U \times G$. The composition $\varphi'_U \circ \varphi \circ \varphi_U$ appears as the map from $U \times G$ to itself that sends any given pair (x, g) to the pair $(x, \mathfrak{h}_U(x)g)$ where $\mathfrak{h}_U$ maps U into G. The image via φ_U of the curvature of A over U is defined by the $\mathfrak{lie}(G)$-valued 2-form $(F_A)_U$, and the image via φ'_U of the curvature of φ^*A is defined by that of the $\mathfrak{lie}(G)$-valued 2-form $(F_{\varphi^*A})_U$. In particular, these two $\mathfrak{lie}(G)$-valued 2-forms are determined one from the other from the formula

$$(F_{\varphi^* A})_U = \mathfrak{h}_U (F_A)_U \mathfrak{h}_U{}^{-1}.$$

Given that the trace obeys $\mathrm{tr}(\mathfrak{h}\mathfrak{m}\mathfrak{h}^{-1}) = \mathrm{tr}(\mathfrak{m})$, the preceding relation between curvatures implies that the two s_p forms are identical.

Two conclusions follow immediately:

- *Principal* G*-bundles* P *and* P′ *are not isomorphic if there exists a representation* ρ *of* G *and an even integer* p *such that* $s_p(P_\rho) \neq s_p(P'_\rho)$ *in* $H^{2p}_{\text{De Rham}}(M)$.
- *A principal* G*-bundle does not have a flat connection if there exists a representation* ρ *of* G *and an even integer* p *such that* $s_p(P_\rho) \neq 0$.

Any given version of s_p is an example of a *characteristic class* for the bundle P.

14.4 Characteristic classes: Part 2

The construction described in the previous two sections of this chapter can be generalized along the following lines: Fix $n \geq 1$ and let $f\colon \mathbb{M}(n; \mathbb{C}) \to \mathbb{C}$ denote any smooth function with the property that $f(\mathfrak{g}\mathfrak{m}\mathfrak{g}^{-1}) = f(\mathfrak{m})$. Such a function is said to be *ad-invariant*. Examples are $\mathrm{tr}(\mathfrak{m}^p)$ for p a positive integer.

Suppose that f is an ad-invariant function. Fix $t \geq 0$ and write the Taylor's expansion of $f(t\mathfrak{m})$ as

$$f_0 + f_1(\mathfrak{m})t + f_2(\mathfrak{m})t^2 + f_3(\mathfrak{m})t^3 + \dots.$$

Each f_k is also an ad-invariant function on $\mathbb{M}(n; \mathbb{C})$, but one that is a homogeneous polynomial of degree p in the entries of $\mathfrak{m}$. This is to say that f_k is the composition

$$f_k = \mathfrak{f}_k \circ \Delta_k$$

where the notation is as follows: First, $\Delta_k\colon \mathbb{M}(n; \mathbb{C}) \to \bigotimes^k \mathbb{M}(n; \mathbb{C})$ sends $\mathfrak{m}$ to its k-fold tensor product $\mathfrak{m} \otimes \cdots \otimes \mathfrak{m}$. Meanwhile, $\mathfrak{f}_k$ is a linear map $\mathfrak{f}_k\colon \bigotimes^k \mathbb{M}(n; \mathbb{C}) \to \mathbb{C}$ that is ad-invariant. This is to say that $\mathfrak{f}_k(\mathfrak{m}_1 \otimes \cdots \otimes \mathfrak{m}_k) = \mathfrak{f}_k(\mathfrak{g}\mathfrak{m}_1\mathfrak{g}^{-1} \otimes \cdots \otimes \mathfrak{g}\mathfrak{m}_k\mathfrak{g}^{-1})$ for any $\mathfrak{g} \in G$. An example is the map that sends $(\mathfrak{m}_1 \otimes \cdots \otimes \mathfrak{m}_k)$ to $\mathrm{tr}(\mathfrak{m}_1 \cdots \mathfrak{m}_k)$. The function f_k is completely determined by $\mathfrak{f}_k$.

Here are two important examples of ad-invariant functions:

- $c(t\mathfrak{m}) = \det(\iota + \frac{i}{2\pi} t\mathfrak{m})$
- $ch(t\mathfrak{m}) = \mathrm{tr}\left(\exp\left(\frac{i}{2\pi} t\, \mathfrak{m}\right)\right)$.

What follows is a basic result concerning the vector space of ad-invariant functions.

Theorem 14.3 (the ad-invariant function theorem) *The vector space of real analytic, ad-invariant functions that are homogeneous of a given positive degree* p *is the* $\mathbb{C}$*-linear span of the set* $\{tr(m^{k_1})\cdots tr(m^{k_q}) : k_1+\cdots+k_q = p\}$.

Appendix 14.1 contains a proof of this assertion. By way of an example, c(t m) and ch(t m) given above can be written as

- $c(m) = 1 + \frac{i}{2\pi} t\, tr(m) - \frac{1}{8\pi^2}(tr(m)^2 - tr(m^2))t^2 + \cdots$.
- $ch(m) = 1 + \Sigma_{k\geq 1}\frac{1}{k!}(\frac{i}{2\pi})^k tr(m^k)t^k$.

With the preceding as background, let $P \to M$ denote a principal G-bundle and let A denote a given connection on P. Let $\rho: G \to Gl(n; \mathbb{C})$ denote a homomorphism of G. Suppose in what follows that f is an ad-invariant function on $\mathbb{M}(n; \mathbb{C})$. Then f and ρ can be used to construct the section $f(\rho, A)$ of $\bigoplus_{1\leq k\leq \dim(M)} \wedge^{2k}T^*M$, this being a differential form that can be thought of as $f(\rho_*(F_A))$.

To give a precise definition, write $f(tm) = f_0 + t f_1(m) + \frac{1}{2}t^2 f_2(m) + \cdots$ where each $f_k(\cdot)$ is a ad-invariant, homogeneous polynomial of degree k in the entries of the matrix. Granted this expansion, define

$$f(\rho_*(F_A)) = 1 + f_1(\rho_*(F_A)) + f_2(\rho_*(F_A)) + \cdots$$

where $f_k(\rho_*(F_A))$ is the 2k-form on M that is defined as follows: View $\rho_*F_A \otimes\cdots\otimes \rho_*F_A$ as a section of $(P \times_{ad(\rho)} (\bigotimes^k \mathbb{M}(n; \mathbb{C}))) \otimes (\bigotimes^k(\wedge^2 T^*M)$. Then antisymmetrize the 2-form part to obtain a section of $(P \times_{ad(\rho)} (\otimes^k \mathbb{M}(n; \mathbb{C})))$ $\otimes \wedge^{2k}T^*M$. To complete the definition, recall that f_k is determined by a function $\mathfrak{f}_k$ from $\bigotimes^k \mathbb{M}(n; \mathbb{C})$ to $\mathbb{C}$ by the rule $f_k = \mathfrak{f}_k \circ \Delta_k$. Take $\mathfrak{f}_k$ and apply it along each fiber to obtain a $\mathbb{C}$-valued section of $\wedge^{2k}T^*M$. If allowed to modify previously introduced notation, I denote the result as $\mathfrak{f}_k(\rho_*F_A \wedge\cdots\wedge \rho_*F_A)$.

By way of an example to make sure things are understood, suppose that $\rho_*(F_A)$ is written on some open set in M as $m_1\omega_1 + m_2\omega_2$ where ω_1 and ω_2 are 2-forms, and m_1 and m_2 are sections of $P \times_{ad(\rho)} \mathbb{M}(n; \mathbb{C})$. Then

- $f_1(\rho_*(F_A)) = \mathfrak{f}_1(m_1)\,\omega_1 + \mathfrak{f}_1(m_2)\,\omega_2$.
- $f_2(\rho_*(F_A)) = \mathfrak{f}_2(m_1, m_1)\,\omega_1 \wedge \omega_1 + (\mathfrak{f}_2(m_1, m_2) + \mathfrak{f}_2(m_2, m_1))\,\omega_1 \wedge \omega_2$
  $+ \mathfrak{f}_2(m_2, m_2)\,\omega_2 \wedge \omega_2$.

Note in particular that only the symmetric part of $\mathfrak{f}_2$ appears. This is a general phenomenon: Only the completely symmetric part of any given $\mathfrak{f}_k$ appears in the formula for $f_k(\rho_*(F_A))$. I leave it to you to convince yourself that this is the case. As a consequence, it is sufficient to consider only symmetric, ad-invariant functions when defining characteristic classes.

The arguments given in the previous section of this chapter concerning the case when $f(m) = tr(m^k)$ can be repeated with only cosmetic modifications to prove the following:

- *Each* $f_k(\rho_*(F_A))$ *is a closed form.*
- *The cohomology class of* $f_k(\rho_*(F_A))$ *does not depend on the chosen connection* A. *This class thus depends only on* P *and* ρ. *It is denoted in what follows by* $f_k(P_\rho)$.
- *If* P *and* P′ *are isomorphic principal bundles, then the cohomology classes* $f_k(P_\rho)$ *and* $f_k(P'_\rho)$ *are the same.*

Each $f_k(P_\rho)$ is also an example of a characteristic class.

The last three sections of this chapter give examples of bundles with nonzero characteristic classes.

It is worth keeping in mind that characteristic classes behave naturally with respect to pull-back by smooth maps between manifolds. To elaborate, suppose that X is a given smooth manifold and $\pi\colon P \to X$ is a principal G-bundle. Suppose that $n \geq 1$ and that $\rho\colon G \to Gl(n; \mathbb{C})$ is a representation. Use ρ to define a characteristic class $f_k(P_\rho)$. Now let M denote a second manifold and $\psi\colon M \to X$ a smooth map. The characteristic class $f_k((\psi^*P)_\rho) \in H^*_{De\ Rham}(M)$ is the pull-back in De Rham cohomology via ψ of the corresponding class $f_k(P_\rho) \in H^*_{De\ Rham}(X)$. This is to say that the cohomology class of $f_k((\psi^*P)_\rho)$ is represented by the pull-back via ψ of a closed 2p-form on X that represents the cohomology class of $f_k(P_\rho)$. Here is why: Fix a connection A on P. Then ψ^*A is a connection on ψ^*P and its curvature 2-form is given by $F_{\psi_*A} = \psi^*F_A$.

As a parenthetical remark, note that any given map $\psi\colon M \to X$ induces by pull-back a linear map $\psi^*\colon H^p_{De\ Rham}(X) \to H^p_{De\ Rham}(M)$. This is defined as follows: Let z denote a given class in $H^p_{De\ Rham}(X)$. Choose a closed p-form, ω, to represent $\mathfrak{z}$. Then $\psi^*z \in H^p_{De\ Rham}(M)$ is represented by $\psi^*\omega$. This pull-back of De Rham cohomology is well defined because the pull-back of a closed form is closed and the pull-back of an exact form is exact. This is to say that $d\psi^* = \psi^*d$.

14.5 Characteristic classes for complex vector bundles and the Chern classes

Let $\pi\colon E \to M$ denote a complex vector bundle; thus the fiber of E is $\mathbb{C}^n$ for some positive integer n. It is customary to associate characteristic classes directly to E rather than going through the construction of a principal bundle that has E as an associated bundle. (For example, one could use the principal bundle $P_{Gl(E)}$ of frames in E.) This is done as follows: Fix a covariant derivative ∇ for sections of E, and then introduce from Chapter 12.4 the exterior covariant derivate d_∇ and define the curvature F_∇ from d_∇^2. Let f denote an ad-invariant function on $\mathbb{M}(n; \mathbb{C})$. The covariant derivative ∇ and an integer k determine the corresponding closed 2k-form $f_k(F_\nabla)$. Here again, this 2k form is obtained from the section $\otimes^k F_\nabla$ of $(\otimes^k \mathrm{End}(E)) \otimes (\otimes^k (\wedge^2 T^*M))$ by applying the antisymmetrization homomorphism to obtain a section of $(\otimes^k \mathrm{End}(E)) \otimes \wedge^{2k} T^*M$ and then applying the linear map $\mathfrak{f}_k$ defined by f to obtain the desired $\mathbb{C}$-valued 2k-form.

The arguments given in the preceding sections of this chapter can be repeated with only notational changes to see that $f_k(F_\nabla)$ is a closed 2k-form on M; and that it's cohomology class does not depend on the covariant derivative. The latter is denoted by $f_k(E)$. Note that if E and E′ are isomorphic vector bundles, then $f_k(E)$ is the same as $f_k(E')$. The cohomology class $f_k(E)$ is an example of a *characteristic class* for the vector bundle E.

The principal bundle and vector bundle notions of these closed forms are essentially the same. To elaborate, suppose that P is a principal G-bundle, that n is a positive integer, and that $\rho: G \to Gl(n; \mathbb{C})$ is representation. Let E denote the associated bundle $P \times_\rho \mathbb{C}^n$. Then $f_k(P_\rho)$ and $f_k(E)$ are identical.

Characteristic classes for vector bundles pull-back via smooth maps just as those for principal bundles, and for essentially the same reason. To elaborate, suppose that X is a smooth manifold, and that $\pi: E \to X$ a complex vector bundle of dimension n. Let M denote another smooth manifold and let $\psi: M \to X$ denote a smooth map. Chapter 6.11 explains how to define the pull-back bundle $\psi^*E \to M$. Then any given characteristic class of ψ^*E is the pull-back via $\psi^*: H^*_{\mathrm{De\ Rham}}(X) \to H^*_{\mathrm{De\ Rham}}(M)$ of the corresponding characteristic class for E.

This last observation has the following consequence: Let M denote a given manifold and E a given complex vector bundle of dimension n. As noted in Chapter 6.11, if m is a sufficiently large integer, then there is a map, ψ, from M to the complex Grassmannian $Gr_{\mathbb{C}}(m; n)$ and a vector bundle isomorphism between E and the pull-back via ψ of the tautological, rank n vector bundle over $Gr_{\mathbb{C}}(m; n)$. Thus, all of E's characteristic classes are pulled back by ψ. This can be interpreted as follows: Characteristic classes are pull-backs of "universal characteristic classes" that live in the cohomology of all sufficiently large m versions of $Gr_{\mathbb{C}}(m; n)$.

In the case when $f(\mathfrak{m}) = c(\mathfrak{m}) = \det(\iota + \frac{i}{2\pi}\mathfrak{m})$, the corresponding characteristic class f_k is denoted by $c_k(E)$ and it is called the *k'th Chern class* of E. The Chern classes are useful by virtue of the fact that $c_k(E) = 0$ if the fiber dimension of E is less than k. This is so because function $\mathfrak{m} \to \det(\iota + \frac{i}{2\pi}\mathfrak{m})$ is a polynomial in the coefficients of $\mathfrak{m}$ of degree equal to the fiber dimension. For example, if E has fiber dimension 1, then only $c_1(E)$ can be nonzero. If E has fiber dimension 2, then only $c_1(E)$ and $c_2(E)$ can be nonzero.

To say more about the utility of the Chern classes, I first remind you that the wedge product of forms endows $H^*_{\mathrm{De\ Rham}}(M) = \otimes_{1 \le k \le \dim(M)} H^k_{\mathrm{De\ Rham}}(M)$ with the structure of an algebra. This algebra is defined as follows: Let α_1 and

α_2 denote two cohomology classes, say with $\alpha_1 \in H^p_{DeRham}(M)$ and with $\alpha_2 \in H^q_{De\ Rham}(M)$. To see about their product, let ω_1 and ω_2 denote any two closed forms that represent these classes. The class $\alpha_1\ \alpha_2$ is represented by the form $\omega_1 \wedge \omega_2$. To see that this form is closed, note that

$$d(\omega_1 \wedge \omega_2) = d\omega_1 \wedge \omega_2 + (-1)^p \omega_1 \wedge d\omega_2.$$

To see that the class of $\omega_1 \wedge \omega_2$ does not depend on the chosen representative, consider that if $\omega_1' = \omega_1 + d\beta$, then

$$\omega_1' \wedge \omega_2 = \omega_1 \wedge \omega_2 + d(\beta \wedge \omega_2).$$

The algebra so defined on $H^*_{De\ Rham}$ is such that $\alpha_1\alpha_2 = (-1)^{pq}\alpha_2\alpha_1$. What was said above about Chern classes can be restated as follows: The set of all characteristic cohomology classes is generated as a subalgebra in $H^*_{De\ Rham}(M)$ by the Chern classes. This is to say that any characteristic class that is defined by a real analytic, ad-invariant function on $\mathbb{M}(n; \mathbb{C})$ is a sum of products in $H^*_{De\ Rham}$ of these Chern classes. The appendix to the chapter explains why this is so.

The classes that arise using $ch(\mathfrak{m}) = tr(\exp(\frac{i}{2\pi}\mathfrak{m}))$ result in what are called *Chern character classes*.

14.6 Characteristic classes for real vector bundles and the Pontryagin classes

Characteristic classes for bundles $E \to M$ with fiber $\mathbb{R}^n$ can also be defined without directly referencing a principal bundle. Here is how this is done: First construct the complexification of E as done in Chapter 6.3. By way of a reminder, this is the rank n complex bundle, $E_{\mathbb{C}}$, whose underlying real bundle is $(M \times \mathbb{R}^2) \otimes E$. This bundle has an almost complex structure, j, that comes from the almost complex structure

$$j_0 = \begin{pmatrix} 0 & -1 \\ 1 & 0 \end{pmatrix} \text{ on } \mathbb{R}^2.$$

To elaborate, j is first defined on decomposable vectors so as to send a pair $z \otimes \mathfrak{e}$ with $z \in \mathbb{R}^2$ and $\mathfrak{e} \in E$ to $j_0 z \otimes \mathfrak{e}$, and it is then extended to the whole of $(M \times \mathbb{R}^2) \otimes E$ by linearity. With $\mathbb{R}^2$ viewed as $\mathbb{C}$, then j_0 acts as multiplication by i; and the $\mathbb{C}$-action on $E_{\mathbb{C}}$ is such that $\sigma \in \mathbb{C}$ sends the decomposable element $z \otimes \mathfrak{e}$ to $\sigma z \otimes \mathfrak{e}$.

A covariant derivative, ∇, for sections of E induces a covariant derivative for sections of $(M \times \mathbb{R}^2) \otimes E$ that commutes with the action of j. The induced covariant derivative acts so as to send a decomposable section, $z \otimes \mathfrak{s}$

to $z \otimes \nabla\mathfrak{s} + dz \otimes s$. Because ∇ commutes with j, it defines a covariant on $E_{\mathbb{C}}$. This induced covariant derivative is denoted by $\nabla_{\mathbb{C}}$. The characteristic classes of $E_{\mathbb{C}}$ as defined by any ad-invariant function on $\mathrm{End}(\mathbb{C}^n)$ are viewed as characteristic classes of the original $\mathbb{R}^n$-bundle E.

An important example of this construction gives the Pontryagin classes. The k'th Pontryagin class of E is the 4k-dimensional chomology class,

$$p_k(E) = (-1)^k c_{2k}(E_{\mathbb{C}}).$$

Note in this regard that the Chern classes $c_k(E_{\mathbb{C}})$ vanish if k is odd; thus there is no motivation to use the odd k Chern classes of $E_{\mathbb{C}}$ to define characteristic classes of E. Here is why these odd k classes vanish: Fix a fiber matrix on E. This done, one can find a covariant derivative on E that comes from a connection on the principal bundle $P_{O(E)}$ of orthonormal frames in E. Let f now denote an ad-invariant function on $\mathrm{End}(\mathbb{C}^n)$. I claim that the odd k versions of $f_k(F_\nabla)$ of the corresponding covariant derivative on $E_{\mathbb{C}}$ are all zero. This is because $-\mathfrak{m}$ can be written as $g\mathfrak{m}g^{-1}$ for some $g \in SO(n)$. To see why this is, remark that $\mathfrak{m}$ has some 2j purely imaginary eigenvalues, these in complex conjugate pairs. Taking the real eigenvectors and the real and complex parts of the complex eigenvectors finds an orthonormal basis for $\mathbb{R}^n$ of the form

$$\{u_1, \ldots, u_{n-2j}, v_1, \ldots, v_{2j}\}$$

with the property that $\mathfrak{m}u_q = 0$ for $q \in \{1, \ldots, n-2j\}$, and $\mathfrak{m}v_{2q-1} = a_q\, u_{2q}$ and $\mathfrak{m}v_{2q} = -a_q\, u_{2q-1}$ for $q \in \{1, \ldots, j\}$ with $a_q > 0$. This understood, let g denote the matrix in SO(n) that switches v_{2q-1} with v_{2q} for each $q \in \{1, \ldots, j\}$ and either fixes the span of $\{u_q\}_{q=1,\ldots,n-2j}$ if j is even, or fixes $\{u_q\}_{2\le q\le n-2j}$ and sends u_j to $-u_j$ if j is odd. This matrix g has the desired properties.

14.7 Examples of bundles with nonzero Chern classes

What follows are some examples where the Chern classes are nonzero.

Example 14.4 This is a principal U(1)-bundle over S^2. View S^2 as the unit sphere in $\mathbb{R}^3$. Reintroduce the Pauli matrices τ_1, τ_2, and τ_3, from Chapter 6.4. Use them as in Chapter 6.4 to define the complex, rank 1 bundle $E_1 \subset S^2 \times \mathbb{C}^2$ to be the set of pairs

$$(x = (x^1, x^2, x^3), v) \textit{ that obey } x^j \tau_j v = iv.$$

The bundle E_1 has a Hermitian metric, this coming from the Hermitian metric on S^2. This is to say that the norm of $(x, v) \in E_1$ is $|v|$. Let $U_E \to S^2$ denote the bundle of unitary frames in E_1. As noted in Chapter 10.3, the space U_E when viewed only as a smooth manifold is just S^3.

The projection Π: $S^2 \times \mathbb{C}^2 \to E$ at $x \in S^2$ is $\frac{1}{2}(1 - ix^j\tau_j)$. A covariant derivative on E_1 is given as follows: View a section, $\mathfrak{s}$, of E_1 as a map, also denoted by $\mathfrak{s}$, from S^2 to $\mathbb{C}^2$ with the property that $\mathfrak{s}(x) \in E_1|_x$ at each $x \in S^2 \times \mathbb{C}^2$. Granted this view, set $\nabla\mathfrak{s} = \Pi d\mathfrak{s}$. As explained in Chapter 12.5, the curvature of this covariant derivative is

$$F_\nabla = \Pi d\Pi \wedge d\Pi\Pi.$$

To be more explicit, note that $d\Pi = -\frac{i}{2}dx^j\tau_j$. Given that the Pauli matrices multiply as the quaternions ($\tau_1\tau_1 = -1$, $\tau_1\tau_2 = -\tau_3$, etc), it follows that

$$d\Pi \wedge d\Pi = \frac{1}{4}dx^i \wedge dx^j\ \varepsilon^{ijk}\tau_k,$$

where ε^{ijk} is completely antisymmetric under interchange of any of its indices; and it obeys $\varepsilon^{123} = 1$. To see what this is at a given $x \in S^2$, fix an oriented, orthonormal frame (x, u, v) for $\mathbb{R}^3$, thus (u, v) are an orthonormal frame for $TS^2|_x$. This done, then the matrix-valued 2-form $d\Pi \wedge d\Pi$ evaluates on (v, u) to give $\frac{1}{2}\varepsilon^{ijk}v^iu^j\tau_k = \frac{1}{2}x^k\tau_k$. Since $x^k\tau_k$ acts as multiplication by i on E_1, it follows that $F_\nabla(u, v)$ is just multiplication by $\frac{i}{2}$. Thus, the Chern form $c_1(E_1; \nabla)$ evaluates on the frame, (v, u) to give $\frac{i}{2\pi}\left(\frac{i}{2}\right) = -\frac{1}{4\pi}$. Note that this is independent of $x \in S^2$. As a consequence, the integral of $c_1(E_1, \nabla)$ over S^2 is -1 as S^2 has area 4π. This tells us that the first Chern class of E is nonzero in $H^2_{\text{De Rham}}(S^2)$. Indeed, were $c_1(E_1) = 0$, then $c_1(E_1; \nabla)$ could be written as $d\alpha$ for some 1-form α. Keep in mind here that there is a well-defined notion of the integral of an n-form on an n-dimensional, compact and oriented manifold, and this integral is zero if and only if the n-form is the exterior derivative of an $(n-1)$-form. Appendix 14.2 discusses this subject of integration on manifolds for those needing a refresher course.

The preceding calculation tells us that E_1 is not isomorphic to $S^2 \times \mathbb{C}$. The calculation also tells us that the principal U(1)-bundle U_E has no flat connections.

Example 14.5 The tangent bundle to S^2 can be viewed as the subbundle in $S^2 \times \mathbb{R}^3$ that consists of the pairs (x, v) such that $x \cdot v = 0$. Here, '·' denotes the dot product. The orthogonal projection Π: $S^2 \times \mathbb{R}^3 \to TS^2$ sends a pair (x, v) to $(x, v - x\,(x \cdot v))$. As explained in Chapter 11.3, this gives a covariant derivative, ∇. Recall that ∇ is defined by viewing a section $S^2 \times \mathbb{R}^3$, and so a section of TS^2 as a map from S^2 to $\mathbb{R}^3$ that lands in TS^2 at each point. This understood, ∇, sends such a map, $\mathfrak{s}$, to $\nabla\mathfrak{s} = \Pi d\mathfrak{s} = d\mathfrak{s} + x dx^T\mathfrak{s}$. The curvature of this covariant derivative is

$$F_\nabla = dx \wedge dx^T.$$

To see what this is, fix $x \in S^2$, and then fix a pair (u, v) of orthonormal vectors in $\mathbb{R}^3$ such that (x, u, v) is an oriented, orthonormal basis for $\mathbb{R}^3$. Then $F_\nabla(u, v)$ is the endomorphism

$$F_\nabla(u, v)^{ij} = u^i v^j - v^i u^j.$$

Meanwhile, TS^2 has an almost complex structure, j, that acts on any given point (x, v) to give $(x, x \times v)$, where '×' here denotes the cross product on $\mathbb{R}^3$. Let $T_{1,0} \to S^2$ denote the corresponding complex, rank 1 vector bundle. Let $x \in S^2$ and let (u, v) be as above. Then $T_{1,0}|_x$ is spanned by $z = u - iv$. A calculation will verify that ∇ commutes with j and so defines a covariant derivative, denoted here by ∇ also, on the space of sections of $T_{1,0}$. The curvature of this connection sends z to $(F_\nabla)^{ij} z^j = -iz^i$. Thus, F_∇ acts as multiplication by $-i$. This implies that $c_1(T_{1,0}; F_\nabla)$ is $\frac{1}{2\pi}$ times the standard area 2-form on S^2, and so its integral over S^2 is equal to 2. This implies that $T_{1,0}$ is neither the trivial $\mathbb{C}$-bundle nor isomorphic to E_1.

Example 14.6 What follows is an application of pull-back to the computation of Chern classes. Let $\Sigma \subset \mathbb{R}^3$ denote a compact, embedded surface. Use $x \to n(x) \in \mathbb{R}^3$ to denote the unit length, outward pointing normal vector to Σ at x. The tangent space to Σ can be viewed as the subbundle of $\Sigma \times \mathbb{R}^3$ consisting of the pairs (x, u) such that $n(x) \cdot u = 0$. As noted in Chapter 6.5, this bundle has the almost complex structure j that acts to send a point (x, u) to the point $(x, n(x) \times u)$. Use $T_{1,0}\Sigma$ to denote the resulting rank 1 complex vector bundle. This bundle is, in fact, the pull-back from S^2 of the bundle $T_{1,0}$ from the previous example. Indeed, it is precisely $n^*T_{1,0}$ where $n: \Sigma \to S^2$ is the map $x \to n(x)$. This said, it follows that $c_1(T_{1,0}\Sigma) = n^*c_1(T_{1,0})$.

The following proposition is needed to exploit this last observation.

Proposition 14.7 *Suppose that* X *is a compact, oriented, n-dimensional manifold and suppose that* M *is another such manifold.*

- *Suppose that* $\psi: M \to X$ *is a smooth map. There exists an integer* p *with the following property: Let* ω *denote a n-form on* X. *Then* $\int_M \psi^*\omega = p\int_X \omega$. *This integer p is said here to be the degree of* ψ.
- *Homotopic maps from* M *to* X *have the same degree.*

The proof of this proposition is given in Appendix 14.3.

To use this proposition, orient Σ so that (n(x), u, v) is an oriented frame for $T\mathbb{R}^3$ at a point x if (u, v) is an oriented frame for $T\Sigma$ at x. This is the case, for example if u is any given nonzero vector and $v = n \times u$. Use this orientation with the preceding proposition to define the notion of the degree of a map from Σ to S^2. Appendix 14.3 describes an embedding of Σ into $\mathbb{R}^3$ whose version of the map $x \to n(x)$ has degree equal to $1 - \text{genus}(\Sigma)$. Granted that such is the case, it follows from the proposition that any 2-form on Σ that represents the corresponding version of $c_1(T_{1,0}\Sigma)$ integrates to zero when Σ is the torus, and it integrates to $2 - 2g$ when Σ is a surface of genus equal to g; this being a surface as depicted in Chapter 1.3 with g holes.

As I now argue, this last conclusion holds for *any* version of $T_{1,0}\Sigma$ whose defining almost complex structure, j, has the following property: If $p \in \Sigma$ is any given point and $u \in T\Sigma|_p$ is any given nonzero vector, then $(u, j{\cdot}u)$ is an oriented frame for $T\Sigma$.

Let j denote an almost complex structure on $T\Sigma$ of the sort just described and let $T_{1,0}\Sigma$ denote the corresponding complex 1-dimensional vector bundle. I use the next lemma to prove this claim that the integral over Σ of any representative of $c_1(T_{1,0}\Sigma)$ is equal to $2 - 2g$.

Lemma 14.8 *Let* M *denote a compact, manifold and* π: $E_{\mathbb{R}} \to M$ *a real vector bundle with even dimensional fiber. Suppose that* $t \to j_t$ *is a smoothly varying, 1-parameter family of almost complex structures on* $E_{\mathbb{R}}$, *this parametrized by* $t \in [0, 1]$. *For each such* t, *let* E^t *denote the complex vector bundle that is defined from* $E_{\mathbb{R}}$ *via defined by* j_t. *Then each* $t \in [0, 1]$ *version of* E^t *is isomorphic to* E^0 *as a complex vector bundle. In particular, the Chern classes do not change as* t *varies.*

Given the example in Appendix 14.3, the proof of the claim about the integral of $c_1(T_{1,0}\Sigma)$ follows from the preceding lemma with a proof of the following assertion: Any two complex structures on $T\Sigma$ that define the same orientation for $T\Sigma$ are connected by a path of complex structures. This fact is proved momentarily. What follows is the proof of the lemma.

Proof of Lemma 14.8 Change the parametrization if necessary so that the map $t \to j_t$ is constant near $t = 0$ and also near $t = 1$. Let $\mathfrak{p}$: $(0, 1) \times M \to M$ denote the projection. The smooth family $t \to j_t$ defines an almost complex structure on $\mathfrak{p}^*E_{\mathbb{R}}$ such that the resulting complex vector bundle, E, restricts to any given $t \in (0, 1)$ version of $\{t\} \times M$ as the bundle E^t. Fix a connection, A on E. Parallel transport by A along the fibers of $\mathfrak{p}$ defines a complex linear isomorphism between $E^{(0)}$ and any $t \in (0, 1)$ version of $E^{(t)}$.

Now consider the assertion that any two almost complex structures on $T\Sigma$ that give the same orientation are connected by a path of almost complex structures. To start, suppose that j is a given almost complex structure on $T\Sigma$. Introduce $T_{1,0j}\Sigma$ to denote j's version of $T_{1,0}\Sigma$. This is, by definition, the subbundle of the complexification $T\Sigma \otimes \mathbb{C}$ on which j acts as multiplication by i. Let j' denote another almost complex structure. Let p denote a point in Σ, and $e \in T_{1,0j}\Sigma|_p$. Then $j'{\cdot}e$ is a vector in $T\Sigma \otimes \mathbb{C}$, and as such, it can be written as $j'e = \alpha\, e + \beta\, \bar{e}$ with α and β complex numbers. Act by j' on this expression for $j'{\cdot}e$ to see that

$$-e = (\alpha^2 - |\beta|^2)\, e + (\alpha + \bar{\alpha})\beta\, \bar{e}\,.$$

Indeed, this follows because $j'^2 = -1$ and because $j'\,\bar{e}$ is the complex conjugate of $j'e$. This last equation requires that $\alpha = \pm i(1 + |\beta|^2)^{1/2}$. Moreover, the + sign must be taken in order that (v, j′v) define the correct orientation. Each point in Σ has a corresponding β, and the assignment of β to the point in Σ defines a section over Σ of $\mathrm{Hom}(T_{1,0j}; T_{0,1j})$.

The converse of this observation is also true: Any given section, β, of the bundle $\mathrm{Hom}(T_{1,0j}; T_{0,1j})$ defines an almost complex structure, j_β, where j_β acts on $e \in T_{1,0j}$ by the rule $j_\beta e = i(1 + |\beta|^2)^{1/2}\, e + \beta\, \bar{e}$. To see j_β as an endomorphism of $T\Sigma$, write e as $u_1 - iu_2$ with u_1 and u_2 vectors in $T\Sigma$. Likewise, write β as $\beta_1 + i\beta_2$ with β_1 and β_2 real. Then

$$j_\beta u_1 = ((1 + |\beta|^2)^{1/2} - \beta_2)u_2 + \beta_1 u_1 \ \text{ and}$$
$$j_\beta u_2 = -((1 + |\beta|^2)^{1/2} + \beta_2)u_1 - \beta_1 u_2.$$

With the preceding understood, write $j' = j_\beta$ as above. Then the 1-parameter family $\{t \to j_{(1-t)\beta}\}_{t\in[0.1]}$ exhibits a path of almost complex structures that starts when $t = 0$ at j and ends when $t = 1$ at j′.

Example 14.9 Let Σ denote a compact, oriented surface. This means that $\wedge^2 TM$ has a distinguished section up to multiplication by a *positive* real number. See Chapter 10.3. Construct a complex, rank 1 vector bundle $E \to \Sigma$ as follows: Choose $k \geq 1$ and a set of k distinct points $\Lambda = \{p_1, \ldots, p_k\} \subset \Sigma$. Assign to each point $p \in \Lambda$ an integer m_p. Fix disjoint coordinate patches around the points in Λ. Given $p \in \Lambda$, let $U_p \subset \Sigma$ denote the corresponding coordinate patch, and let $\psi_p\colon U_p \to \mathbb{R}^2$ a coordinate chart sending p to 0. Assume that the differential of ψ_p is orientation preserving in the sense that it sends the distinguished section of $\wedge^2 TM$ to a *positive* multiple of $\frac{\partial}{\partial x_1} \wedge \frac{\partial}{\partial x_2}$. I use (r, θ) for standard polar coordinates on $\mathbb{R}^2$. Thus $x_1 = r\cos\theta$ and $x_2 = r\sin\theta$.

The bundle E is defined over the cover of M given by the set $U_0 = \Sigma - \Lambda$ and $\{U_p\}_{p\in\Lambda}$. Over each such set, E is defined to be the product bundle. The only intersections are between U_0 and U_p for each $p \in \Lambda$. The ψ_p image of $U_0 \cap U_p$ is the subset $\mathbb{R}^2 - \{0\}$ of $\mathbb{R}^2$. This understood, define the bundle E so that when $x \in U_0 \cap U_p$, then the point $(x, v) \in U_p \times \mathbb{C}$ corresponds to the point $(x, g_{0,p}(x)v) \in U_0 \times C$ where $g_{0,p}(x)$ is such that if $x = \psi_p(r, \theta)$ then $g_{0,p} = e^{im_p\theta}$. I claim that the integral over Σ of $c_1(E)$ is equal to $-\Sigma_{p\in\Lambda}\, m_p$.

To verify this claim, I will fix a convenient covariant derivative for E whose corresponding first Chern form is easy to integrate. To start, fix a smooth function, χ, from $[0, \infty)$ to $[0, 1]$ which is equal to 0 near 0 and equal to 1 on $[1, \infty)$. I now view χ as the function on $\mathbb{R}^2$ that sends (r, θ) to $\chi(r)$.

The covariant derivative, ∇, on E is defined as follows: Identify E's restriction to U_p with $U_p \times \mathbb{C}$ so as to view a section of E on U_p as a map $x \to \mathfrak{s}_p(x)$ from U_p to $\mathbb{C}$. Use ψ_p to view $\nabla\mathfrak{s}$ on U_p as a $\mathbb{C}$-valued 1-form, $(\nabla\mathfrak{s})_p$, on $\mathbb{R}^2$. Then ∇ is defined so that

$$(\nabla_{\mathfrak{s}})_p = d\mathfrak{s}_p + i\, m_p\, \mathfrak{s}_p\, \chi\, d\theta.$$

Identify E's restriction to U_0 as $U_0 \times \mathbb{C}$ so as to view $\mathfrak{s}$ over U_0 as a map, $x \to \mathfrak{s}_0(x)$ from U_0 to $\mathbb{C}$. By way of a reminder, $\mathfrak{s}_0$ over $U_0 \cap U_p$ is given in terms of $\mathfrak{s}_p$ by the formula

$$\mathfrak{s}_0(x) = e^{im_p\theta}\, \mathfrak{s}_p(x)$$

where θ is the angular coordinate of $\psi_p(x)$. View $\nabla\mathfrak{s}$ over U_0 as a $\mathbb{C}$-valued 1-form, this denoted by $(\nabla\mathfrak{s})_0$. Then ∇ is defined so that

$$(\nabla_{\mathfrak{s}})_0 = d\mathfrak{s}_0 - \textstyle\sum_{p\in\Lambda} i\, m_p\, \mathfrak{s}_0 \psi_p^*((1-\chi)d\theta).$$

You can check that $(\nabla\mathfrak{s})_0 = e^{im_p\theta}(\nabla\mathfrak{s})_p$ at any given point $x \in U_0 \cap U_p$. In particular, ∇ as defined is a bonafide covariant derivative.

The preceding formulae for ∇ show that $F_\nabla = d_\nabla^2 = 0$ except in the sets U_p. In any such set, the formula above for $(\nabla\mathfrak{s}_p)$ implies that $(\psi_p^{-1})^* F_\nabla$ is the $Gl(1; \mathbb{C})$-valued 2-form on $\mathbb{R}^2$ given in polar coordinates by

$$(F_\nabla)_p = i m_p\, \chi'\, dr \wedge d\theta.$$

Here, $\chi' = \frac{d}{dr}\chi$. It follows that $c_1(E; \nabla)$ has support only in the sets U_p, and in any such set, it pulls back via ψ_p^{-1} to the 2-form $-\frac{1}{2\pi} m_p\, \chi' dr\, d\theta$. As a consequence, its integral over U_p is equal to $-m_p$. Thus, its integral over Σ is $-\sum_{p\in\Lambda} m_p$.

Example 14.10 Fix $n \geq 1$ and recall that $\mathbb{CP}^n$ is the Grassmannian of 1-dimensional complex vector spaces in $\mathbb{C}^{n+1}$ as defined in Chapter 6.8. An alternate view of $\mathbb{CP}^n$ as S^{2n+1}/S^1 described in Example 10.3 and Example 10.8. By way of a reminder, view S^{2n+1} as the unit sphere in $\mathbb{C}^{n+1}$ and view S^1 as the unit sphere in $\mathbb{C}$. Have S^1 act on $\mathbb{C}^{n+1}$ so that any given $\lambda \in S^1$ sends any given $(n+1)$-tuple $(z_1, \ldots, z_{n+1})$ to $(\lambda z_1, \ldots, \lambda z_{n+1})$. This action preserves S^{2n+1}. The action is free, and the quotient is the space $\mathbb{CP}^n$. As explained in Chapters 10.5 and 10.6, the projection π: $S^{2n+1} \to \mathbb{CP}^n$ defines a principal $S^1 = U(1)$ bundle. As the Lie algebra of $U(1)$ is the imaginary complex numbers, a connection on this bundle is an i-valued 1-form on S^{2n+1} that is pulled back to itself by the action of any $\lambda \in S^1$ and pairs with the vector field

$$v = i\,\Sigma_{1\le k\le 1}\left(z_k\frac{\partial}{\partial z_k} - \bar{z}_k\frac{\partial}{\partial \bar{z}_k}\right)$$

to give i. Note that the vector field v spans the kernel of π_* and is such that if f is any given function on S^{2n+1} and $z = (z_1, \ldots, z_{n+1}) \in S^{2n+1}$, then $\frac{d}{dt} f(e^{it}z_1, \ldots, e^{it}z_{n+1})$ at $t = 0$ is equal to $v(f)|_z$.

Granted what was just said, the i-valued 1-form

$$A = \frac{1}{2}\,\Sigma_{1\le k\le n+1}(\bar{z}_k\, dz_k - z_k d\bar{z}_k)$$

from $\mathbb{C}^{n+1}$ is a connection when viewed as a 1-form on S^{2n+1}. The pull-back via the projection $\pi: S^{2n+1} \to \mathbb{CP}^n$ of the curvature 2-form F_A is the restriction to S^{2n+1} of the form

$$\pi^*F_A = \Sigma_{1\le k\le n+1}\, d\,\bar{z}_k \wedge dz_k$$

on $\mathbb{C}^{n+1}$. Note that this form on S^{2n+1} annihilates the vector v since $\pi^*F_A(v, \cdot)$ is the 1-form $d(\sum_{1\le k\le n+1} |z_k|^2)$ and $\sum_{1\le k\le n+1} |z_k|^2 = 1$ on S^{2n+1}. The resulting 2-form that defines the first Chern class of this line bundle pulls back to S^{2n+1} as the restriction from $\mathbb{C}^{n+1}$ of the form $\omega = \frac{i}{2\pi}\,\Sigma_{1\le k\le n+1} d\,\bar{z}_k \wedge dz_k$.

To see that the first Chern class is nonzero, it is enough to note that the (2n+1)-form $\frac{i}{2\pi}\, A \wedge \omega^k = \frac{i}{2\pi} A \wedge \omega \wedge \cdots \wedge \omega$ is a constant multiple of the volume form on S^{2n+1}. Thus, its integral is nonzero. This can happen if and only if the k'th power of the Chern class $c_1(E)$ is nonzero in $H^{2k}_{\text{De Rham}}(\mathbb{CP}^n)$. Thus, $c_1(E) \neq 0$. Indeed, suppose that $c_1(E)^k = 0$ in $H^{2k}_{\text{De Rham}}(\mathbb{CP}^n)$. Then the form $(\frac{i}{2\pi})^k F_A \wedge \cdots \wedge F_A$ can be written as $d\alpha$ with α a $(2k-1)$-form on $\mathbb{CP}^n$. Were such the case, then $\omega^k = d(\pi^*\alpha)$. The integral of $\frac{i}{2\pi} A \wedge \omega^k$ would then equal that of $\frac{i}{2\pi} A \wedge d(\pi^*\alpha)$ which is

$$-d\left(\frac{i}{2\pi} A \wedge \pi^*\alpha\right) + \frac{i}{2\pi} dA \wedge \pi^*\alpha.$$

However, $dA = \pi^*F_A$ and so the integral of $\frac{i}{2\pi} A \wedge \omega^k$ over S^{2n+1} would equal the integral of $\pi^*\omega \wedge \pi^*\alpha$. However the latter is $\pi^*(\omega \wedge \alpha)$, which is to say that it is the pull-back of a degree $(2n+1)$-form on $\mathbb{CP}^n$. All such forms are zero since $\dim(\mathbb{CP}^n) = 2n$.

Example 14.11 This example considers principal SU(2)-bundles over an oriented 4-dimensional Riemannian manifold. Let M denote the manifold. The bundle will be specified by an open cover with suitable transition functions on the overlapping regions between sets of the cover. To start, fix a nonempty, finite set $\Lambda \subset M$. For each $p \in \Lambda$, fix an open set U_p that contains p and that has an orientation-preserving diffeomorphism $\psi_p : U_p \to \mathbb{R}^4$ sending p to 0. I use $y = (y_1, y_2, y_3, y_4)$ for the Euclidean coordinates of $\mathbb{R}^4$. To say that M is oriented is to say that the bundle $\wedge^4 T^*M$ has a given nowhere zero section (see Chapter 10.3). To say that ψ_p is orientation preserving is to say that the pull-back via ψ_p of the 4-form $dy_1 \wedge dy_2 \wedge dy_3 \wedge dy_4$ from $\mathbb{R}^4$ is a positive multiple of this section.

Choose these charts so that the U_p is disjoint from $U_{p'}$ when $p \neq p'$. Let $U_0 = M - \Lambda$. The only transition functions to consider are those for $U_0 \cap U_p$ for $p \in \Lambda$. Such a function is given by a map, $g_{0,p}: \mathbb{R}^4 - \{0\} \to SU(2)$, by the rule $g_{0,p}(x) = g_p(\psi_p(x))$. To specify g_p, I first choose an integer, m_p. Set $g: \mathbb{R}^4 - \{0\} \to SU(2)$ to be the map that sends a point $y \in \mathbb{R}^4$ to the matrix

$$|y|^{-1} \begin{pmatrix} y_4 + iy_3 & iy_1 - y_2 \\ iy_1 + y_2 & y_4 - iy_3 \end{pmatrix}.$$

Now set $g_p = g^m$ where $g^m = g \cdots g$ if $m > 0$, or $g^{-1} \cdots g^{-1}$ if $m < 0$, or the identity matrix ι if $m = 0$.

Let $\rho: SU(2) \to Gl(2; \mathbb{C})$ denote the defining representation. I claim that the Chern class $c_2(P_\rho)$ is nonzero, and that it evaluates on M's fundamental class as $\Sigma_{p \in \Lambda}\, m_p$. To do this, I need to choose a convenient connection. For this purpose, I take the function χ from Example 14.9. The connection A is defined by giving, for each $p \in \Lambda$, a $\mathfrak{lie}(SU(2))$-valued 1-form $\mathfrak{a}_p$ on U_p, and also a $\mathfrak{lie}(SU(2))$-valued 1-form $\mathfrak{a}_0$ on U_0. These must be such that $\mathfrak{a}_p = g_{0p}^{-1} dg_{0p} + g_{0p}^{-1} \mathfrak{a}_0\, g_{0p}$ on $U_0 \cap U_p$. This understood, I take $(\psi_p^{-1})^* \mathfrak{a}_p$ to be the $\mathfrak{lie}(SU(2))$-valued 1-form on $\mathbb{R}^4 - \{0\}$ given by

$$(\psi_p^{-1})^* \mathfrak{a}_p = \chi(|y|)\, g^{-m_p} d(g^{m_p}).$$

Meanwhile, I take $\mathfrak{a}_0 = 0$ except on U_p, and here

$$(\psi_p^{-1})^* \mathfrak{a}_0 = \chi(|y| - 1) d(g^{m_p}) g^{-m_p}.$$

Since the trace of a matrix in $\mathfrak{lie}(SU(2))$ is zero, the 2-form that defines the second Chern class is

$$\frac{1}{8\pi^2} \operatorname{tr}(F_A \wedge F_A).$$

I argue momentarily that the second Chern class is nonzero if $\Sigma_{p \in \Lambda}\, m_p$ is nonzero. This follows from the assertion that the integral of the preceding form over M is $-\Sigma_{p \in \Lambda}\, m_p$. This shows, in particular, that this construction of principal bundles gives nonisomorphic bundles when the respective sums differ.

To see what the integral of this Chern form is, note first that the form in question is zero except in U_p for $p \in \Lambda$. Moreover, the support of this form is contained in the inverse image via ψ_p of the unit radius ball about the origin in $\mathbb{R}^4$. This understood, it is enough to compute, for each p, the following integral

$$\frac{1}{8\pi^2}\int\limits_{|y|\le 1} \mathrm{tr}(F_{\mathfrak{a}_p}\wedge F_{\mathfrak{a}_p})$$

where $F_{\mathfrak{a}_p} = d\mathfrak{a}_p + \mathfrak{a}_p \wedge \mathfrak{a}_p$. The computation that follows finds this integral equal to $-m_p$.

To see about this integral, I will employ the following fact: Let G denote a Lie group and $\mathfrak{a}$ denote a $\mathfrak{lie}(G)$-valued 1-form defined on some open set in any given manifold. Introduce $F_{\mathfrak{a}} = d\mathfrak{a} + \mathfrak{a} \wedge \mathfrak{a}$. Then

$$\mathrm{tr}(F_{\mathfrak{a}}\wedge F_{\mathfrak{a}}) = d\left(\mathrm{tr}\left(\mathfrak{a}\wedge d\mathfrak{a} + \frac{2}{3}\,\mathfrak{a}\wedge\mathfrak{a}\wedge\mathfrak{a}\right)\right).$$

This identity will be proved in Chapter 14.9. Take the $\mathfrak{a} = \mathfrak{a}_p$ version of this identity and integrate by parts to see that the integral at the end of the previous paragraph is equal to

$$\frac{1}{8\pi^2}\int\limits_{|y|=1}\mathrm{tr}\left(\mathfrak{a}_p\wedge d\mathfrak{a}_p + \frac{2}{3}\mathfrak{a}_p\wedge\mathfrak{a}_p\wedge\mathfrak{a}_p\right).$$

A computation finds the integrand equal

$$-\frac{1}{3}\mathrm{tr}(g^{-m_p}d(g^{m_p})\wedge g^{-m_p}\,d(g^{m_p})\wedge\, g^{-m_p}\,d(g^{m_p})).$$

This in turn is the pull-back via the map from S^3 to SU(2) given by $y \to g^{m_p}|_y$ of the form

$$-\frac{1}{3}\mathrm{tr}(g^{-1}dg\wedge g^{-1}dg\wedge g^{-1}dg)$$

on SU(2). This latest 3-form is invariant under both left and right multiplication. This said, its evaluation at the identity $\iota \in$ SU(2) finds it equal to -4 times the volume form on SU(2) that comes by identifying SU(2) with the unit radius sphere in $\mathbb{R}^4$.

Granted all of this, it then follows that the integral of $\frac{1}{8\pi^2}\mathrm{tr}(F_A\wedge F_A)$ over U_p is given by the product of three factors. The first is $\frac{1}{8\pi^2}$. The second is -4 times the volume of the 3-sphere, which is $-8\pi^2$. The third is the degree of the map $y \to g^{m_p}|_y$ from S^3 to SU(2), or what is the same thing, the integral over S^3 of the pull-back via this map of the $\frac{1}{2\pi^2}$ times the volume form. The first two factors give an overall factor of -1, and so the contribution from U_p to the integral of the form that gives the second Chern class is just -1 times the pull-back of $\frac{1}{2\pi^2}$ times the pull-back volume form by the map $y \to g^{m_p}|_y$. As explained momentarily, the latter number is m_p. Granted this, then the integral of the second Chern class does have the predicted value $-\Sigma_{p\in\Lambda}\, m_p$.

14.8 The degree of the map $g \to g^m$ from SU(2) to itself

The map $y \to g^m|_y$ from S^3 to SU(2) where

$$g(y) = |y|^{-1}\begin{pmatrix} y_4 + iy_3 & iy_1 - y_2 \\ iy_1 + y_2 & y_4 - iy_3 \end{pmatrix}$$

becomes the map $g \to g^m$ from SU(2) to SU(2) when S^3 is identified in the usual way with SU(2). This map has degree m. More to the point, what is proved next is that the integral of the pull-back of the volume form on SU(2),

$$\Omega = \frac{1}{12}\,\mathrm{tr}(g^{-1}dg \wedge g^{-1}dg \wedge g^{-1}dg),$$

by this map is m times the volume of S^3.

To see this, start with the case $m = 0$. The map in this case is $g \to \iota$, the constant map to the identity. Thus, the map pulls back the volume form as zero. In the case $m = 1$, the map is $g \to g$ which is the identity map, so the pull-back of the volume form is itself. To see about $m = -1$, note that the pull-back of $g^{-1}dg$ by the map $g \to g^{-1}$ is $-dgg^{-1}$. As a consequence, this map pulls back Ω as

$$(-1)^3\frac{1}{12}\mathrm{tr}(dgg^{-1} \wedge dgg^{-1} \wedge dgg^{-1})$$

which is seen to be $-\Omega$ using the cyclic property of the trace. Note in this regard that if $\mathfrak{a}$ is a matrix-valued p-form and $\mathfrak{b}$ is a matrix-valued q-form, then

$$\mathrm{tr}(\mathfrak{a}\mathfrak{b}) = (-1)^{pq}\,\mathrm{tr}(\mathfrak{b}\mathfrak{a}).$$

It follows from this and the fact that $g^{-m} = (g^m)^{-1}$ that it is sufficient for the purposes at hand to show that the map $g \to g^m$ for m positive pulls Ω back to a form whose integral over S^3 is m times that of Ω. To do so, suppose that this has been established for all integers $m \in \{1, \ldots, n\}$. To consider the case $m = n + 1$, write the map $g \to g^n$ as gh. Then the map $g \to gh$ pulls back $g^{-1}dg$ as $h^{-1}g^{-1}dgh + h^{-1}dh$. This is a matrix-valued 1-form. Multiplying matrices and using the aforementioned cyclic property of the trace finds that the map $g \to g^{n+1}$ pulls back Ω as

$$\frac{1}{12}\,\mathrm{tr}(g^{-1}dg \wedge g^{-1}dg \wedge g^{-1}dg) + \frac{1}{12}\,\mathrm{tr}(h^{-1}dh \wedge h^{-1}dh \wedge h^{-1}dh)$$
$$+\frac{1}{4}\,\mathrm{tr}(g^{-1}dg \wedge g^{-1}dg \wedge dhh^{-1}) + \frac{1}{4}\,\mathrm{tr}(dhh^{-1} \wedge dhh^{-1} \wedge g^{-1}dg).$$

Now the two left-most terms above are Ω and $h^*\Omega$, respectively, so the sum of their integrals over SU(2) is $n + 1$ times the volume of SU(2). As for the two right-most terms, note that $d(g^{-1}) = -g^{-1}dgg^{-1}$ and $d(h^{-1}) = -h^{-1}dhh^{-1}$. Therefore, the sum of the two right-most terms above can be written as

$$d\left(-\frac{1}{4}\mathrm{tr}(g^{-1}dg \wedge dhh^{-1})\right).$$

Thus, the integral of the two right-most terms over SU(2) is zero.

14.9 A Chern–Simons form

The identity asserted in Chapter 14.8 can be proved by directly computing both sides. What follows is another approach that can be used for other purposes. Let t denote the coordinate on [0, 1]. Then $\hat{A} = g^{-1}dg + tg^{-1}ag$ is a connection on the principal bundle $([0, 1] \times U) \times G$. Its curvature 2-form is $F_{\hat{A}} = t da + dt \wedge a + t^2 a \wedge a$. This understood, the 4-form on $[0, 1] \times U$ given by $\mathrm{tr}(F_{\hat{A}} \wedge F_{\hat{A}})$ is closed. This form can be written in the manner introduced in Chapter 12.2 as $dt \wedge \alpha_3 + \alpha_4$ where

$$\alpha_3 = 2\mathrm{tr}(ta \wedge da + t^2 a \wedge a \wedge a),$$

and where α_4 has no factor of dt. At any given $t \in [0, 1]$, the 4-form α_4 is $\mathrm{tr}(F_{\hat{A}(t)} \wedge F_{\hat{A}(t)})$ where $\hat{A}(t)$ is the connection on the principal $U \times G$ on U given by $\hat{A}(t) = g^{-1}dg + t\, g^{-1}ag$. For example, $\hat{A}(0)$ is the product, flat connection and $\hat{A}(1) = g^{-1}dg + g^{-1}ag$. In any event, since $\mathrm{tr}(F_{\hat{A}} \wedge F_{\hat{A}})$ is a closed form, it follows that $\frac{\partial}{\partial t}\alpha_4 = d\alpha_3$. Integrate this identity with respect to $t \in [0, 1]$ to see that

$$\mathrm{tr}(F_{\hat{A}(1)} \wedge F_{\hat{A}(1)}) = \mathrm{tr}(F_{\hat{A}(0)} \wedge F_{\hat{A}(0)}) + d\left(\mathrm{tr}\left(a \wedge da + \frac{2}{3}\, a \wedge a \wedge a\right)\right).$$

This last equality gives the identity in Chapter 14.8 by virtue of the fact that $F_{\hat{A}(0)} = 0$ and $F_{\hat{A}(1)} = da + a \wedge a$.

The 3-form

$$-\frac{1}{8\pi^2}\mathrm{tr}\left(\hat{a} \wedge d\,\hat{a} + \frac{2}{3}\hat{a} \wedge \hat{a} \wedge \hat{a}\right)$$

is an example of *Chern–Simons* form.

Appendix 14.1 The ad-invariant functions on $\mathbb{M}(n; \mathbb{C})$

What follows first is a proof of the ad-invariant function theorem (Theorem 14.3) from Chapter 14.4. Recall that this theorem asserts that the real analytic, ad-invariant functions on $\mathbb{M}(n; \mathbb{C})$ are generated as an algebra by the functions from the set $\{m \to tr(m^k)\}_{k\geq 1}$. A few words are then said about the assertion made at the end of Chapter 14.5 to the effect that the algebra of characteristic classes of a given vector bundle are generated by its Chern classes.

Proof of the ad-invariant function theorem Note to start that any ad-invariant function on the vector space $\mathbb{M}(n; \mathbb{C})$ must be a symmetric function of the eigenvalues of the matrix. The reason being that an open dense set of matrices are diagonalizable over $\mathbb{C}$. (To be diagonalizable means that the matrix is conjugate to a diagonal matrix.) You can find a proof of this fact in almost any undergraduate book on linear algebra.

As there are n eigenvalues, and granted the observation of the preceding paragraph, it is enough to know that the set $\{\Sigma_{1\leq k\leq n} \lambda_j^q\}_{q=0,1,\ldots n}$ of functions of the coordinates $(\lambda_1, \ldots, \lambda_n)$ for $\mathbb{C}^n$ generates the algebra of symmetric, analytic functions on $\mathbb{C}^n$. To see that such is the case, view the coordinates $\lambda = (\lambda_1, \ldots, \lambda_n)$ as the roots of the polynomial $z \to h_\lambda(z) = \prod_k (z - \lambda_k)$. The coefficients of this n'th order polynomial are symmetric functions of the roots. This is to say that when $h_\lambda(z)$ is written out as

$$h_\lambda(z) = z^n - a_1 z^{n-1} + \cdots + (-1)^n a_n,$$

then each a_j is a symmetric function of the roots. For example, $a_1 = \Sigma_k \lambda_k$ and $a_n = \prod_k \lambda_k$. If two polynomials have the same coefficients, then they have the same roots, so this implies that the functions $\{a_1, \ldots, a_n\}$ of the roots generate the algebra of symmetric functions of the coordinates on $\mathbb{C}^n$. Granted this, consider now the function $u \to g(u)$ on $\mathbb{C}$ given by $g(u) = \prod_j (1 + \lambda_j u) = a_n u^n + \cdots + a_1 u + 1$. As is clear, the coefficients of $g(u)$ are just this set of symmetric functions $\{a_1, \ldots, a_n\}$ of the roots. It follows as a consequence that the coefficients of the power series expansion of

$$\ln(g(u)) = \Sigma_j \ln(1 + \lambda_j u)$$

provide a generating set for the algebra of symmetric functions of the coordinates of $\mathbb{C}^n$. This is because the k'th order term in the expansion of $g(u)$ consists of sums of product of the set $\{a_j : j < k\}$ plus a nonzero multiple of a_k. This said, note that the power series for $\ln(g(u))$ is

$$\ln(g(u)) = \Sigma_{m\geq 1} (-1)^{m+1}\, m^{-1} (\Sigma_{1\leq j\leq n} \lambda_j^m)\, u^m.$$

The preceding formula shows that the coefficients are proportional to the symmetric functions $\{\Sigma_j \lambda_j^m\}_{m\geq 0}$. This is what was claimed by the theorem.

The final remarks in this appendix concern the assertion from Chapter 14.5 that the algebra of characteristic classes of a given vector bundle are generated by its Chern classes. Given what was said above in the proof of the ad-invariant function theorem, this assertion is an immediate consequence of the following two observations: First, the coefficients of the polynomial $u \to c(u) = \Pi_j(1 - \frac{i}{2\pi}\lambda_j u)$ generate the algebra of symmetric functions on $\mathbb{C}^n$. Second, the coefficients of this polynomial are the very same functions (viewed as ad-invariant functions on $\mathbb{M}(n; \mathbb{C})$) that define the Chern classes.

Appendix 14.2 Integration on manifolds

This appendix reviews the basics of integration on compact manifolds. The following proposition summarizes the story.

Proposition 14.12 *Let* M *denote a compact, oriented,* n*-dimensional manifold. There is a well-defined notion of the integral of an* n*-form on* M*. An* n*-form that is not identically zero and defines the given orientation where it is nonzero has positive integral. On the other hand, the integral of the exterior derivative of an* (n−1)*-form is zero.*

An example involving spheres is given after the proof of this proposition.

What follows is an immediate corollary.

Corollary 14.13 *Let* M *denote a compact, oriented,* n*-dimensional manifold. Then the De Rham cohomology in dimension* n *has dimension at least* 1*. Indeed, any* n*-form whose integral is nonzero projects with nonzero image to* $H^n_{\text{De Rham}}(M)$.

Proof of Corollary 14.13 As there are no nonzero (n+1)-forms, the De Rham cohomology in dimension n is the vector space $C^\infty(M; \wedge^n T^*M)/\text{im}(d)$. This understood, the corollary follows from the proposition's assertion that the integration of the exterior derivative of any given (n−1)-form is zero.

As it turns out, an n-form can be written as the exterior derivative of an (n−1)-form *if and only if* its integral is zero. This implies that $H^n_{\text{De Rham}}(M)$ has dimension 1.

Proof of Proposition 14.12 The proof has four parts. The first part reviews integration over domains in $\mathbb{R}^n$. Part 2 states three important properties of the integral. Part 3 uses what is said in Part 1 to give a definition of the integral of an n-form that employs a cover of M by coordinate charts. Part 4 proves that the value of the integral as defined in Part 3 does not depend on the chosen coordinate chart cover. Part 5 proves that a nontrivial n-form that defines the orientation where it is nonzero has positive integral, and that the integral of the exterior derivative of an (n−1)-form is zero.

Part 1: This part gives a definition of the integral of a compactly supported function in $\mathbb{R}^n$. The integral here is the n-dimensional version of the 1-variable calculus definition via the "midpoint approximation" that you might remember from your second semester of calculus.

To start, suppose that f is a smooth function on $\mathbb{R}^n$ which is zero outside of some bounded set. What follows defines the integral of f. Because the set where f is nonzero is bounded, there exists $R > 0$ such that this set is contained in the n-dimensional cube where the coordinates $(x_1, \ldots, x_n)$ satisfy $-R \leq x_i \leq R$ for each $i \in \{1, \ldots, n\}$. Fix $N \geq 1$ and introduce Λ_N to denote the set of vectors of the form $k = (k_1, \ldots, k_n)$ where each $i \in \{1, \ldots, n\}$ version of k_i is drawn from the set $\{1, \ldots, N\}$. This set Λ_N has N^n elements. For each $k \in \Lambda_N$, introduce f_k to denote the value of f at the point whose coordinate entries obeys

$$x_i = R\left(\frac{2k_i - 1}{N} - 1\right) \text{ for each } i \in \{1, \ldots, n\}.$$

As explained next, the following limit exists:

$$(*) \qquad \lim_{N\to\infty} (2R)^n N^{-n} \textstyle\sum_{k\in\Lambda_N} f_k.$$

This limit is defined to be the integral of the function f. I denote the latter by

$$\int_{R^n} f d^n x.$$

To see that this limit exists, consider the difference between the N and $N + L$ version for any given integer $L > 1$. For $k \in \Lambda_k$, let $\square_{R,k} \subset \mathbb{R}^n$ denote the n-dimensional cube where the coordinates are such that each entry obeys

$$R\frac{2k_i - 2 - N}{N} \leq x_i \leq R\frac{2k_i - N}{N}.$$

The center of $\square_{R,k}$ is the point that defines f_k. The union of these cubes is $\times_n [-R, R)$, this the cube that contains the support of f. Meanwhile, pair-wise distinct cubes are disjoint. The replacement of N by $N + L$ when evaluating

$$(2R)^n N^{-n} \textstyle\sum_{k\in\Lambda_N} f_k$$

involves the evaluation of f at $(2L)^n$ points in any given version of $\square_{R,k}$. Given that f is differentiable, the value of f at any such point differs from f_k by no more than a $c_f RN^{-1}$, where c_f is the supremum as x ranges in $\times_n [-R, R]$ of $|df|$. This being the case, it follows that the $N + L$ version of the sum written above differs from the N version by at most $c_f RN^{-1}$. This last fact proves that the sequence of such sums, as indexed by N, is a Cauchy sequence and so has a unique limit. As noted, this limit is defined to be the integral of f.

Part 2: Three properties of the integral of f are used in what follows.

- Positivity: $\int_{\mathbb{R}^n} f d^n x > 0$ *if* f *is nonnegative and strictly greater than zero at some point.*
- The integral of a divergence is zero: $\int_{\mathbb{R}^n} f d^n x = 0$ *if* f *can be written as* $\Sigma_{1\le i\le n} \frac{\partial v_i}{\partial x_i}$ *where* $(v_1, \ldots, v_n)$ *are compactly supported functions on* $\mathbb{R}^n$.
- Change of variable formula: *Suppose that* $V \subset \mathbb{R}^n$ *is an open set and let* $\psi: V \to \mathbb{R}^n$ *denote an embedding whose image contains the set where* f *is nonzero. Then*

$$\int_{\mathbb{R}^n} (\psi^* f)|\det(\psi_*)| d^n x = \int_{\mathbb{R}^n} f \, d^n x.$$

The claim made by the first bullet is, I trust, self-evident. To see about the second bullet, it is enough to prove that the integral of each term in the sum is zero. I consider the case $\frac{\partial v_1}{\partial x_1}$. The arguments for the other terms differ only in notation. To start the argument, fix N large. Write a given vector $k \in \Lambda_N$ as a pair (k_1, k') with $k_1 \in \{1, \ldots, n\}$ and $k' \in \times_{n-1} \{1, \ldots, N\}$. Define $x = (x_i, x_2, \ldots, x_n)$ to be the center point of the n-dimensional cube $\square_{R,k}$, thus the point in $\mathbb{R}^n$ whose i'th coordinate is given by $x_i = R(2k_i - 1)/N$. Write $x = (x_1, x')$ with $x' = (x_1, \ldots, x_n)$. Now recall the definition of the partial derivative to write

$$\left(\frac{\partial v_1}{\partial x_1}\right)(x_1, x') = \frac{N}{2R}\left(v_1\left(\frac{2R}{N}k_1, x'\right) - v_1\left(\frac{2R}{N}\left(k_1 - 1\right), x'\right)\right) + \mathfrak{e}$$

where $|\mathfrak{e}|$ is bounded by an N-independent multiple of $\frac{1}{N}$. With the preceding understood, note next that the sum (∗) can be ordered so that one fixes k' and then sums over all values of k_1 before going to another value of k'. Keeping this in mind, it follows that the sum of the various k_1 versions of the expression above for $\frac{\partial v_1}{\partial x_1}$ cancel but for the error term $\mathfrak{e}$. For any given fixed k', there are at most N such error terms, so their sum is bounded by some k' and N independent number. As there are N^{n-1} distinct vectors k' to consider, it follows that the totality of these error terms contribute to (∗) at most some N-independent multiple of $R^n N^{-n}$ times this factor N^{n-1}; thus at most an N-independent multiple of $R^n N^{-1}$. The $N \to \infty$ limit of these error terms is therefore zero.

The change of variable can be seen by using (∗) with the following observation: Suppose that $x \in \mathbb{R}^n$ is in the domain of ψ. Fix $r > 0$ and let $\square$ denote a cube of side length r with center at x. Then the volume of $\psi(\square)$ differs from $|\det(\psi_*|_x)|$ by a term whose size is bounded by an x-dependent multiple of r^{n+1}. This is seen by writing a first order Taylor's expansion for ψ near x, thus writing $\psi(x + y) = \psi(x) + \psi_*|_x \cdot y + \mathfrak{r}_x(y)$ where $|\mathfrak{r}_x(y)|$ is bounded by an x-independent multiple of $|y|^2$. The claim about the volume of $\psi(\square)$ follows from the fact that a linear transformation multiplies volumes by the absolute value of its determinant.

Part 3: Fix a finite, open cover, $\mathfrak{U}$, of M such that each set $U \subset \mathfrak{U}$ comes with a coordinate chart map $\varphi_U: U \to \mathbb{R}^n$ that is a diffeomorphism onto an open subset. Let $\{\chi_U\}_{U \in \mathfrak{U}}$ denote a subordinate partition of unity. See Appendix 1.2 if you are hazy about such things. Let Ω denote a nowhere zero n-form that defines the given orientation for TM. If $U \in \mathfrak{U}$, write $\varphi_U^* \Omega$ as $\sigma_U\, dx_1 \wedge \cdots \wedge dx_n$ with $(x_1, \ldots, x_n)$ denoting the Euclidean coordinates for $\mathbb{R}^n$. Here, σ_U is a nowhere zero function on the image of φ_U. Let ε_U denote the sign of σ_U.

Let Ξ denote any given n-form. The integral of Ξ over X is denoted by $\int_M \Xi$ and it is defined as follows: If $U \in \mathfrak{U}$, write $\varphi_U^* \Xi$ as $z_U\, dx_1 \wedge \cdots \wedge dx_N$ with z_U a function on the image of φ_U. Granted this notation, the integral of Ξ is

$$(**) \qquad \Sigma_{U \in \mathfrak{U}}\, \varepsilon_U \int_{\mathbb{R}^n} ((\varphi_U^{-1})^* \chi_U) z_U d^n x.$$

Part 4: What follows explains why this definition associates to any given n-form a number that is independent of the chosen coordinate chart cover of M. Suppose that U' is a second coordinate chart cover. There is a version of the integral above for U', this defined by a partition of unity $\{\chi'_{U'}\}_{U' \in \mathfrak{U}'}$. I have to show that

$$\Sigma_{U \in \mathfrak{U}} \varepsilon_U \int_{\mathbb{R}^n} ((\varphi_U^{-1})^* \chi_U) z_U d^n x = \Sigma_{U' \in \mathfrak{U}'}\, \varepsilon_{U'} \int_{\mathbb{R}^n} ((\varphi_{U'}^{-1})^* \chi'_{U'}) z_{U'} d^n x.$$

Using the fact that $\Sigma_{U \in \mathfrak{U}}\, \chi_U = 1$ and $\Sigma_{U' \in \mathfrak{U}'}\, \chi'_{U'} = 1$, write the respective left- and right-hand expressions above as

$$\Sigma_{U \in \mathfrak{U}, U' \in \mathfrak{U}'}\, \varepsilon_U \int_{\mathbb{R}^n} ((\varphi_U^{-1})^* (\chi_U \chi'_{U'})) z_U d^n x \textit{ and}$$
$$\Sigma_{U \in \mathfrak{U},\, U' \in \mathfrak{U}'}\, \varepsilon_{U'} \int_{\mathbb{R}^n} ((\varphi_{U'}^{-1})^* (\chi_U \chi'_{U'})) z_{U'}\, d^n x.$$

This understood, it is sufficient to verify the following: Fix any given $(U, U') \in \mathfrak{U} \times \mathfrak{U}'$ with $U \cap U' \neq \emptyset$. Then the corresponding integrals,

$$\varepsilon_U \int_{\mathbb{R}^n} ((\varphi_U^{-1})^* (\chi_U \chi'_{U'})) z_U d^n x \textit{ and } \varepsilon_{U'} \int_{\mathbb{R}^n} ((\varphi_{U'}^{-1})^* (\chi_U \chi'_{U'})) z_{U'} d^n x,$$

agree. To see that this is the case, introduce the coordinate transition function $\psi = \varphi_U \circ (\varphi_{U'}^{-1})$. The latter is a diffeomorphism that maps the region in $\mathbb{R}^n$ where the right-hand integrand is nonzero to that where the left-hand integrand is nonzero. I use $\det(\psi_*)$ to denote the determinant of the differential of ψ, this a nowhere zero function on the domain of ψ. These are relevant by virtue of the fact that $z_{U'} = \det(\psi_*) \psi^* z_U$ and $\varepsilon_{U'} = \text{sign}(\det(\psi_*)) \varepsilon_U$. Meanwhile,

$(\varphi_{U'}^{-1})^*(\chi_U\chi'_{U'}) = \psi^*((\varphi_{U'}^{-1})^*(\chi_U\chi'_{U'}))$. This understood, let f denote the integrand on the left-hand side above. The integrand on the right-hand side above is the function $|\det(\psi_*)|\ \psi^*f$. Granted this last observation, then the desired equality between the integrals above follows from the *change of variables property* given in Part 2 above.

Part 5: Consider the integral of n-form that can be written as $\Xi = f\Omega$ where Ω is a nowhere zero n-form that defines the given orientation, and where f is a nonnegative function that is greater than zero at some point. To see that the integral of Ξ is nonzero, it is enough to note the following: Let U denote a given coordinate chart from the cover U that is used in Part 3 to compute the integral of Ξ, and let $\varphi_U: U \to \mathbb{R}^n$ denote the associated embedding. Then the function z_U that appears in U's contribution to the integral of Ξ is equal to $\varepsilon_U\ ((\varphi_U^{-1})^*f)$. This the case, then the claimed positivity follows from the *positivity property* stated in Part 2.

Suppose instead that $\Xi = d\alpha$ where α is an $(n-1)$-form. Let $\mathfrak{U}$ denote a given coordinate chart cover that is used in Part 3 to define the integral of Ξ. Write the $(n-1)$-form α as $\Sigma_{U\in\mathfrak{U}}\ d(\chi_U\alpha)$. This done, it follows that the integral of Ξ can be written as

$$\sum\nolimits_{U\in\mathfrak{U}} \int_U d(\chi_U\alpha)$$

because the form $d(\chi_U\alpha)$ is nonzero only in U. As such, its pull-back to $\mathbb{R}^n$ via φ_U^{-1} has support in a bounded set. Write this pull-back using the standard coordinate basis for the $(n-1)$-forms on $\mathbb{R}^n$ as

$$\alpha_U = v_1 dx^2 \wedge \cdots \wedge dx^n - v_2 dx^1 \wedge dx^3 \wedge \cdots \wedge dx^n + \cdots + (-1)^{n-1} v_n dx^1 \wedge \cdots \wedge dx^{n-1}.$$

Having done this, it follows that $d(\chi_U\alpha) = \Sigma_{1\le i\le n}\frac{\partial v_i}{\partial x_i}$. Granted the latter, then the vanishing of the integral of $d\alpha$ follows from the *divergence property* from Part 2.

The example of S^n:

The examples in Chapter 14.7 involve integrals of n-forms on the sphere S^n. What follows computes a few. To set the stage, view S^n as the set of points $x = (x_1, \ldots, x_{n+1}) \in \mathbb{R}^{n+1}$ with $|x| = 1$. The form

$$\Omega = x_1 dx_2 \wedge \cdots \wedge dx_n - x_2 dx_1 \wedge dx_3 \wedge \cdots \wedge dx_{n+1} + \cdots + (-1)^n x_{n+1} dx_1 \wedge \cdots \wedge dx_n$$

on $\mathbb{R}^n$ restricts to TS^n as a nowhere zero n-form. To see this, remark that the $x_{n+1} > 0$ part of S^n has a coordinate chart whose inverse is the map from interior of the ball in $\mathbb{R}^n$ that sends

$$U = (U_1, \ldots, U_n) \to \left(U_1, \ldots, U_n, (1 - |U|^2)^{1/2}\right).$$

The pull-back of Ω using this map is the n-form

$$\left(1 - |U|^2\right)^{-1/2} du_1 \wedge \cdots \wedge du_n.$$

Thus, Ω is nonzero where $x_{n+1} > 0$. A similar calculation finds it nonzero for $x_{n+1} > 0$. You can check points on the equator by repeating the previous observations with x_{n+1} replaced by the other coordinate functions.

In the case of S^2, the form Ω can be written using the spherical coordinates (θ, φ) as $\Omega = \sin\theta \, d\theta \wedge d\varphi$. You can either use the latter formula, or the representation above as $(1 - |u|^2)^{-1/2} du_1 \wedge du_2$ to see that its integral is 4π.

Appendix 14.3 The degree of a map

This appendix first proves the proposition in Chapter 14.7 that defines the degree of a map. The appendix then computes the degree for a map from a certain surface in $\mathbb{R}^3$ to S^2 of the sort that is described by Example 14.6.

What follows is Proposition 14.7 from Chapter 14.7.

Proposition 14.7 *Suppose that* X *is a compact, oriented, n-dimensional manifold and suppose that* M *is another such manifold*

- *Suppose that* $\psi: M \to X$ *is a smooth map. There exists an integer* p *with the following property: Let* ω *denote a n-form on* X. *Then* $\int_M \psi^*\omega = p \int_X \omega$. *This integer* p *is, by definition, the degree of* ψ.
- *Homotopic maps from* M *to* X *have the same degree.*

Proof of Proposition 14.7 The assertion in the second bullet is proved first. To start this proof, remark that maps ψ_0 and ψ_1 from M to X are homotopic if there exists a smooth map from $[0, 1] \times M$ to X that equals ψ_0 on some neighborhood of $\{0\} \times M$ and is equal to ψ_1 on some neighborhood of $\{1\} \times M$. Meanwhile, as ω is an n-form and X is n-dimensional, it follows that $d\omega = 0$. This being the case, Chapter 12.2 explains why $\psi_1^*\omega - \psi_0^*\omega$ is the exterior derivative of an $(n-1)$-form on M. It follows as a consequence that $\psi_1^*\omega$ and $\psi_0^*\omega$ have the same integral.

Turn now to the proof of the claim made by the first bullet. The proof has six steps.

Step 1: The proof benefits from a review of something said earlier. To start, suppose that Y is a smooth, oriented n-dimensional Riemannian manifold. Let $y \in Y$ denote a given point. Chapter 9.3 describes a certain sort of coordinate chart map that is defined on a small radius, metric ball centered on y. These are the Gaussian coordinates. Let U denote a ball on which these coordinates are defined, and let $\varphi: U \to \mathbb{R}^n$ denote the corresponding coordinate chart map. As explained by the proposition in Chapter 9.3, the metric on Y pulls back via φ^{-1} to a Riemannian metric on the image of φ in $\mathbb{R}^n$ that is given by a symmetric, positive definite $n \times n$ matrix valued function whose (i, j) entry has the form $g_{ij} = \delta_{ij} + K_{ij}(x)$ where $|K_{ij}| \le c_0 |x|^2$ with c_0 denoting a constant. Here, δ_{ij} is 1 if $i = j$ and 0 otherwise.

The metric also determines a preferred section of $\wedge^n T^*Y$, this I denote by Ω_Y. It is the unique, unit normed section that defines the given orientation. If $\{e^1, \ldots, e^n\}$ is an oriented, orthonormal frame for T^*M at a given point, then $\Omega_Y = e^1 \wedge \cdots \wedge e^n$. It follows from this that Ω_Y pulls back via the inverse of the Gaussian coordinate chart map as an n-form that can be written as

$$(\varphi^{-1})^*\Omega_Y = (1 + \kappa)dx^1 \wedge \cdots \wedge dx^n$$

where $|\kappa| \le c_0 |x|^2$. Here, c_0 is another constant.

If $O \subset Y$ is an open set, then the *volume* of O is, by definition, the integral of Ω_Y over O. If O is a ball of radius r centered on y, then it follows from the depiction just given of Ω_Y that the volume of O can be written as $\frac{1}{n+1}\mu_n(1 + \mathfrak{z}(r))\, r^n$ where μ_n depends only on n, and where $|\mathfrak{z}| \le c_0 r^2$. Here again, c_0 is independent of r.

Step 2: It proves useful to invoke a special case of Sard's theorem (Theorem 1.3; Chapter 1 states the full version). Fix a Riemannian metric on X to measure distances and norms. Use Ω_X to denote the unique n-form on X with norm 1 that defines the orientation. Let $C_X \subset X$ denote the set of points with the following property: If $x \in C_X$, then $\psi^{-1}(x)$ contains a point where the determinant of ψ's differential is zero. A point in C_X is said to be a *critical value* of ψ.

Theorem 14.14 (the n-to-n Sard's theorem) *Given* $\varepsilon > 0$, *there exists an open set* $V_\varepsilon \subset X$ *that contains* C_X *and is such that* $\int_{V_\varepsilon} \Omega_X$ *is less than* ε.

Proof of Theorem 14.14 Fix also a Riemannian metric on M to measure distances and norms. Let p denote a given point in M and let $U \subset M$ denote a Gaussian coordinate chart centered at p. Introduce $\varphi: U \to \mathbb{R}^n$ to denote a corresponding Gaussian coordinate chart map that send p to the origin also. Let $x \in \psi(p)$ and let $U' \subset X$ denote a Gaussian coordinate chart centered at x. Introduce φ' to denote a Gaussian coordinate chart embedding from U' onto an open set in $\mathbb{R}^n$ that sends x to the origin. Use $\psi_{U',U}$ to denote the composition of $\varphi' \circ \psi \circ \varphi^{-1}$, this mapping some open neighborhood of $0 \in \mathbb{R}^n$ into another neighborhood of 0 in $\mathbb{R}^n$. Fix $r > 0$ so that the ball of radius r centered at the origin in $\mathbb{R}^n$ is contained in the domain of $\psi_{U',U}$. As explained momentarily, the image of this ball via $\psi_{U,U'}$ is contained in an open set, O_r, whose volume obeys

$$(*) \qquad \text{volume} = (O_r) \leq c(\lambda_p + r)r^n$$

where λ_p is the absolute value of the smallest eigenvalue of the differential of $\psi_{U,U}$ at the origin in $\mathbb{R}^n$. Meanwhile, c denotes a number that is independent of r and $p \in M$. Accept this claim for the moment to complete the proof of the n-to-n Sard's theorem.

To complete the proof, choose $r > 0$ but small enough so that the following is true: Fix any point $p \in M$. A ball of radius r centered at p is contained in a Gaussian coordinate chart centered at p, and the image via ψ_r of this ball is contained in a Gaussian coordinate chart centered at $\psi(p)$. Note that this is the case for all sufficiently small r by virtue of the fact that M is compact and ψ is continuous. Now, let C_M denote the set of critical points of ψ. These are the points where ψ's differential is not an isomorphism. Let $\vartheta_r \subset C_M$ denote a maximal set of points in C_M with the property that distinct points in ϑ_r have distance at least $\frac{1}{4}r$ apart. Let N_r denote the number of elements in ϑ_r. This set has at most $c_1 r^{-n}$ members as can be seen by the following argument: The set of balls in M with centers on the points in ϑ_r and radius $\frac{1}{16}r$ are disjoint. Let Ω_M denote the n-form with norm 1 that defines the orientation. It follows from what is said in Step 1 that the volume of each such ball is no less than $c_1^{-1}r^n$ where $c_1 \geq 1$ is a constant that is independent of the point in question or r. Granted this last fact, it follows that volume of the union of the radius $\frac{1}{4}r$ balls centered at the points of ϑ_r is at least $c_1^{-1}\, r^n\, N_r$. This must be less than the volume of M. Thus, $N_r \leq c_1 r^{-n}$. Granted the latter bound, remark next that the union of the balls of radius r centered on the points in ϑ_r covers C_M. Thus, the image of this union via ψ covers C_X. It follows from $(*)$ that the image of these balls is contained in an open set on which the integral of Ω_X is bounded by $c\, r^{n+1}N_r$, and thus by $cc_1 r$.

Turn now to the proof of $(*)$. View the differential of $\psi_{U'U}$ at the origin as a linear transformation from the vector space $\mathbb{R}^n$ to itself. Let A denote this linear transformation. I can find orthonormal bases $\{e^1, \ldots, e^n\}$ and $\{e'^1, \ldots, e'^n\}$ for $\mathbb{R}^n$ so that A acts as an upper triangular matrix with (n, n) entry equal to the absolute value of A's smallest eigenvalue. Thus $Ae^k = \Sigma_{j \geq k} A^{kj} e'^j$ and $A_{n,n} = \lambda_p$ with λ_p as in $(*)$. This has the following implication: Fix $r > 0$ and let $\square_r$ denote the n-dimensional cube where each coordinate has absolute value less than ρ. Then A maps $\square_r$ in to an n-dimensional rectangle where the coordinates $(x_1, \ldots, x_n)$ are constrained as follows: If $k < n$, then $|x_k|$ is no greater $(n + 1 - k)|A|\, r$ and $|x_n| \leq \lambda_p r$. As a consequence, the volume of $A(\square_r)$ is less than the volume of this cube, hence less than $c_0\, \lambda_p\, r^n$ with c_0 independent of r and the point in p.

Granted the preceding, use Taylor's theorem with remainder to write $\psi_{U',U}$ near the origin as the map $x \to Ax + \mathfrak{e}$ where $|\mathfrak{e}| \leq c_0\, r$ with c_0 independent of r and p. Given what was just said about rectangles, this implies that $\psi_{U',U}$ maps the ball of radius r centered at the origin in $\mathbb{R}^n$ into the rectangle where the coordinates $(x_1, \ldots, x_k)$ are constrained so that $|x_k| \leq (n + 1 - k)(|A| + c_1 r)\, r$ for $k < n$ and $|x_n| < c_1\,(\lambda_p + r)\, r$. Here, c_1 is independent of r and of p. This implies what is asserted in (*).

Step 3: This step asserts and then proves an analog of the n-to-n Sard's theorem, this a bound for the integral of $\psi^*\Omega_X$ over the set $\psi^{-1}(V_\varepsilon)$.

Lemma 14.15 *Given* $\delta > 0$, *there exists* ε *such that* $\int_{\psi^{-1}(V_\varepsilon)} \psi^*\Omega_X < \delta$.

Proof of Lemma 14.15 Let $x \in X$ and let $U' \subset X$ denote a Gaussian coordinate chart centered at p. Introduce φ' to denote the Gaussian coordinate chart embedding from U onto an open set in $\mathbb{R}^n$. Write $((\varphi')^{-1})^*\Omega_X$ as $z\, dx_1 \wedge \cdots \wedge dx_n$ with $\mathfrak{z}$ a strictly positive function on the image of φ. Suppose $p \in \psi^{-1}(x)$. Fix a corresponding Gaussian coordinate chart $U \subset M$ containing p, and let $\varphi\colon U \to \mathbb{R}^n$ denote the associated embedding. Use $\psi_{U',U}$ to denote the composition of $\varphi' \circ \psi \circ \varphi^{-1}\colon \varphi(U' \cap \psi^{-1}(U')) \to \varphi'(U')$. It follows from what is said in Chapter 5.5 that $(\varphi^{-1})^*\psi^*\Omega_X$ has the form

$$\det(\psi_{U',U*})\psi_{U,U'}{}^*(z)dx^1 \wedge \cdots \wedge dx^n$$

where $(\psi_{U',U})_*$ denotes the differential of $\psi_{U,U'}$ and $\det(\cdot)$ denotes the determinant of the indicated $n \times n$ matrix. Given $r > 0$, let W_r denote the set of points in M with distance less than r from C_M. Since C_M is the set of points where ψ_* is not an isomorphism, the determinant of $(\psi_{U',U})_*$ vanishes on the φ image of C_M. This and the preceeding formula for $(\varphi^{-1})^*\psi^*\Omega_X$ imply that there is an r and p-independent constant, c, such that $|\psi^*\Omega| < cr$ at all points W_r. This last observation has the following consequence: Given $\delta > 0$, there exists r_δ such that if $r < r_\delta$, then

$$\left| \int_{W_r} \psi^*\Omega_X \right| \leq \frac{1}{2}\delta.$$

Keeping the latter inequality in mind, note next that each point in $M - W_r$ has a neighborhood that is embedded by ψ. This follows from the inverse function theorem (Theorem 1.1) since ψ's differential is invertible at each such point. Since W_r is open, so $M - W_r$ is closed and compact. This being the case, there exists N_r such that $M - W_{r/2}$ contains a collection of at most N_r Gaussian coordinate chart maps that cover $M - W_r$ and that are mapped in a 1–1 fashion by ψ as an embedding onto their image. Let $\mathfrak{U}_r$ denote such a collection of sets. The formula for $(\varphi^{-1})^*\psi^*\Omega_X$ tells us the following: If $U \subset \mathfrak{U}$, then

$$\int_{U} \psi^{*}(\Omega_X) = \varepsilon_U \int_{\psi(U)} \Omega_X$$

where $\varepsilon_U \in \{\pm 1\}$ is the sign of the determinant of the differential of $\psi_{U',U}$. This last equality has the following consequence: Fix $\varepsilon > 0$ so as to define V_ε using the n-to-n Sard's theorem. Then

$$\left| \int_{\psi^{-1}(V_\varepsilon) \cap (M - W_R)} \psi^{*}(\Omega_X) \right| \leq N_r \varepsilon.$$

With the preceding understood, take $\varepsilon < \frac{1}{2} N_r^{-1} \delta$ to deduce what is asserted by the lemma.

Step 4: Now let ω denote a given n-form on X. What is said in Steps 2 and 3 lead to the following: Given $\delta > 0$, there exists $\varepsilon > 0$ such that

- $\left| \int_X \omega - \int_{X - V_\varepsilon} \omega \right| \leq \delta,$
- $\left| \int_M \psi^{*} \omega - \int_{M - \psi^{-1}(V_\varepsilon)} \psi^{*} \omega \right| \leq \delta.$

Indeed, this follows by writing ω as $f\,\Omega_X$ with $f\colon X \to \mathbb{R}$ a smooth function.

To proceed from here, use the fact that $X - V_\varepsilon$ is compact to find a finite collection $\mathfrak{U}'_\varepsilon$ of Gaussian coordinate charts centered at points in $X - V_{\varepsilon/2}$ with the following properties: First, the collection of charts covers $X - V_\varepsilon$. Second, if $U' \subset \mathfrak{U}'_\varepsilon$ is any given chart, then $\psi^{-1}(U')$ consists of a finite, disjoint collection, $\mathfrak{U}_{U'}$, of open sets that are each contained in a Gaussian coordinate chart which is mapped by ψ as an embedding onto its image in X. Fix a subordinate partition of unity for the open cover of X consisting of the sets in $\mathfrak{U}'_\varepsilon$ and the extra set V_ε. If $U' \in \mathfrak{U}'_\varepsilon$, I use $\chi_{U'}$ to denote the corresponding partition function. Note in this regard that $\sum_{U \in U'_\varepsilon} \chi_{U'}$ is equal to 1 on $X - V_\varepsilon$.

With $\mathfrak{U}'_\varepsilon$ as just described, it follows now that

$$\int_{M = \psi^{-1}(V_\varepsilon)} \psi^{*} \omega = \sum_{U' \in \mathfrak{U}'_\varepsilon} \sum_{U \in \mathfrak{U}_{U'}} \int_{U} \psi^{*} \omega.$$

Write ω again as $f\ \Omega_X$ and invoke the formula for $(\varphi^{-1})^{*} \psi^{*} \Omega_X$ in Step 3 to write the preceding formula as

$$\int_{M = \psi^{-1}(V_\varepsilon)} \psi^{*} \omega = \sum_{U' \in \mathfrak{U}'_\varepsilon} \left(\sum_{U \in \mathfrak{U}_{U'}} \varepsilon_U \int_{U'} \chi_{U'} \omega_X \right).$$

Granted this last equality, and the fact that δ can be taken as small as desired, the proposition follows if the sum $\Sigma_{U \in \mathfrak{U}_{U'}}$ does not depend on the chosen set $U' \in \mathfrak{U}'_\varepsilon$.

Step 5: To see about the sum in question, fix a point $x \in X-V_\varepsilon$ and a Gaussian coordinate chart, U′, centered at x. Let φ′ denote the corresponding Gaussian coordinate chart map. Fix $r > 0$ but much smaller than the distance between x and V_ε, and let $B' \subset X$ denote the ball of radius r, centered at x. The map φ′ identifies B′ with the ball of radius r about the origin in $\mathbb{R}^n$. Now, fix a smooth function σ on $\mathbb{R}^n$ that equals 1 on the ball of radius $\frac{1}{2}r$ and equals zero on the complement of the ball of radius $\frac{3}{4}r$. Use this function to map B′ to S^n as follows: This map sends $x \in B'$ to the point on the unit sphere in $\mathbb{R}^{n+1}$ with coordinate

$$\left(\sigma^2|x|^2 + (1-2\sigma)^2\right)^{-1/2}(\sigma x, 1-2\sigma) \in \mathbb{R}^n \times \mathbb{R}.$$

Denote this map by $\mathfrak{f}_x$. Note that there is a small ball centered on the north pole in S^n whose inverse image via $\mathfrak{f}_x$ is a small ball centered on 0 in $\mathbb{R}^n$. Use $D \subset S^n$ to denote this ball about the north pole in S^n. Note also that a neighborhood of the boundary of the ball of radius r is mapped by $\mathfrak{f}_x$ to the south pole in S^n.

The map $\mathfrak{f}_x$ can be used to construct a map, $\mathfrak{F}_x$, from X to S^n. The map $\mathfrak{F}_x$ sends the complement of B'_x to the south pole, and it sends a point $p \in B'_x$ to $\mathfrak{f}_x(\varphi'(p))$.

Let Ω_{S^n} denote an n-form on S^n that gives the orientation that is consistent with $\mathfrak{f}_x$, and has norm 1 as measured with the round metric. Fix next a nonnegative function, u, on S^n which is nonzero on the complement of D, and is such that the integral of $u\,\Omega_{S^n}$ is equal to 1. Choose u so as to be invariant under rotations of S^n with axis the north pole. The change of variable formula for integration tells us that the integral over X of the n-form $\mathfrak{F}_x^*(u\,\Omega_{S^n})$ is equal to 1. This understood, it follows from what is said in Step 4 that the integral over M of $\psi^*(\mathfrak{F}_x^*(u\,\Omega_{S^n}))$ is equal to $\Sigma_{U\in\mathfrak{U}_{U'}}\varepsilon_U$.

Step 6: Recall that the *second* bullet of Proposition 14.7 as stated at the start of this appendix has already been proved. Granted this, and granted the conclusions of Step 5, the assertion that the sum $\Sigma_{U\in\mathfrak{U}_{U'}}\varepsilon_U$ is independent of U′ follows if it can be shown that the maps $\mathfrak{F}_x \circ \psi$ and $\mathfrak{F}_{x'} \circ \psi$ are homotopic in the case that x and x′ are any two points in X. In particular, this will be the case if $\mathfrak{F}_x$ and $\mathfrak{F}_{x'}$ are homotopic. Note, by the way, that the question of a homotopy between $\mathfrak{F}_x$ and $\mathfrak{F}_{x'}$ makes no reference to either M or the map ψ. In any event, the fact that the maps $\mathfrak{F}_x$ and $\mathfrak{F}_{x'}$ are homotopic follows from what is said by Proposition 8.4 about geodesics. To see how this comes about, choose a smooth path, say a geodesic, from x to x′. Let γ denote this geodesic. Fix an oriented, orthonormal frame at x, and then parallel transport the frame vectors using the Levi-Civita connection to obtain a trivialization of the orthonormal frame bundle of TX along γ. Compose this frame with a smoothly varying, t-dependent rotation if needed so that the resulting 1-parameter family of frames gives the respective frames at x and x′ that are used to define the Gaussian coordinate charts and thus the maps $\mathfrak{F}_x$ and $\mathfrak{F}_{x'}$. Let $\{e_1|_{\gamma(t)}, \ldots, e_n|_{\gamma(t)}\}$ denote the frame at the point $\gamma(t)$. Use this frame for γ^*T^*X to define a map, ϑ, from $\mathbb{R} \times \mathbb{R}^n$ into M as follows: The map sends a given point $t \in R$ and $x \in \mathbb{R}^n$ to the point $\exp_{\gamma(t)}(\Sigma_{1\le i\le n} x_i e_i|_{\gamma(t)})$. Proposition 8.4 tells us that this map ϑ is smooth.

Given that X is compact, there exists $r_X > 0$ with the following significance: Let $B' \subset \mathbb{R}^n$ denote the ball of radius r_M. Then the restriction of ϑ to $\mathbb{R} \times B'$ embeds each $t \in R$ version of $\{t\} \times B'$ so as to give the inverse map of a Gaussian coordinate chart map centered at $\gamma(t)$. Let $\{\varphi_t\}_{t\in\mathbb{R}}$ denote the corresponding, 1-parameter family of Gaussian coordinate chart maps. Then these constitute a smoothly varying family of maps, and this understood, it follows that the family $\{\mathfrak{F}_{\gamma(t)}\}_{t\in\mathbb{R}}$ varies smoothly with t also.

With the proposition stated at the outset of this appendix in hand, consider by way of a sample calculation, the proof of the following claim from Example 14.6: Fix $g \ge 0$. There is a surface Σ in $\mathbb{R}^3$ of genus g with the property that the association to any given point of the corresponding outward pointing normal vector defines a degree $1 - g$ map from the surface to S^2.

The surface I have in mind is the one that is described in Chapter 1.3 as the level set of a function $f: \mathbb{R}^3 \to \mathbb{R}$ that has the form $f(x, y, z) = z^2 - h(x, y)$ with h chosen to have the following properties:

- $h(x, y) \le 0$ *where* $|x|^2 + |y|^2 \ge 1$.
- *For each* $k = \{1, \ldots, g\}$, *require* $h(x, y) \le 0$ *where* $|x - \frac{k}{2g}|^2 + |y|^2 \le r^2$.
- $h(x, y) > 0$ *otherwise.*
- $dh = \frac{\partial}{\partial x}h dx + \frac{\partial}{\partial y}h dy \ne 0$ *at any point where* $h = 0$, *thus on the circle of radius 1 about the origin, and on the* g *circles of radius* r *about the respective points in the set* $\{(x = \frac{k}{2g},\ y = 0)\}_{k=1,\ldots,g}$.

Here, r is positive, but less than $\frac{1}{100g}$. I now add $g + 1$ extra conditions on the function h:

- $\frac{\partial^2 h}{\partial y^2} < 0$ *at* $(1, 0, 0)$.
- $\frac{\partial^2 h}{\partial y^2} < 0$ *at* $\{(\frac{k}{2g} - r,\ 0,\ 0)\}_{1\le k\le g}$.

These points comprise half of the set of points on the x-axis where the value of h makes the transition from negative to positive.

To say more about the significance of these points, note that the image of a point $(x, y, z) \in \Sigma$ via the map n is the unit vector

$$n = \left(-\frac{\partial h}{\partial x}, -\frac{\partial h}{\partial y}, 2z\right)\left(\left|\frac{\partial h}{\partial x}\right|^2 + \left|\frac{\partial h}{\partial y}\right|^2 + 4|z|^2\right)^{-1/2}.$$

It is a direct consequence of this depiction of the map n that the point $(1, 0, 0) \in S^2$ has precisely $g+1$ inverse images via n in Σ, these being the points given above, thus $(1, 0, 0)$ and the points in the set $\{(\frac{k}{2g} - r, 0, 0)\}_{1\le k\le g}$. The condition on the second derivatives implies that the differential of the map n is an isomorphism at each of these points. However, the sign of its determinant is positive at the point (1, 0, 0) and negative at the other points. In any event, the fact that the differential is an isomorphism has, via Chapter 1.1's inverse function theorem, the following implication: There is a small radius disk centered at the point (1, 0, 0) in S^2 whose inverse image via n has g + 1 disjoint, inverse images, each mapped diffeomorphically by n onto the original disk. One of these g + 1 inverse images contains (1, 0, 0) and the rest are distinguished as follows: Each point from the set $\{(\frac{k}{2g} - r, 0, 0)\}_{1 \le k \le g}$ is contained in one and only one of these inverse images. I use $D \subset S^2$ to denote this disk.

Now, let Ω denote the standard area form on S^2, this the constant section of $\wedge^2 S^2$ that gives the orientation and has norm 1 as defined by the round metric. Fix a smooth function u: $S^2 \to [0, 1]$ that is nonzero only in D and is such that u Ω has integral equal to 1. Given that n maps each component of $n^{-1}(D)$ diffeomorphically onto D, and given what was said in the preceding paragraph about the sign of the determinant of the differential of n at (1, 0, 0) and the points that comprise the set $\{(\frac{k}{2g} - r, 0, 0)\}_{1 \le k \le g}$, it follows from the change of variable formula for the integral as given in Appendix 14.1 that the integral of $n^*(u\,\Omega)$ over Σ is 1 − g. Since the integral of u Ω is 1, it follows by definition that n has degree 1 − g.

Additional reading

- *Characteristic Classes*, John Milnor, Princeton University Press, 1974.
- *From Calculus to Cohomology: De Rham Cohomology and Characteristic Classes*, I. H. Madsen and Tornehave Jxrgen, Cambridge University Press, 1977.
- *Curvature and Characteristic Classes*, J. L. Dupont, Springer, 1978.
- *The Geometry of Differential Forms*, Shigeyuki Morita, Teruko Nagase and Katsumi Nomizu, American Mathematical Society, 2001.
- *Differential Forms in Algebraic Topology*, Raoul Bott and Loring W. Tu, Springer, 1982.
- *Differential Topology*, Victor Guillemin and Alan Pollack, Prentice-Hall, 1974.
- *An Introduction to Manifolds*, Loring W. Tu, Springer, 2007.
- *Topology from the Differentiable Viewpoint*, John W. Milnor, Princeton University Press, 1997.

15 Covariant derivatives and metrics

A very large part of differential geometry exploits various relations between certain sorts of covariant derivatives and metrics, especially metrics on the tangent bundle. This chapter introduces and discusses some of these fundamental relations.

15.1 Metric compatible covariant derivatives

Suppose that $\pi\colon E \to M$ is a vector bundle with fiber $\mathbb{R}^n$. Assume that E also has a fiber metric. By definition, the metric is a positive definite, symmetric section of the bundle $\mathrm{Hom}(E \otimes E; \mathbb{R})$. The inner product of vectors $v, u \in E|_x$ is denoted for the time being by $\langle v, u\rangle$; and the norm of $v \in E$ is denoted $|v|$. Suppose that $K \to M$ is some other vector bundle on M. The metric induces a vector bundle homomorphism from the bundle $(E\otimes K)\otimes E$ to K; thus a section of $\mathrm{Hom}((E\otimes K) \otimes E; K)$. This homomorphism is defined first on the decomposable elements in $(E\otimes K)\otimes E$ and it is then extended by linearity to the whole of $(E \otimes K) \otimes E$. The action of the homomorphism on a decomposable element $(v\otimes \mathfrak{k})\otimes u$ is given by $\langle v, u\rangle\, \mathfrak{k}$. There is a corresponding section of $\mathrm{Hom}(E \otimes (E \otimes K); K)$, this sending $v \otimes (u \otimes \mathfrak{k})$ to $\langle v, u\rangle\, \mathfrak{k}$. These metric induced elements in $\mathrm{Hom}((E \otimes K) \otimes E; K)$ and $\mathrm{Hom}(E \otimes (E \otimes K); K)$ are both also denoted for now by $\langle\ ,\ \rangle$.

Let ∇ now denote a covariant derivative on sections of E. The covariant derivative ∇ is said to be *metric compatible* when the following is true: Let $\mathfrak{s}, \mathfrak{s}'$ denote any given pair of sections of E. Then $\langle \mathfrak{s}, \mathfrak{s}'\rangle$ is a function on M, and its exterior derivative obeys

$$d\langle \mathfrak{s}, \mathfrak{s}'\rangle = \langle \nabla\mathfrak{s}, \mathfrak{s}'\rangle + \langle \mathfrak{s}, \nabla\mathfrak{s}'\rangle.$$

Note that taking $\mathfrak{s} = \mathfrak{s}'$ implies that

$$d|\mathfrak{s}|^2 = 2\langle \mathfrak{s}, \nabla \mathfrak{s}\rangle.$$

To see what this means in a less abstract setting, suppose that $U \subset M$ is an open set and that $\{e^b\}_{1\leq b\leq n}$ is a basis of sections for E over U. If ∇ is any given covariant derivative, then its action on a basis vector e^b can be written as

$$\nabla e^b = \sum_{1\leq a\leq n} e^a \otimes \mathfrak{a}^{ab}$$

where each element from the set $\{\mathfrak{a}^{ab}\}_{1\leq a,b\leq n}$ is a 1-form on U. This the case, the $n \times n$ matrix of 1-forms $\mathfrak{a}$ with the components $\{\mathfrak{a}^{ab}\}_{1\leq a,b\leq n}$ defines the covariant derivative on U. If the basis is orthonormal and if ∇ is metric compatible, then $\mathfrak{a}^{ab} = -\mathfrak{a}^{ba}$, which is to say that $\mathfrak{a}$ is a $\mathfrak{lie}(SO(n))$-valued matrix of 1-forms.

What follows is another, equivalent way to state the metric compatibility condition. To start, suppose that $K \to M$ is any vector bundle and ∇ is a covariant derivative on K. A section, $\mathfrak{s}$, of K is said to be *covariantly constant* if $\nabla\mathfrak{s} = 0$. A covariant derivative on E is metric compatible if and only if the metric, when viewed as a section of $K = \mathrm{Hom}(E \otimes E; \mathbb{R}) = E^*\otimes E^*$, is covariantly constant with respect to covariant derivative on sections of $E^* \otimes E^*$ that is induced by ∇.

By way of reminder, if ∇ is any given covariant derivative for sections of E, then it induces one for sections of E^*, and for sections of all multiple tensor powers and direct sum bundles whose factors are E and/or E^*. This induced derivative can be defined in two equivalent ways. The first is to view ∇ as coming from a connection on the principal bundle $P_{Gl(E)}$ of frames for E, and then using the fact that E^* and multiple tensor products whose factors are E and/or E^* are associated vector bundles to $P_{Gl(E)}$ using representations of the general linear group. This point of view is described in Chapter 11.5.

An equivalent definition first defines ∇ on sections of E^* as follows: Let $\mathfrak{w}$ denote any given section. If $\mathfrak{s}$ is a section of E, then the evaluation of $\mathfrak{w}$ on $\mathfrak{s}$ is a function on M, this denoted by $\mathfrak{w}\cdot\mathfrak{s}$. This understood, then $\nabla\mathfrak{w}$ is defined by the requirement that

$$d(\mathfrak{w}\cdot\mathfrak{s}) = (\nabla\mathfrak{w})\cdot\mathfrak{s} + \mathfrak{w}\cdot(\nabla\mathfrak{s})$$

where $\nabla\mathfrak{w}$, a section of $E^* \otimes T^*M$, is viewed in the left-most term on the right side above as a section of $\mathrm{Hom}(E; T^*M)$ and $\mathfrak{w}$ in the right-most term is viewed as a section of $\mathrm{Hom}(E \otimes T^*M; T^*M)$. Granted the latter definition of ∇ on E^*, the induced covariant derivative is defined on sections of multiple tensor powers by defining it first on decomposable sections, thus those of the form $\mathfrak{u}_1 \otimes \mathfrak{u}_2 \otimes \cdots$; and then extending this definition by linearity. The definition of ∇ on a decomposable element is such that

$$\nabla(\mathfrak{u}_1 \otimes \mathfrak{u}_2 \otimes \cdots) = \nabla\mathfrak{u}_1 \otimes \mathfrak{u}_2 \otimes \cdots + \mathfrak{u}_1 \otimes \nabla\mathfrak{u}_2 \otimes \cdots + \text{etc.}$$

Metric compatible covariant derivatives exist in all cases. In fact, any given such covariant derivative corresponds to a connection on the bundle $P_{O(E)}$ of orthonormal frames in E. Conversely, any connection on $P_{O(E)}$ defines a metric compatible covariant derivative. To see why this is, go to an open set $U \subset M$ where there is an isomorphism φ_U: $E|_U \to U \times \mathbb{R}^n$. As explained in Chapter 7.1, such a set U can be found with an isomorphism φ_U that gives a fiberwise an isometry with respect to the metric on E and the Euclidean metric on $\mathbb{R}^n$. Assume in what follows that such is the case here for φ_U. Let $\mathfrak{s}$ denote a section of E. Then $\varphi_U(\mathfrak{s})$ appears as the section $x \to (x, \mathfrak{s}_U(x))$ of $U \times \mathbb{R}^n$ where $\mathfrak{s}_U$ maps U to $\mathbb{R}^n$. Meanwhile, $\varphi_U(\nabla\mathfrak{s})$ appears as the section $x \to (x\,(\nabla\mathfrak{s})_U)$ of the bundle $(U \times \mathbb{R}^n) \otimes T^*M|_U$ where $(\nabla\mathfrak{s}_U)$ can be written as $d\mathfrak{s}_U + \mathfrak{a}_U\mathfrak{s}_U$ with $\mathfrak{a}_U$ a $Gl(n;\mathbb{R})$-valued 1-form on U. Since φ_U is an isometry, the inner product of sections $\mathfrak{s}$ and $\mathfrak{s}'$ on U gives the function $\mathfrak{s}_U^T\mathfrak{s}_U'$, and its exterior derivative is therefore

$$d\mathfrak{s}_U^T\mathfrak{s}_U' + \mathfrak{s}_U^T d\mathfrak{s}_U'.$$

This can be written as

$$(\nabla\mathfrak{s}_U)^T\mathfrak{s}_U' + \mathfrak{s}_U^T(\nabla\mathfrak{s}')_U - \mathfrak{s}_U^T(\mathfrak{a}_U^T + \mathfrak{a}_U)\mathfrak{s}_U'.$$

This implies that ∇ is metric compatible if and only if $\mathfrak{a}_U = -\mathfrak{a}_U^T$. This is to say that $\mathfrak{a}_U$ is a $\mathfrak{lie}(SO(n))$-valued 1-form. If such is the case, then ∇ comes from a connection on $P_{O(E)}$. What is essentially the same calculation shows that a connection on $P_{O(E)}$ supplies E with a metric compatible covariant derivative.

Now suppose that ∇ is a metric compatible connection, and let F_∇ denote its curvature 2-form, this a section of the associated bundle $(P_{O(E)} \times_{ad} \mathfrak{lie}(SO(n))) \otimes \wedge^2 T^*M$. Said in colloquial terms, F_∇ is a 2-form on M with values in the skew-symmetric endomorphisms of E. Note in this regard that the metric defines what is meant by skew-symmetric: An endomorphism $\mathfrak{m}$ of E is skew-symmetric if and only if $\langle\,\cdot\,, \mathfrak{m}(\cdot)\rangle = -\langle\mathfrak{m}(\cdot), \cdot\,\rangle$.

What follows is one additional observation in the case that $E = T^*M$ with M oriented. A metric on TM induces one on T^*M (and vice versa). Thus, in either case, M has a canonical, nowhere zero section of the real line bundle $\wedge^n T^*M$, this the section with length 1 (as defined by the metric) and giving the chosen orientation. Here $n = \dim(M)$. This section is denoted here by μ_M. To be more explicit, suppose that $U \subset M$ is an open set and that $\{e^1, \ldots, e^n\}$ is an orthonormal frame for T^*M over U. Then this canonical section of μ_M of $\wedge^{\dim(M)} T^*M$ is $\mu_M = e^1 \wedge \cdots \wedge e^n$. If ∇ is a metric compatible connection on T^*M, then this section is covariantly constant; thus $\nabla\mu_M = 0$. To see this, write $\nabla e^b = \sum_{1\le a\le n} e^a\, \mathfrak{a}^{ab}$ as before. Then

$$\nabla(e^1 \wedge \cdots \wedge e^n) = (e^a \wedge e^2 \wedge \cdots \wedge e^n) \otimes \mathfrak{a}^{a1} + (e^1 \wedge e^a \wedge \cdots \wedge e^n) \otimes \mathfrak{a}^{a2} + \cdots;$$

and each term in this sum is zero. Indeed, $e^a \wedge e^2 \wedge \cdots \wedge e^n = 0$ unless $a = 1$, but $\mathfrak{a}^{11} = 0$ since $\mathfrak{a}$ is skew symmetric. The remaining terms vanish for analogous reasons. The n-form μ_M is said to be the metric's *volume form*. Appendix 14.2 explains how to integrate the form μ_M over any given open set in M with compact closure. This integral is in all cases a positive number. It is deemed the *volume* of the set in question. The integral of μ_M over M is guaranteed to be finite when M is compact. The value of the latter integral is said to be the *volume of* M.

An analogous notion of metric compatibility exists for covariant derivatives on a complex vector bundle with a Hermitian metric. To elaborate, recall from Chapter 6.1 that a complex vector bundle, E, with fiber $\mathbb{C}^n$ can be viewed as a real vector bundle with fiber $\mathbb{R}^{2n}$ and an almost complex structure, j. The real bundle is denoted by $E_{\mathbb{R}}$. The almost complex structure is a section of $\mathrm{End}(E_{\mathbb{R}})$ that obeys $j^2 = -\iota$. The fiber of E at any given $x \in M$ is by definition the span of the complex eigenvectors of j on $E_{\mathbb{R}}|_x$ with eigenvalue i. As discussed in Chapter 7.2, a Hermitian inner on E comes from a positive definite inner product, $\langle \, , \rangle$ on $E_{\mathbb{R}}$ with the property that $\langle v, ju\rangle = -\langle jv, u\rangle$. This induced Hermitian inner product for E is denoted also by $\langle \, , \rangle$. A covariant derivative for sections of E comes from a covariant derivative for sections of $E_{\mathbb{R}}$ with the following property: Let $\mathfrak{s}$ denote any given section of $E_{\mathbb{R}}$. Then $\nabla(j\,\mathfrak{s}) = j\nabla\mathfrak{s}$. Such a covariant derivative on sections of $E_{\mathbb{R}}$ will give a metric compatible covariant derivative on sections of E if and only if it is a metric compatible covariant derivative on sections of $E_{\mathbb{R}}$. Metric compatible covariant derivates for sections of E are in 1–1 correspondence with connections on the principal U(n)-bundle $P_{U(E)}$ of unitary frames for E.

The curvature, F_∇, in this case defines a section of $(P_{U(E)} \times_{ad} \mathfrak{lie}(U(n))) \otimes \wedge^2 T^*M$, which is to say that it is a 2-form on M with values in the anti-Hermitian endomorphisms of E.

15.2 Torsion free covariant derivatives on T^*M

As always, let M denote a smooth manifold. The exterior derivative gives a completely canonical way to obtain a section of $\wedge^2 T^*M$ from one of T^*M. However, if T^*M has a covariant derivative, then the exterior covariant derivative as defined in Chapter 12.4 gives a second way to obtain such a section. By way of a reminder, suppose that ∇ is a covariant derivative on sections of T^*M and that α is section of T^*M, thus a 1-form on M. Then $\nabla\alpha$ is a section of $T^*M \otimes T^*M$ and antisymmetrizing after multiplying by -1 gives a section of $\wedge^2 T^*M$. This section of $\wedge^2 T^*M$ is also denoted by $d_\nabla\alpha$.

The *torsion* of the covariant derivative ∇ as applied to the 1-form α measures the extent to which $d_\nabla\alpha$ and $d\alpha$ differ: $d_\nabla\alpha - d\alpha$. What follows is an important observation: The two form $d_\nabla\alpha - d\alpha$ can be written as

$$d_\nabla \alpha - d\alpha = T_\nabla \alpha$$

where T_∇ is a section of the bundle $\mathrm{Hom}(T^*M; \wedge^2 T^*M)$. This is to say that T_∇ takes no derivatives. Here is why: Suppose that f is a smooth function. Then

$$d(f\alpha) = df \wedge \alpha + f d\alpha \quad \textit{and} \quad d_\nabla(f\alpha) = df \wedge \alpha + f d_\nabla\alpha\,.$$

Thus $T_\nabla(f\alpha) = f\, T_\nabla\alpha$ and this can happen only if no derivatives are involved. Indeed, this follows as an application of Lemma 11.1 with $E = T^*M$, $E' = \wedge^2 T^*M$ and with $\mathcal{L} = d_\nabla - d$.

A covariant derivative on T^*M is said to be *torsion free* when $T_\nabla = 0$. If $T_\nabla = 0$, then d_∇ and d agree not just on 1-forms, but on p-forms for all p. This follows directly from the Leibnitz rule that both obey, namely $d_\nabla(\alpha \wedge \beta) = d_\nabla\alpha \wedge \beta + (-1)^{\deg(\alpha)} \alpha \wedge d_\nabla\beta$ and likewise $d(\alpha \wedge \beta) = d\alpha \wedge \beta + (-1)^{\deg(\alpha)} \alpha \wedge d\beta$.

There exist in all cases, torsion free covariant derivatives. To construct such a thing, first fix a covariant derivative ∇. The task is to find a section, $\mathfrak{a}$, of the bundle $\mathrm{End}(T^*M) \otimes T^*M$ such that $\nabla' = \nabla + \mathfrak{a}$ is torsion free. This is done by first writing the bundles $\mathrm{End}(T^*M) \otimes T^*M$ and $\mathrm{Hom}(T^*M; \wedge^2 T^*M)$ in an alternate form so as to view $\mathfrak{a}$ and T_∇ as sections of the *same* vector bundle. Once this is done, it is a simple matter to choose $\mathfrak{a}$ so that $d_{\nabla'}$ has zero torsion. To start, use the identification of $\mathrm{End}(T^*M)$ with the bundle $T^*M \otimes TM$ to view $\mathfrak{a}$ as a section of the bundle $(T^*M \otimes TM) \otimes T^*M$. Meanwhile T_∇ is a section of $\mathrm{Hom}(T^*M; \wedge^2 T^*M)$ which is equivalent to being a section of $(\wedge^2 T^*M) \otimes TM$. As such, T_∇ can also be viewed as a section of $(T^*M \otimes T^*M) \otimes TM$, albeit one that is antisymmetric with respect to the endomorphism that interchanges the two T^*M factors. To continue, note that the bundle $(T^*M \otimes T^*M) \otimes TM$ is isomorphic to $(T^*M \otimes TM) \otimes T^*M$. The desired isomorphism takes any given decomposable element $(u \otimes u') \otimes v$ to the element $(u \otimes v) \otimes u'$. Granted all of these identifications, then $\mathfrak{a}$ and T_∇ are seen as sections of the same vector bundle. Viewed in this way take $\mathfrak{a}$ to equal -1 times T_∇ to obtain a torsion free covariant derivative.

Repeat the construction just given for the case $T_\nabla = 0$ to verify the following basic fact: Suppose that ∇ is a torsion free covariant derivative on T^*M. Then so is $\nabla + \mathfrak{a}$ if and only if $\mathfrak{a}$ is a section of $(T^*M \otimes TM) \otimes T^*M$ that is *symmetric* with respect to the automorphism of $(T^*M \otimes TM) \otimes T^*M$ that interchanges the T^*M factors.

15.3 The Levi-Civita connection/covariant derivative

Suppose now that M has a Riemannian metric. Then its orthonormal frame bundle has a very special connection, this being the *Levi-Civita* connection. This is the connection described in

Theorem 15.1 (The Levi-Civita theorem) *There is one and only one connection on the orthonormal frame bundle that induces a torsion free covariant derivative on* T^*M. *Said differently, there is one and only one covariant derivative for sections of* T^*M *that is both metric compatible and torsion free.*

What follows are some comments to put this theorem in perspective. First, the bundle T^*M is an associated vector bundle to the bundle of frames in TM. This said, a connection on this principal bundle induces a covariant derivative on the space of sections of T^*M. The bundle T^*M also inherits a canonical fiber metric from the metric on TM. There are any number of ways to see this. One way to see this is to write T^*M as the just described associated bundle to the bundle, $P_{Gl(M)}$, of frames in TM. As such, it is given as $P_{Gl(M)} \times_\rho \mathbb{R}^n$, where ρ is the representation of $Gl(n; \mathbb{R})$ on $\mathbb{R}^n$ that has a matrix g act so as to send a vector v to $(g^{-1})^T$. Use the given metric to define the principal bundle $P_{O(n)}$ of orthonormal frames in TM. This identifies $P_{Gl(M)}$ as the principal bundle given by $P_{O(n)} \times_{def} Gl(n; \mathbb{R})$ where def is the defining representation of O(n) in $Gl(n; \mathbb{R})$. This identifies T^*M with TM because the representations of O(n) on $\mathbb{R}^n$ that have g act on a vector $v \in \mathbb{R}^n$ to give gv and $(g^{-1})^T v$ are one and the same representations of O(n). This identification gives T^*M the fiber metric that the theorem is referring to.

Here is another way to see this identification between TM and T^*M: The metric is a section of $Hom(TM \otimes TM; \mathbb{R})$ and this bundle is canonically isomorphic to $Hom(TM; T^*M)$. Viewed in this light, the metric defines an isomorphism from TM to T^*M. The induced metric on T^*M is defined by the requirement that the latter isomorphism define an isometry.

Proof of the Levi-Civita theorem (Theorem 15.1) I find the only proof I know long on index manipulation and short on intuition. In any event, choose an orthornormal frame for T^*M over some coordinate chart $U \subset M$. Use $\{e^1, \ldots, e^n\}$ to denote the chosen frame. Then ∇e^i can be written as

$$(*) \qquad \nabla e^j = e^i \otimes \mathfrak{a}_k^{ji} e^k$$

where repeated indices come implicitly with a summation over the indexing set $\{1, \ldots, n\}$. This covariant derivative is metric compatible if and only if

$$\mathfrak{a}_k^{ij} = -\mathfrak{a}_k^{ji} \textit{ for all } 1 \le i \le j \le n.$$

Meanwhile, write

$$de^j = \frac{1}{2} w^{\,j}_{ik} e^i \wedge e^k,$$

where $w^{\,j}_{ik} = -w^{\,j}_{ik}$ for all indices $1 \le i, k \le n$. Thus, ∇ is torsion free if and only if

$$a^{ij}_k - a^{kj}_i = w^{\,j}_{ik}.$$

These conditions on a are satisfied by

$$a^{ij}_k = -\frac{1}{2}(w^{\,i}_{jk} - w^{\,j}_{ik} + w^{\,k}_{ji}).$$

To see that this is the only solution, remark that if there were two, then their difference, a^{ij}_k, would have to obey both $a^{ij}_k = -a^{ji}_k$ and $a^{ij}_k = a^{kj}_i$. This is possible if and only if all a^{ij}_k are zero. To see why, call the first equality the *antisymmetry* condition and the second the *symmetry* condition. Then note that alternating applications of first the *symmetry* condition and then the *antisymmetry* condition lead to the following sequence of equalities

$$a^{ij}_k = a^{kj}_i = -a^{jk}_i = -a^{ik}_j = a^{ki}_j = a^{ji}_k = -a^{ij}_k.$$

As the first and last terms are identical but for the sign, so each term must be zero.

15.4 A formula for the Levi-Civita connection

Equation (∗) above says all that is needed to define the covariant derivative since it tells us how to take covariant derivatives of any given basis of orthonormal sections of T^*M that are defined over an open set in M. Indeed, given such a basis $\{e^i\}_{1 \le i \le n}$, their covariant derivatives are defined by the rule

$$\nabla e^j = e^i \otimes a^{ij}_k e^k$$

and if $\mathfrak{w} = s_i e^i$ is any other section of T^*M, then

$$\nabla \mathfrak{w} = e^i \otimes ds_i + s_i a^{ij}_k e^j \otimes e^k.$$

Here as before (and in what follows), repeated indices come with an implicit summation over the indexing set $\{1, \ldots, n\}$. To take the covariant derivative of a vector field, introduce the dual basis $\{\mathfrak{v}_a\}$ of T^*M, this defined so that the pairing is $e^i \cdot \mathfrak{v}_a = 1$ or 0 depending on whether $i = a$ or $i \ne a$ respectively. Then ∇a_a is defined so that

$$0 = d(e^i \cdot \mathfrak{v}_a) = \nabla e^i \cdot \mathfrak{v}_a + e^i \cdot \nabla \mathfrak{v}_a.$$

This finds

$$\nabla \mathfrak{v}_a = -\mathfrak{a}_k^{ia} \mathfrak{v}_i \otimes e^k.$$

Suppose now that U is an open set in M and that $\psi: U \to \mathbb{R}^m$ is a coordinate chart. Here, I use m for the dimension of M. The Riemannian metric pulls back via ψ as the section

$$g_{ij} dx^i \otimes dx^j$$

of $T^*\mathbb{R}^m \otimes T^*\mathbb{R}^m$.

Proposition 15.2 *The Levi-Civita covariant derivative* ∇ *is given by* $\nabla dx^i = -\Gamma^i_{jk}\, dx^j \otimes dx^k$ *where* $\Gamma^i_{jk} = \frac{1}{2} g^{in}(\partial_j g_{nk} + \partial_k g_{nj} - \partial_n g_{jk})$.

Proof of Proposition 15.2 Noting that $\Gamma^i_{jk} = \Gamma^i_{kj}$, it follows that $d_\nabla(dx^i) = \Gamma^i_{jk}\, dx^j \wedge dx^k = 0$. This verifies that ∇ is torsion free. Consider now $\nabla(g_{np}\, dx^n \otimes dx^p)$. If this is zero, then ∇ is also metric compatible. Write out this expression to see that such is the case:

$$\nabla(g_{np} dx^n \otimes dx^p) = (\partial_k g_{np} - g_{rp}\Gamma^r_{nk} - g_{nr}\Gamma^r_{pk})(dx^n \otimes dx^p) \otimes dx^k.$$

The coefficient of $(dx^n \otimes dx^p) \otimes dx^k$ on the right-hand side is

$$\partial_k g_{np} - \frac{1}{2}(\partial_n g_{pk} + \partial_k g_{pn} - \partial_p g_{nk}) - \frac{1}{2}(\partial_p g_{nk} + \partial_k g_{np} - \partial_n g_{pk}).$$

This sum is zero. This being the case, the Levi-Civita covariant derivative has the form asserted by the proposition.

Recall from Chapter 8.2 that this same Γ^i_{jk} appears in the definition of a geodesic. Indeed, the geodesic equation from Chapter 8.2 is the assertion that a map $t \to \gamma(t)$ from an interval $I \subset \mathbb{R}$ is a geodesic if and only if $\dot{\gamma}$ is *covariantly constant* along γ. This term has the following meaning: If $\dot{\gamma}$ is viewed as a section over I of γ^*M, then its covariant derivative as defined using the pull-back of ∇ is zero.

15.5 Covariantly constant sections

As an aside, note that the term *covariantly constant* is used in greater generality. To elaborate, suppose that $E \to M$ is a vector bundle and ∇ is a connection on E. Recall from Chapter 15.1 that a section, $\mathfrak{s}$, of E is said to be *covariantly* *constant* if $\nabla\mathfrak{s} = 0$. Of course, if $\nabla\mathfrak{s} = 0$, then $d_\nabla\nabla\mathfrak{s}$ must also vanish, and as $d_\nabla\nabla\mathfrak{s} = F_\nabla\mathfrak{s}$, this requires $F_\nabla\mathfrak{s}$ to vanish: The curvature 2-form, a section of $End(E) \otimes \wedge^2 T^*M$, must annihilate $\mathfrak{s}$. This, in general, is fairly strong constraint on the curvature, for it asserts that if v and u are any two vector fields on M, then the endomorphism $F_\nabla(u, v)$ of E annihilates $\mathfrak{s}$.

If the span of the linear space of covariantly constant sections of E is constant over an open set in M, then this span defines a subbundle, $E' \subset E$. The covariant derivative acts as zero on sections of E', and so it induces a covariant derivative of sections of E' with zero curvature. This gives a flat connection on the bundle $P_{GL(E')}$.

Now suppose that $I \subset \mathbb{R}$ is an interval and that $\gamma: I \to M$ a smooth map, the image is a path in M. A section $\mathfrak{s}$ of γ^*E is said to be *parallel transported along* γ, or just *parallel along* γ, when $\mathfrak{s}$ is covariantly constant with respect to the pull-back covariant derivative, $\gamma^*\nabla$. This is to say that $(\gamma^*\nabla)\mathfrak{s} = 0$. To see what this means, fix any given isomorphism $\varphi: \gamma^*E \to I \times V$ where V is $\mathbb{R}^n$ or $\mathbb{C}^n$ as the case may. Write $\varphi(\mathfrak{s})$ as the section $t \to (t, \mathfrak{s}_\varphi(t))$, where $\mathfrak{s}_\varphi$ maps I to V. Likewise, write $\varphi((\gamma^*\nabla)\mathfrak{s})$ as the section $t \to (t, ((\gamma^*\nabla)\mathfrak{s})_\varphi dt)$ of the bundle $(I \times V) \otimes T^*I$ over the interval. In general $((\gamma^*\nabla)\mathfrak{s})_\varphi$ has the form $\frac{d}{dt}\mathfrak{s}_\varphi + \mathfrak{a}_\varphi\mathfrak{s}_\varphi$ where $\mathfrak{a}_\varphi$ is a map from I to Gl(V). The section $\mathfrak{s}$ is covariantly constant (equivalent terms are 'parallel' and 'parallel transported') when

$$\frac{d}{dt}\mathfrak{s}_\varphi + \mathfrak{a}_\varphi\mathfrak{s}_\varphi = 0.$$

Granted this, it follows from what is said in Chapter 8.3's vector field theorem about solving systems of ordinary differential equations that there exists, in all cases, a covariantly constant basis of sections of γ^*E. Suppose that $\{\mathfrak{s}_1, \ldots, \mathfrak{s}_n\}$ is such a basis. Define a new isomorphism from γ^*E to $I \times V$ by declaring that its inverse be given by the map that sends a given point $(t, (v_1, \ldots, v_n))$ to the point $\sum_{1\le k\le n} v_k\mathfrak{s}_k|_t$. Let ψ denote this basis. It then follows by virtue of the fact that $(\gamma^*\nabla)\mathfrak{s}_k = 0$ for all k that the $((\gamma^*\nabla)\mathfrak{s})_\psi = \frac{d}{dt}\mathfrak{s}_\varphi$. Indeed, this is because $\mathfrak{s}$ when written in terms of the basis $\{\mathfrak{s}_k\}$ if it has the form $\sum_k v_k(\cdot)\,\mathfrak{s}_k$ where each $v_k(\cdot)$ is a function on I. Then

$$(\gamma^*\nabla)\mathfrak{s} = \left(\sum\nolimits_k \left(\frac{d}{dt}v_k\right)\mathfrak{s}_k\right) \otimes dt + \sum\nolimits_k v_k(\gamma^*\nabla)\mathfrak{s}_k.$$

Thus, the trivialization of γ^*E given by the covariantly constant basis makes $\gamma^*\nabla$ look just like the ordinary derivative. This is a reason to prefer a basis of covariantly constant sections.

15.6 An example of the Levi-Civita connection

Suppose that M is an n-dimensional submanifold in $\mathbb{R}^N$ for some $N \geq n$. This makes TM into a subbundle of the product bundle $M \times \mathbb{R}^N$. The Euclidean metric on $\mathbb{R}^N$ supplies a Riemannian metric for TM. To be explicit, fix any point $x \in M$. Then the inner product of vectors v and v′ in $TM|_x$ is their Euclidean inner product as vectors in $\mathbb{R}^N$ that happen to lie in the subspace $TM|_x$. Use g in what follows to denote this metric.

The identification of TM as a subbundle of $M \times \mathbb{R}^N$ also gives a covariant derivative for sections of TM. To see how this comes about, let $\Pi: M \times \mathbb{R}^N \to M \times \mathbb{R}^N$ denote the bundle homomorphism that is given at each $x \in M$ by the orthogonal projection onto the vector subspace $TM|_x \subset \mathbb{R}^N$. Write a section, v, of the product bundle $M \times \mathbb{R}^N$ as $x \to (x, v(x))$ where $v: M \to \mathbb{R}^N$ is a smooth map. A section of TM (i.e. a vector field on M) is such a section $x \to \mathfrak{v}(x) = (x, v(x))$ such that $v: M \to \mathbb{R}^N$ lands in $TM|_x$ at each x. Thus, $\Pi v = v$ for a section of TM.

The bundle $TM \otimes T^*M$ can likewise be viewed as a subbundle inside the bundle $(M \times \mathbb{R}^N) \otimes T^*M$. This is because it is spanned by the decomposable elements of the form $\mathfrak{v} \otimes \alpha$ where α is a 1-form and $\mathfrak{v}(x) = (x, v(x))$ with $v: M \to \mathbb{R}^n$ is such that $\Pi v = v$ at each x. If $x \to \mathfrak{v}(x) = (x, v(x))$ is a section of $M \times \mathbb{R}^N$ (whether or not $v(x) \in TM|_x$), then $x \to (x, dv|_x)$ defines a section of $(M \times \mathbb{R}^N) \otimes T^*M$. Here, dv is the vector of 1-forms that is obtained by taking the exterior derivative of each component of the vector valued map v. This understood, suppose that $x \to \mathfrak{v}(x) = (x, v(x))$ defines a section of TM. Set $\nabla\mathfrak{v}$ to be the section of $TM \otimes T^*M$ given by $x \to (x, \Pi dv|_x)$. Here, Π is defined on sections of the bundle $(M \times \mathbb{R}^N) \otimes T^*M$ by its action on the decomposable elements.

The covariant derivative just defined is the Levi-Civita covariant derivative for the metric just defined. To see why this is, it is enough to check that this covariant derivative and the Levi-Civita covariant derivative have the same affect on any given vector field. Let $\nabla^\diamond$denote for the moment the Levi-Civita covariant derivative. The difference, $\nabla - \nabla^\diamond$, is a section, $\mathfrak{a}$, of $\text{End}(TM) \otimes T^*M$. The task is to prove that $\mathfrak{a}$ is zero at each point. To do this for a given point x, the strategy is to choose a coordinate basis for M near x which makes the comparison relatively easy to do. Since $\mathfrak{a}$ is a tensor, the assertion that $\mathfrak{a}|_x = 0$ does not depend on the chosen coordinate chart. To reiterate, the strategy for this proof exploits the fact that the assertion to be proved is coordinate independent. By the way, this same strategy should be considered first for any calculation of any coordinate independent statement in differential geometry.

To start the argument, use the fact that the Euclidean metric on $\mathbb{R}^N$ is translation and rotation invariant to make a translation and rotation of $\mathbb{R}^N$

so that x is the origin and so that the splitting $\mathbb{R}^N = \mathbb{R}^n \times \mathbb{R}^{N-n}$ has the first factor tangent to M at x. Thus, $TM|_x$ is the factor $\mathbb{R}^n \subset \mathbb{R}^n \times \mathbb{R}^{N-n}$. The first n coordinates can serve now as coordinates on M near x. This is to say that M near x is the image of a neighborhood of the origin in $\mathbb{R}^n$ via a map, φ, that sends $y = (y_1, \ldots, y_n)$ to the N-tuple

$$\varphi(y) = (y_1, \ldots, y_n, f_1, \ldots, f_{N-n})$$

where $f = (f_1, \ldots, f_{N-n})\colon \mathbb{R}^n \to \mathbb{R}^{N-n}$ is such that $f_k = 0$ at $y = 0$ and also

$$df_k = \sum\nolimits_{1\le a\le n}\left(\frac{\partial}{\partial y_a} f_k\right) dy^a = 0 \;\; at \;\; y = 0.$$

The metric g appears in these coordinates as $g_{ab}dy^a dy^b$ where $g_{ab}|_y$ is the Euclidean inner product at $\varphi(y) \in \mathbb{R}^N$ between $\varphi_* \frac{\partial}{\partial y_a}$ and $\varphi_* \frac{\partial}{\partial y_b}$. Here, $\varphi_* \frac{\partial}{\partial y_a}$ is the push-forward to $\mathbb{R}^N$ via the embedding φ of the coordinate vector field $\frac{\partial}{\partial y_a}$:

$$\varphi_* \frac{\partial}{\partial y_a} = \frac{\partial}{\partial x_a} + \sum\nolimits_{1\le b\le N-n} \frac{\partial f_k}{\partial y_a} \frac{\partial}{\partial x_{n+k}}.$$

Thus, the metric coefficient g_{ab} is $g_{ab} = \delta_{ab} + \partial_a f_k \partial_b f_k$ where $\delta_{ab} = 1$ if $a = b$ and $\delta_{ab} = 0$ otherwise.

This formula can be used to compute the Levi-Civita covariant derivative of the coordinate 1-form basis $\{dy_a\}_{1\le a\le n}$. Recall that I use $\nabla^\diamond$ to denote the Levi-Civita covariant derivative. Then $\nabla^\diamond dy_a$ can be written as $\nabla^\diamond dy_a = -\Gamma^a_{bc}\, dy_b \otimes dy_c$ where

$$\Gamma^a_{bc} = \frac{1}{2} g^{ad}\left(\frac{\partial}{\partial y_b} g_{dc} + \frac{\partial}{\partial y_c} g_{db} - \frac{\partial}{\partial y_d} g_{bc}\right).$$

I use in this last equation and subsequently the convention where $\{g^{ab}\}_{1\le a,b\le n}$ denotes the entries of the inverse of the matrix, g, with entries $\{g_{ab}\}_{1\le a,b\le n}$; the position of the index as superscript or subscript is meant to convey the distinction between g and the inverse matrix g^{-1}.

Here is the first place where the choice of coordinates is exploited: The metric coefficient g_{ab} is the constant δ_{ab} to order $|y|^2$. Thus, $\Gamma^a_{bc} = O(|y|)$. In particular, $\Gamma^a_{bc}|_{y=0} = 0$. This is to say that $\nabla^\diamond dy^a = 0$ at $y = 0$. As a consequence

$$\nabla^\diamond \frac{\partial}{\partial y_a} = 0 \;\; at \;\; y = 0$$

because $\nabla \frac{\partial}{\partial y_a} = \Gamma^b_{ac} \frac{\partial}{\partial y_b} \otimes dy^c$. To elaborate, this last equation follows from the relation between the covariant derivatives on TM and T*M; they determine each other using the rule

$$d(\alpha(\mathfrak{v})) = (\nabla^\diamond \alpha)(\mathfrak{v}) + \alpha(\nabla^\diamond \mathfrak{v}).$$

Here, α is any given 1-form, $\mathfrak{v}$ is any given vector field, and $\alpha(\mathfrak{v})$ denotes the function that is obtained by pairing the 1-form with the vector field.

The next task is to compute $\nabla\frac{\partial}{\partial y_a}$ at $y = 0$. If this is also zero, then $\nabla = \nabla^\diamond$ as claimed. To prove this equality, note first that the definition of ∇ is such that

$$\nabla\frac{\partial}{\partial y_a} = \Pi\left(d\left(\frac{\partial}{\partial x_a} + \sum\nolimits_{1\le h\le N-n} \frac{\partial f_k}{\partial y_a}\frac{\partial}{\partial x_{n+k}}\right)\right)$$
$$= \Pi\left(\sum\nolimits_{1\le h\le N-n}\left(\frac{\partial^2 f_k}{\partial y_d \partial y_a}\frac{\partial}{\partial x_{n+k}}\right)\otimes dy^d\right).$$

Meanwhile, the projection Π at any given point near x is the projection in $\mathbb{R}^N$ onto the span at the given point of $\{\varphi_* \frac{\partial}{\partial y_a}\}$. In particular, it takes a vector $v = v_c\frac{\partial}{\partial x_c} + v_{k+n}\frac{\partial}{\partial x_{n+k}}$ to

$$\Pi v = g^{cb}\left(v_b + \sum\nolimits_{1\le h\le N-n} v_k \frac{\partial f_k}{\partial y_b}\right)\left(\frac{\partial}{\partial x_c} + \sum\nolimits_{1\le a\le N-n}\frac{\partial f_k}{\partial y_c}\frac{\partial}{\partial x_{n+k}}\right).$$

What with these last two equations, it follows that

$$\nabla\frac{\partial}{\partial y_a} = g^{cb}\sum\nolimits_{1\le h\le N-n}\frac{\partial f_k}{\partial y_b}\frac{\partial^2 f_k}{\partial y_d \partial y_a}\left(\frac{\partial}{\partial x_c} + \sum\nolimits_{1\le h\le N-n}\frac{\partial f_k}{\partial y_c}\frac{\partial}{\partial x_{n+k}}\right)\otimes dy^d.$$

Note in particular that the expression on the right-hand side vanishes at $y = 0$ because the first derivatives of the functions f_k all vanish at $y = 0$. Thus, the covariant derivatives ∇ and $\nabla^\diamond$ are indeed equal at x.

15.7 The curvature of the Levi-Civita connection

Suppose first that $E \to M$ is a vector bundle with fiber $\mathbb{R}^n$, that g is a fiber metric on E, and that ∇ is a metric compatible connection on E. This means that g is a covariantly constant section of the bundle $E^* \otimes E^*$, thus $\nabla g = 0$. To see what this entails, suppose that $U \subset M$ is an open set and $\{e^b\}_{1\le b\le n}$ is an orthonormal basis of sections of E^*. Thus, the metric g can be written as

$$g = e^b \otimes e^b.$$

Here and in what follows, repeated indices are implicitly summed over the indexing set $\{1,\ldots,n\}$. Meanwhile, ∇e^b can be written as $\nabla e^b = e^a \otimes \mathfrak{a}^{ab}$ where each $a,b \in \{1,\ldots,n\}$ version of $\mathfrak{a}^{ab}$ is a 1-form on U. The collection $\{\mathfrak{a}^{ab}\}_{1\le a,b\le n}$ can be viewed as the coefficients of a 1-form, $\mathfrak{a}$, on U with values in the vectors space of $n \times n$ antisymmetric matrices.

The condition $d_\nabla \nabla g = 0$ says that the curvature 2-form F_∇ annihilates the section g. To understand what this implies, write $(F_\nabla){\cdot}e^b = e^a \otimes (F_\nabla)^{ab}$. The condition $d_\nabla \nabla g = 0$ says that $(F_\nabla)^{ab} = -(F_\nabla)^{ba}$; this asserts what was said previously in Chapter 15.1: The curvature of a metric compatible connection is a 2-form on M with values in the bundle of antisymmetric endomorphism of E.

Now specialize to the case when E is the tangent bundle of M, and ∇ is a torsion free connection. Let $\{e^b\}$ denote an orthonormal basis of 1-forms over some open set U in M. Write $\nabla e^b = e^a \otimes \mathfrak{a}^{ab}$ and $(F_\nabla)e^b = e^a \otimes (F_\nabla)^{ab}$ where $(F_\nabla)^{ab} = d\mathfrak{a}^{ab} + \mathfrak{a}^{ac} \wedge \mathfrak{a}^{cb}$. Write the 2-form indices of F_∇ explicitly:

$$(F_\nabla)^{ab} = \frac{1}{2}(F_\nabla)^{ab}_{cd} e^c \wedge e^d$$

where $(F_\nabla)^{ab}_{cd} = -(F_\nabla)^{ab}_{dc}$. Given that $d^2_\nabla e^b = e^a \otimes (F_\nabla)^{ab}$, and given the torsion free condition, which is that d_∇ acts as d, it follows that $(F_\nabla)^{ab} \wedge e^b = 0$. This says that

$$(F_\nabla)^{ab}_{cd}\, e^c \wedge e^d \wedge e^b = 0.$$

Here as before (and subsequently), repeated indices are implicitly summed over the indexing set $\{1, \ldots, n\}$. What is written here says neither more nor less than

$$(F_\nabla)^{ab}_{cd} + (F_\nabla)^{ac}_{db} + (F_\nabla)^{ad}_{bc} = 0.$$

If ∇ is the Levi-Civita connection, then it is also the case that $(F_\nabla)^{ab}_{cd} = -(F_\nabla)^{ba}_{cd}$, and this condition with the equation written above imply that

$$(F_\nabla)^{ab}_{cd} = (F_\nabla)^{cd}_{ab}.$$

This is to say that $(F_\nabla)^{ab}_{cd}$ is *symmetric* with respect to interchanging its first two and last two pairs of indices. Since it is *antisymmetric* in each pair, the curvature of the Levi-Civita connection can be viewed as a section of the bundle of symmetric endomorphisms of $\wedge^2 T^*M$. Viewed in this way, F_∇ is called the *curvature operator.*

The curvature of the Levi-Civita covariant derivative is called the *Riemann curvature tensor.* It is written with respect to a given orthonormal frame for T^*M as $(F_\nabla)^{ab}_{cd} = R_{abcd}$. As noted above, it enjoys the following symmetries:

- $R_{abcd} = -R_{bacd}$,
- $R_{abcd} = -R_{abdc}$,
- $R_{abcd} = R_{cdab}$,
- $R_{abcd} + R_{adbc} + R_{acdb} = 0$,
- $\nabla_e R_{abcd} + \nabla_c R_{abde} + \nabla_d R_{abec} = 0$.

The last one is the Bianchi identity $d_\nabla F_\nabla = 0$. With regards to the notation of the last line, the Riemann tensor is viewed as the section

$$\text{Riem} = \frac{1}{4} R_{abcd}(e^a \wedge e^b) \otimes (e^c \wedge e^d)$$

of $(\wedge^2 T^*M) \otimes (\wedge^2 T^*M)$, and, as such, it has the covariant derivative ∇Riem, this a section of $((\wedge^2 T^*M) \otimes (\wedge^2 T^*M)) \otimes T^*M$ which is written as

$$\nabla\text{Riem} = \frac{1}{4} \nabla_e R_{abcd}((e^a \wedge e^b) \otimes (e^c \wedge e^d)) \otimes e^e.$$

With $\nabla e^a = e^b\, \mathfrak{a}^{ab}$, this means

$$\nabla_e R_{abcd} = \partial_e R_{abcd} + \mathfrak{a}^{af} R_{fbcd} + \mathfrak{a}^{bf} R_{afcd} + \mathfrak{a}^{cf} R_{abfd} + \mathfrak{a}^{df} R_{abcf},$$

where ∂_e indicates the directional derivative of the function R_{abcd} in the direction of the vector field e_e.

The components of the Riemann curvature tensor are used to define a symmetric section of $T^*M \otimes T^*M$, this being the *Ricci tensor*; its components are

$$\text{Ric}_{ac} = \textstyle\sum_b R_{abcb}.$$

Thus, the tensor is $\text{Ric}_{ac}\, e^a \otimes e^c$. The trace of the Ricci tensor (its inner product with the metric) is called the *scalar curvature*:

$$R = \textstyle\sum_a \text{Ric}_{aa} = \sum_{ac} \text{Ric}_{acac}.$$

The Bianchi identities imply in part that Ricci tensor obeys

$$\nabla_e \text{Ric}_{ac} - \nabla_c \text{Ric}_{ae} + \nabla_b R_{abec} = 0 \quad \textit{and} \quad \nabla_c \text{Ric}_{ac} - \frac{1}{2} \nabla_a R = 0.$$

The symmetric section of $T^*M \otimes T^*M$ with components

$$\text{Ric}_{ab} - \frac{1}{2} R\delta_{ab}$$

is called the *Einstein* tensor. Here δ_{ab} is 1 if $a = b$ and 0 otherwise. The section $\delta_{ab}\, e^a \otimes e^b = e^a \otimes e^a$ of $T^*M \otimes T^*M$ is the Riemannian metric. The Ricci tensor, Einstein tensor and the scalar curvature play a central role in modern applications of metric geometry.

The next chapter describes these tensors in various examples.

Additional reading

- *Riemannian Geometry: A Modern Introduction*, Isaac Chavel, Cambridge University Press, 2006.

- *Riemannian Geometry*, Manfredo P. Carmo, Birkäuser, 1992.
- *Riemannian Manifolds: An Introduction to Curvature*, John M. Lee, Springer, 1997.
- *Riemannian Geometry*, Peter Petersen, Springer, 2006.
- *Riemannian Geometry*, Sylvestre Gallot, Dominque Hulin and Jacques Lafontaine, Springer, 2004.

16 The Riemann curvature tensor

This chapter studies first some fundamental examples where the Riemann curvature tensor has an especially simple form. It then briefly discusses some of the ways in which the Riemann and Ricci curvatures are used to study questions in geometry and differential topology.

16.1 Spherical metrics, flat metrics and hyperbolic metrics

The examples that follow have, in some sense, the simplest possible Riemann curvature tensor. In these examples, the components of the Riemann tensor when written with respect to a given orthonormal frame are

$$R_{abcd} = \kappa(\delta_{ac}\delta_{bd} - \delta_{bc}\delta_{ad}).$$

Here, κ is a constant and δ_{ac} is 1 if $a = c$ and 0 otherwise. Let $\{e^a\}_{1\leq a\leq n=\dim(M)}$ denote the orthonormal frame in question. The corresponding section Riem of $\wedge^2 T^*M \otimes \wedge^2 T^*M$ is

$$\text{Riem} = \frac{1}{4}\kappa(e^a \wedge e^b) \otimes (e^a \wedge e^b).$$

(Here, as in the previous chapter, repeated indices are summed.) Viewed as an endomorphism of $\wedge^2 T^*M$, this section acts as κ times the identity endomorphism. Let n denote the dimension of M.

The corresponding Ricci tensor is such that $R_{ac} = (n - 1)\kappa\delta_{ac}$. Stated without reference to an orthonormal frame, this is saying that

$$\text{Ric} = (n - 1)\kappa g.$$

Meanwhile, the scalar curvature is $R = n(n - 1)\kappa$.

16.1.1 Spheres

Fix $n \geq 2$, $\rho > 0$ and embed S^n in $\mathbb{R}^{n+1}$ as the space of vectors in $\mathbb{R}^{n+1}$ with length ρ. The corresponding metric obeys the equation above with $\kappa = \rho^{-2}$. What follows does the computation directly using the formula given in Chapter 8.4.3 for the metric of a submanifold in $\mathbb{R}^{N=n+1}$. Recall that the sphere of radius ρ is invariant under the $SO(n+1)$ group of rotations of $\mathbb{R}^{n+1}$ about the origin. This understood, it is enough to verify the formula for R at the north pole, this where $x_1 = x_2 = \cdots = x_n = 0$ and $x_{n+1} = \rho$. The coordinate embedding in this case has $\phi(y) = (y_1, \ldots, y_n, (\rho^2 - |y|^2)^{1/2})$. The corresponding metric components $\{g_{ab}\}$ are given by

$$g_{ab} = \delta_{ab} + (\rho^2 - |y|^2)^{-1} y_a y_b.$$

As the Christoffel symbols $\{\Gamma^a_{bc}\}$ involve first derivatives of g and the curvature involves first derivatives of the Christoffel symbols, it is enough for the computation of R_{abcd} at $y = 0$ to replace g_{ab} by its second-order Taylor's approximation, this being

$$g^*_{ab} = \delta_{ab} + \rho^{-2} y_a y_b.$$

Likewise, the Christoffel symbols can be replaced by their first-order Taylor's approximations, thus by

$$\Gamma^a_{bc} = \rho^{-2} \delta_{bc} y_a.$$

The curvature 2-form at $y = 0$ is $R^{ab} = d\Gamma^a_{bc} \wedge dy^c = \rho^{-2}\, dy^a \wedge dy^b$. This is to say that R^{ab}_{cd} is equal to $\rho^{-2}\, (\delta_{ac}\delta_{bd} - \delta_{ad}\delta_{bc})$.

16.1.2 Tori

Fix $n \geq 2$ and fix a *lattice* $\Lambda \subset \mathbb{R}^n$, this being a set of basis vectors, $\{\mathfrak{v}_1, \ldots, \mathfrak{v}_n\}$. Do not require any orthonormality. Define $\mathbb{T}_\Lambda$ to be $\mathbb{R}^n/\mathbb{Z}^n$ where an integer n-tuple $(k_1, \ldots, k_n) \in \mathbb{Z}^n$ acts on $x \in \mathbb{R}^n$ as the translation by the vector $2\pi \sum_{1\leq a\leq n} k_a \mathfrak{v}_a$. The quotient is a smooth manifold that is diffeomorphic to $\times_n S^1$. A convenient diffeomorphism is given as follows: Identify S^1 as $\mathbb{R}/2\pi Z$. The diffeomorphism, φ, sends a given point $(t_1, \ldots, t_n) \in \mathbb{R}/2\pi\mathbb{Z}$ to the equivalence class of the vector $\sum_{1\leq a\leq n} t_a\, \mathfrak{v}_a$. As the Euclidean metric on $\mathbb{R}^n$ is translationally invariant, it descends to $\mathbb{T}_\Lambda$ to define a metric there. This is to say that the metric pulls back via ϕ to the metric $g = g_{ab} dt^a dt^b$ where

$$g_{ab} = \mathfrak{v}_a \cdot \mathfrak{v}_b,$$

with $\mathfrak{v}\cdot\mathfrak{v}'$ denoting the Euclidean dot product. Since the Christoffel symbols involve a derivative of g_{ab}, they must all vanish; thus the metric so defined has Riemann curvature tensor equal to zero. Such a metric is said to be *flat*.

Two lattices Λ and Λ' give the same metric (up to a diffeomorphism of $\times_n S^1$) if there exists an integer-valued matrix $\mathfrak{m} \in Gl(n; \mathbb{R})$ with determinant 1 such that $\Lambda = m\Lambda'$.

16.1.3 The hyperbolic space

Define $\mathbb{R}^{1,n}$ to be $\mathbb{R} \times \mathbb{R}^n$ with the *Lorentz* metric

$$\mathfrak{l} = -dt \otimes dt + dx_a \otimes dx_a.$$

Here, I use t for the coordinate on the $\mathbb{R}$ factor and $x = (x_1, \ldots, x_n)$ for the coordinates of the $\mathbb{R}^n$ factor. Although not positive definite, this bilinear form is nonetheless called a "metric". The bilinear form is, however, nondegenerate.

Like its spherical counterpart, this metric has a large symmetry group, known as the *Lorentz group*. This is the subgroup of $Gl(n + 1; \mathbb{R})$ of matrices $\mathfrak{m}$ that preserve the quadratic form $x^2 - t^2$. For example, it contains the SO(n) subgroup that fixes the t-coordinate. It also contains matrices that fix, say $x_2, \ldots, x_n$ and act on the (t, x_1) coordinates as

$$\begin{pmatrix} \cosh(u) & \sinh(u) \\ \sinh(u) & \cosh(u) \end{pmatrix}$$

with $u \in \mathbb{R}$. The subgroup generated by the matrices in SO(n) and the matrices of that fix all but one x_a as just described is the subgroup, L^+, of the Lorentz group that fixes the $t \geq 0$ cone in $\mathbb{R}^1 \times \mathbb{R}^n$.

Fix $\rho > 0$ and let $M \subset \mathbb{R}^1 \times \mathbb{R}^n$ denote the manifold where $x^2 - t^2 = -\rho^2$ and $t > 0$. Although the Lorentz metric is not positive definite, the induced metric on M *is* positive definite. To see this, remark that a tangent vector v to M at a point (t, x) can be written as $v = v_t \frac{\partial}{\partial t} + v_a \frac{\partial}{\partial x_a}$ where (v_t, v_a) are such that $tv_t - x_a v_a = 0$. As $t = (\rho^2 + |x|^2)^{1/2} > 0$ on M, this means that $v_t = (\rho^2 + |x|^2)^{-1/2} x_a v_a$. Meanwhile, $\mathfrak{l}(v, v) = v_a v_a - v_t v_t$, and as $t > 0$ on M, this tells us that

$$\mathfrak{l}(v, v) = v_a v_b (\delta_{ab} - (\rho^2 + |x|^2)^{-1} x_a x_b).$$

In particular, the eigenvalues of this quadratic form are 1 (with multiplicity $n - 1$) and $\rho^2/(\rho^2 + |x|^2)$, this last with multiplicity 1. The metric just described is called the *hyperbolic metric*.

As is explained next, the Riemann curvature tensor for this metric is such that $R_{abcd} = -\rho^2(\delta_{ac}\delta_{bd} - \delta_{ad}\delta_{bc})$. The Ricci tensor has components $R_{ab} = -(n - 1)\rho^{-2}\,\delta_{ab}$, this the negative, constant factor $(n - 1)\rho^{-2}$ times the metric tensor. The scalar curvature is R is equal to the constant $-n(n - 1)\rho^{-2}$.

To prove these assertions, use the fact that the L^+ acts to preserve both M and the Lorentz inner product $\mathfrak{l}$ on $\mathbb{R}^{1,n}$ on M to see that it is enough to prove this assertion at the point $(t = \rho,\ x = 0)$ in M. To consider the curvature at this point, use the coordinates supplied by the embedding $\varphi: \mathbb{R}^n \to M \subset \mathbb{R} \times \mathbb{R}^n$ that sends $y = (y_1, \ldots, y_n)$ to

$$\varphi(y) = ((\rho^2 + |y|^2)^{1/2}, y_1, \ldots, y_n).$$

The metric on M appears in these coordinates as $g_{ab}dy^a dy^b$ where $g_{ab} = \mathfrak{l}(\varphi_* \frac{\partial}{\partial y_a}, \varphi_* \frac{\partial}{\partial y_b})$. Given that

$$\varphi^* \frac{\partial}{\partial y_a} = (\rho^2 + |y|^2)^{-1/2} y_a \frac{\partial}{\partial t} + \frac{\partial}{\partial x_a},$$

it follows that

$$g_{ab} = \delta_{ab} - (\rho^2 + |y|^2)^{-1} y_a y_b.$$

As in the spherical case, it is sufficient for computing R_{abcd} at $y = 0$ to use the second-order Taylor's approximation for g_{ab} and the first-order Taylor's approximation for Γ^a_{bc}. The subsequent calculations are identical to those done in the spherical case save for the factor of -1 between the two terms in the preceding expression for g_{ab}. This leads to the factor of -1 in the expression $R_{abcd} = -\rho^{-2}(\delta_{ac}\delta_{bd} - \delta_{ad}\delta_{bc})$.

16.2 The Schwarzchild metric

What follows is an example of a metric with zero Ricci curvature, but a more complicated Riemann curvature tensor. This metric is called the Schwarzchild metric, and it is a metric on the noncompact manifold $S^2 \times \mathbb{R}^2$. There is a 1-parameter family of such metrics, the parameter being $m > 0$. The metric is written first in a coordinate basis with the coordinates (t, r, θ, φ) where θ and φ are the usual angular coordinates on S^2, where t is an $\mathbb{R}/(2\pi\mathbb{Z})$-valued coordinate, and $r > m$.

$$g = (1 - m/r)4m^2 dt \otimes dt + (1 - m/r)^{-1} dr \otimes dr$$
$$+ r^2(d\theta \otimes d\theta + \sin^2\theta d\varphi \otimes d\varphi).$$

Note in this regard that $d\theta \otimes d\theta + \sin^2\theta\, d\varphi \otimes d\varphi$ is the standard round metric on S^2. The apparent singularity as $r \to m$ is spurious and due to the fact that these coordinates are not valid on $r = m$. Taking $r \to m$ corresponds on

$S^2 \times \mathbb{R}^2$ to going to $S^2 \times \{0\}$, as can be seen using the change of coordinates (t, ρ, θ, ϕ) with $\rho = (1 - m/r)^2$ taking values in $[0, 1)$. This change writes g as

$$4m^2(\rho^2 dt \otimes dt + (1-\rho^2)^{-4} d\rho \otimes d\rho) \\ + m^2(1-\rho^2)^{-2}(d\theta \otimes d\theta + \sin^2\theta d\varphi \otimes d\varphi).$$

The (ρ, t) part is a smooth metric on the interior of the radius 1 disk in $\mathbb{R}^2$, and the (θ, ϕ) part is a smooth, ρ-dependent metric on S^2.

16.3 Curvature conditions

An oft-used strategy in differential geometry is that of finding a canonical metric on a given manifold. If the precise notion of "canonical" asserts a condition that is equivariant under the action of the diffeomorphism group, then the existence or not of a "canonical" metric says something about the smooth structure on M. Conditions that involve the Riemann curvature, the Ricci curvature, or the scalar curvature of the metric are often considered because these objects behave as one might hope under the action of the diffeomorphism group. To elaborate, suppose that g is a metric on M and $\phi: M \to M$ is a diffeomorphism. View the Riemann curvature as a section of $(\wedge^2 T^*M) \otimes (\wedge^2 T^*M)$, view the Ricci curvature tensor as a section of $T^*M \otimes T^*M$, and view the scalar curvature as a function on M. Then the Riemann curvature of ϕ^*g is the pull-back via ϕ of the Riemann curvature of g. Likewise, the Ricci tensor of ϕ^*g is the pull-back via ϕ of the Ricci tensor of g. And, the scalar curvature function of ϕ is the pull-back of the scalar curvature function of g. These assertions follow from the three remarks that follow.

Remark 16.1 Suppose that M and N are manifolds, that $\phi: M \to N$ is a smooth map and that $\pi: E \to N$ is a vector bundle with fiber either $\mathbb{R}^n$ or $\mathbb{C}^n$. As explained in Chapter 5.1, the map ϕ can be used to define pull-back vector bundle $\pi: \phi^*E \to M$, this the subspace in $M \times E$ of points (x, v) with $\phi(x) = \pi(v)$. Let ∇ denote a given covariant derivative on sections of E. Chapter 11.8 explains how ϕ can be used to construct a corresponding covariant derivative on sections of ϕ^*E, this denoted by $\phi^*\nabla$.

Suppose now that $P \to N$ is a principal bundle with fiber some group G and that E is associated to P via a representation ρ from G to some general linear group. Chapter 10.7 explains how the map ϕ is used to define pull-back principal bundle $\phi^*P \to M$. Let A denote a given connection on P. Chapter 11.8 uses ϕ to define the pull-back $\hat{\phi}^*A$ on ϕ^*P. As explained in Chapter 11.5, the connection A defines the covariant derivate ∇_A on Sections of E. Meanwhile

$\hat{\phi}^*A$ defines a covariant derivative on sections of ϕ^*A. You can unwind definitions to see that the latter is $\phi^*\nabla_A$.

Here is one final point: Suppose that E has a fiber metric, $\mathfrak{m}$. Then $\mathfrak{m}$ gives ϕ^*E a fiber metric, this denoted by $\phi^*\mathfrak{m}$. To elaborate, view $\phi^*E \subset M \times E$ as the subspace of pairs (p, v) such that $\phi(p) = \pi(v)$. Viewed in this way, the inner product as defined by $\phi^*\mathfrak{m}$ between pairs (x, v) and (x, v′) in ϕ^*E is $\mathfrak{m}(v, v')$. If a given covariant derivative ∇ for sections of E is metric compatible, then the covariant derivative $\phi^*\nabla$ is metric compatible for the metric $\phi^*\mathfrak{m}$. This is also checked by unwinding all of the various definitions.

Remark 16.2 Specialize to the case when $E = T^*N$. As explained in Chapter 5.3, there is a canonical vector bundle homomorphism between the pull-back bundle $\phi^*(T^*N)$ and T^*M. Don't confuse this with $\hat{\phi}$, for the latter moves the base point (in fact, over to N), while this canonical homomorphism does not. This bundle homomorphism is just the Chain rule in fancy dress. To elaborate, let $x = (x^1, \ldots, x^n)$ denote coordinate functions on an open set in M centered on given point $p \in M$, and let $y = (y^1, \ldots, y^{n'})$ denote coordinate functions on an open set in N centered on $\phi(p)$. Here, $n = \dim(M)$ and $n' = \dim(N)$. The map ϕ sends a given point near p with the coordinates $x = (x^1, \ldots, x^n)$ to the point near $\phi(p)$ with coordinates given by $y^1 = \phi^1(x), \ldots, y^{n'} = \phi^{n'}(x)$.

Meanwhile $(dx^1, \ldots, dx^n)|_x$ is a basis for the fiber of T^*M at the point with coordinates x, while $(dy^1, \ldots, dy^{n'})|_{y=\phi(x)}$ is a basis the fiber of $\phi^*(T^*N)$ at the image point. The aforementioned bundle map from $\phi^*(T^*N)|_x$ to $T^*M|_x$ acts with respect to this basis as

$$dy^k \to \frac{\partial \phi^k}{\partial x^a} dx^a$$

where repeated indices are summed as usual.

Now suppose that $\mathfrak{m} = \mathfrak{m}_{ik} dy^i dy^j$ is a metric on T^*N. Then $\mathfrak{m}_{ik}|_{\phi(\cdot)}\, dy^i dy^k$ defines the metric $\phi^*\mathfrak{m}$ on $\phi^*(T^*N)$. The homomorphism given by the preceding equation defines from $\mathfrak{m}$ a symmetric section $\mathfrak{m}_\phi$, of $T^*M \otimes T^*M$, this given by

$$\mathfrak{m}_{ik} \frac{\partial \varphi^i}{\partial x^a} \frac{\partial \phi^k}{\partial x^b} dx^a \otimes dx^b.$$

Note that this is nonnegative definite, but it won't be an honest positive definite metric unless the homomorphism $dy^k \to \frac{\partial \phi^k}{\partial x^a} dx^a$ from $\phi^*(T^*N) \to T^*M$ is surjective.

Suppose that ∇ is a covariant for T^*N. It is determined near $\phi(p)$ by its action on the basis of sections $(dy^1, \ldots, dy^{n'})$. Write this action as

$$\nabla dy^k = W_j^{ik} dy^i \otimes dy^j.$$

The covariant derivative $\phi^*\nabla$ on $\phi^*(T^*N)$ acts on sections of $\phi^*(T^*N)$ to give sections of $\phi^*(T^*N) \otimes T^*M$. It's action is defined by its action on the sections $(dy^1, \ldots, dy^{n'})|_{\phi(\cdot)}$ near p by the rule

$$(\phi^*\nabla)(dy^k) = \frac{\partial \phi^j}{\partial x^a} W_j^{ik} dy^i \otimes dx^a.$$

Notice in particular that this formula does not give a covariant derivative on T^*M unless there is an inverse to the homomorphism $dy^k \to \frac{\partial \varphi^k}{\partial x^a} dx^a$.

Remark 16.3 Granted all of the above, suppose now that $N = M$ and that ϕ is a diffeomorphism. If this is the case, then there is an inverse homomorphism, this given in the coordinate basis by $dx^a = \frac{\partial \psi^a}{\partial y^k}|_{\phi(x)} dy^k$ where ψ is the inverse to ϕ. I can use the formula above now to define $(\phi^*\nabla)(dx^a)$. This is done by first viewing $dx^a = \frac{\partial \psi^a}{\partial y^k} dy^k$ as a section of $\phi^*(T^*N)$, then using the formula above, and finally using the homomorphism $dy^k \to \frac{\partial \phi^k}{\partial x^a} dx^a$ to view the resulting covariant derivative as a section of T^*M. Here is the final product:

$$(\phi^*\nabla)(dx^a) = \left(\frac{\partial^2 \psi^a}{\partial y^i \partial y^j} + \frac{\partial \psi^a}{\partial y^k} W_j^{ik} \right)\Big|_{\phi(x)} \frac{\partial \phi^i}{\partial x^b} \frac{\partial \phi^j}{\partial x^c} dx^b \otimes dx^c.$$

To finish the story, remark first that this covariant derivative is compatible with the metric $\mathfrak{m}_\phi = \mathfrak{m}_{ik} \frac{\partial \phi^i}{\partial x^a} \frac{\partial \phi^k}{\partial x^b} dx^a \otimes dx^b$. Moreover, if ∇ is torsion free, then so is this covariant derivative. The point being that the condition that ∇ is torsion is signified here by the fact that $W_j^{ik} = W_i^{jk}$. Writing $(\phi^*\nabla)(dx^a)$ as $W_{bc}^a dx^b \otimes dx^c$, you see that W_{bc}^a obeys the corresponding symmetry condition; and so it is also torsion free.

Granted the preceding, what follows are some common, diffeomorphism-invariant conditions that are imposed on the Riemann curvature, Ricci tensor, and scalar curvature.

Vanishing Riemann curvature tensor: This is the condition that the Riemann tensor vanish everywhere on M. Thus, the Levi-Civita covariant derivative is defined by a flat connection.

Positive or negative definite curvature operator: View the Riemann curvature tensor as a symmetric section of $(\wedge^2 T^*M) \otimes (\wedge^2 T^*M)$. As such, it defines a symmetric bilinear form on $\wedge^2 TM$. The condition here is that this form is either negative definite everywhere or positive definite everywhere.

Positive or negative sectional curvature: View the Riemann curvature tensor again as a symmetric bilinear form $\mathrm{Riem}(\cdot, \cdot)$ on $\wedge^2 T^*M$. Let $x \in M$ and let u, v denote a pair of linearly independent vectors in $TM|_x$. The *sectional curvature* of the plane spanned by u and v is, by definition, the number $\mathrm{Riem}(u \wedge v, u \wedge v)$. When the Riemann curvature is written in terms of an orthonormal basis $\{e^a\}$

for T^*M as Riem $= \frac{1}{4} R_{abcd}(e^a \wedge e^b) \otimes (e^c \wedge e^d)$, then the sectional curvature is $R_{abcd}u^a v^b u^c v^d$ where u^a is the pairing of the vector u with the basis element e^a, and likewise v^a is the pairing of v with e^a. The positive or negative sectional curvature condition requires that all such sectional curvatures are respectively positive or negative.

The Einstein equations: These equations ask the Ricci tensor to obey $\text{Ric} = \frac{1}{2} Rg$ where g is the metric, R is the scalar curvature. One can also ask that Ric $= 0$, or that Ric be a positive or negative definite when viewed as a symmetric, bilinear form on TM.

Constant scalar curvature: This demands that the scalar curvature function R be constant over M.

More is said about some of these conditions in the subsequent parts of this chapter.

There are other conditions that are special to certain dimensions (such as 4), or to metrics on manifolds with additional structure imposed such as an almost complex structure. The latter is defined in Chapter 17. (Chapter 17 also introduces a curvature condition that is special to this sort of manifold.)

16.4 Manifolds of dimension 2: The Gauss–Bonnet formula

Suppose that M is a 2-dimensional manifold with a Riemannian metric. There is but one independent component of the Riemann curvature on a 2-dimensional manifold, and this can be taken to be R, the scalar curvature. Indeed, fix an open set $U \subset M$ where there is an orthonormal frame for T^*M on U. Let $\{e_1, e_2\}$ deote this frame. As defined with respect to this frame, the Riemann curvature tensor has $R_{1212} = \frac{1}{2}R$ and the other components of the Riemann tensor are either equal to $\pm R_{1212}$ or zero. The next proposition implies that the function R cannot be just any function on M.

Proposition 16.4 *If* M *is compact and oriented, then*

$$\int_M R\mu_M = 8\pi(1 - \text{genus}(M)).$$

Proof of Proposition 16.4 A Riemannian metric on an oriented surface endows the TM with an almost complex structure. Two observations are needed in order to define this almost complex structure. Here is the first: To say that M is orientable is to say that the line bundle $\wedge^2 T^*M$ admits a nowhere vanishing section. To say that M is oriented is to say that such a section has been chosen up to multiplication by a *positive*, nowhere zero function. The metric on M gives such a section, that with norm 1 at each point. Let μ_M denote the latter. If $\{e^1, e^2\}$ are an orthonormal basis for T^*M at a given point, then μ_M at that point is $e^1 \wedge e^2$. Put μ_M away for the moment.

Here is the second comment: The metric can be viewed as an isomorphism from TM to T^*M. This is the homomorphism that sends any given vector v to the section of T^*M given by $g(v, \cdot)$. Let g^{-1} denote the inverse isomorphism.

Let j denote the almost complex structure whose action on a given vector v is defined by the rule

$$j\cdot v = g^{-1}\cdot(\mu_M(v,\cdot)).$$

Note that $g(jv, u) = \mu_M(v, u) = -g(v, ju)$. This and the fact that g is symmetric implies that $j^2 = -1$. Here is another way to see this: The almost complex structure j acts on an oriented, orthonormal frame, $\{e_1, e_2\}$, for TM at a given point as $je_1 = e_2$ and $je_2 = -e_1$.

It is customary to use $T_{1,0}M$ to denote the resulting rank 1, complex vector bundle that is defined using this almost complex structure on TM.

As explained next, the Levi-Civita covariant derivative on sections of TM induces a covariant derivative on sections of $T_{1,0}M$. Here is a quick way to see this: The metric g is covariantly constant, and thus so is g^{-1}. Note in this regard that if m is any isomorphism between TM and T^*M, then $\nabla m^{-1} = -m^{-1}(\nabla m)\, m^{-1}$. Meanwhile, μ_M is also covariantly constant. This is proved in Chapter 15.1. As a consequence, j is covariantly constant. This last fact implies that $\nabla(jv) = j\nabla v$ when v is a section of TM. This confirms that ∇ commutes with j and so defines a covariant derivative for the resulting rank 1, complex bundle $T_{1,0}M$.

To elaborate on this last point, recall from Chapter 6.3 that $T_{1,0}M$ sits inside the tensor $(TM \otimes_{\mathbb{R}} \mathbb{C})$ as the subspace of vectors of the form $v \otimes 1 - jv \otimes i$. As ∇ commutes with j, it sends a section of this subspace to a section of its tensor product with T^*M.

Here is another way to see the action of ∇ on sections of E: Fix attention on an open set $U \subset M$ where there is an orthonormal basis of sections for TM. Let $\{e_1, e_2\}$ denote such a basis, but one where $\mu_M = e^1 \wedge e^2$. The fact that the connection is metric compatible to write $\nabla e^1 = e^2 \otimes \mathfrak{a}$ and $\nabla e^2 = -e^1 \otimes \mathfrak{a}$ with $\mathfrak{a}$ is a 1-form on U. The covariant derivative of the dual frame is $\nabla e_1 = e_2 \otimes \mathfrak{a}$ and $\nabla e_2 = -e_1 \otimes \mathfrak{a}$. This understood, then $\nabla(je_1) = \nabla e_2 = -e_1 \otimes \mathfrak{a}$ which is $j(\nabla e_1)$, and likewise, $\nabla(je_2) = j(\nabla e_2)$.

The curvature 2-form for this covariant derivative for sections of $T_{1,0}M$ can be written as $i\kappa\, \mu_M$ where κ is a real-valued function on M. The integral over M of $-\frac{1}{2\pi}\kappa$ is the integral over M of the first Chern class of the bundle $T_{1,0}M$. This was computed in Chapter 14.7 to be $2 - 2\text{genus}(M)$. The computation in Chapter 14.7 was done using the construction above for a particular choice of metric on TM and thus connection on $T_{1,0}M$. As the cohomology class of the

Chern forms do not depend on the chosen covariant derivative, this value for the integral must hold for any choice of metric.

To make contact with the function R, note that $T_{1,0}M$ is spanned over the open set U by the section $e = \frac{1}{\sqrt{2}}(e_1 - ie_2)$; and $\nabla e = e \otimes i\mathfrak{a}$. Thus, the curvature 2-form for the covariant derivative on $T_{1,0}M$ is given over U by $i\, d\mathfrak{a} = i\,\kappa\, e^1 \wedge e^2$. On the other hand, writing $d\mathfrak{a}$ as $\kappa\, e^1 \wedge e^2$ identifies the Riemann curvature tensor component R_{1212} with $-\kappa$. Thus, $R = -2\kappa$, and so the integral of $R\,\mu_M$ is $8\pi\,(1 - \text{genus}(M))$.

16.5 Metrics on manifolds of dimension 2

The preceding proposition implies that only the torus can have a Riemannian metric with zero curvature. In the case of a surface with positive genus one can ask if there are metrics where R is constant. As seen previously, there is a metric on S^2 with constant R; the metric that is induced from its standard inclusion into $\mathbb{R}^3$ as the sphere of radius 1 has $R = 2$ and the sphere of radius $\rho > 0$, the latter has $R = 2\rho^{-2}$.

What follows gives an example of an $R = 0$ metric on the torus. Let $\Lambda = \{\mathfrak{v}_1, \mathfrak{v}_2\}$ denote an ordered pair of linearly independent vectors in $\mathbb{R}^2$. Define a $\mathbb{Z}^2$-action on $\mathbb{R}^2$ by sending a given pair of integers, (m, n), to $2\pi(m\mathfrak{v}_1 + n\mathfrak{v}_2)$. As explained earlier, the quotient space $\mathbb{T}_\Lambda = \mathbb{R}^2/\mathbb{Z}^2$ is diffeomorphic to the standard torus $S^1 \times S^1$ with a flat metric. The standard Euclidean metric on $\mathbb{R}^2$ is invariant under translations, so it defines a Riemannian metric on this quotient. The former has $R = 0$, and thus so does the latter.

The next theorem classifies the metrics on compact 2-dimensional manifolds, and in doing so, classifies the metrics with constant scalar curvature. The theorem uses the following terminology: Metrics g and g′ on a given manifold M are said to be *conformal* if $g = f\,g'$ with f a positive function on M. Here, g and g′ are to be viewed as sections of $T^*M \otimes T^*M$. More is said about conformal metrics in Chapter 16.6.

Theorem 16.5 *Let* M *denote a compact, oriented 2-dimensional manifold, and let* g *denote a Riemannian metric on* M. *There exists a positive real function* $f\colon M \to R$ *and a diffeomorphism* $\phi\colon M \to M$ *such that*

- *If* $M = S^2$, *then* $f\,\phi^*g$ *is the spherical metric that is obtained by viewing* S^2 *as the unit sphere in* $\mathbb{R}^3$. *The diffeomorphism* ϕ *is unique up to the action of the group* $Sl(2; \mathbb{C})$ *on the unit sphere in* $\mathbb{R}^3$. *View* S^2 *as* $\mathbb{CP}^1$, *which is to say the space of complex lines in* $\mathbb{C}^2$. *The group* $Sl(2; \mathbb{C})$ *acts as a group of* $\mathbb{C}$*-linear transformations of* $\mathbb{C}^2$ *by definition, and so it acts on the space of complex* *lines. The pull-back of the round metric by a diffeomorphism from* $\mathrm{Sl}(2;\mathbb{C})$ *is conformal to the round metric.*

- *If* $M = S^1 \times S^1$, *then* $f\phi^*g$ *is a metric with zero curvature, in fact, a metric as described above from the identification* $S^1 \times S^1 \to \mathbb{T}_\Lambda$ *for some lattice* $\Lambda \subset \mathbb{R}^2$. *The lattice* Λ *is unique up to the action of* $\mathrm{Sl}(2;\mathbb{Z})$ *on* $\mathbb{R}^2$.
- *If* M *is a surface of genus greater than 1, then* $f\,\phi^*g$ *has constant curvature* $R = -1$. *In fact,* M *is the quotient of the 2-dimensional hyperbolic space (with* $\rho = 1$*) by the action of a discrete subgroup of the Lorentz group. The set of equivalence classes of such hyperbolic metrics can be given a topology such that the complement of a locally finite subspace is a smooth manifold with dimension* $6(\mathrm{genus}(M) - 1)$.

With regards to the third item, the equivalence relation here posits that metrics g and g' are equivalent if and only if there exists a diffeomorphism $\phi\colon M \to M$ such that $\phi^*g = g'$.

The proof of this theorem requires some tools that this book does not discuss, and so the proof is not given. However, the next part of this chapter says a little bit about changes of the form $g \to f\,g$. Meanwhile, Chapter 16.8 discusses the topology of manifolds that admit metrics with constant negative or zero sectional curvature. Both subjects are germaine to the proof of this theorem.

16.6 Conformal changes

Let M be a smooth manifold of any dimension, and suppose that g is a metric on TM. A metric g' is said to be *conformal* to g when $g' = f\,g$ where $f\colon M \to (0, \infty)$ is a positive function on M. Length measured by g' differs from those of g, but not angles. Suppose that $\{e^a\}_{1\le a\le m}$ is an orthonormal frame for T^*M as defined by g over an open set $U \subset M$. Write $h = e^u$ and then $(e')^a = h\,e^a$ is orthonormal for g'. This fact can be used to compute the curvature of g' in terms of derivatives of h and the curvature of g. To do so write $\nabla e^b = e^a \otimes \Gamma^{ab}$ with ∇ being the Levi-Civita covariant derivative for g. Here $\Gamma^{ab} = -\Gamma^{ba}$ and $de^b = -\Gamma^{ab} \wedge e^a$. (The first condition asserts metric compatibility and the second is the torsion free condition.) I also use here and below the convention that repeated indices are implicitly summed over their range of values. Use ∇' to denote the g' Levi-Civita covariant derivative and write $\nabla'(e')^a = (e')^b \otimes (\Gamma')^{ba}$. The equation $de^b = -\Gamma^{ab} \wedge e^a$ implies that

$$d(e')^b = -\Gamma^{ab} \wedge (e')^a + \partial_a h e^a \wedge e^b,$$

where $\{\partial_b h\}$ are the components of dh when written in terms of the basis $\{e^b\}$. Thus, $\partial_b h$ denotes the directional derivative in the direction of the dual vector field to e^b. The left hand side of the last equation must equal $-(\Gamma')^{ab} \wedge (e')^a$ with $(\Gamma')^{ab} = -(\Gamma')^{ba}$. As seen in Chapter 15.2 there is just one way to solve this equation, and here is the solution:

$$(\Gamma')^{ab} = \Gamma^{ab} + \partial_b h h^{-1} e^a - \partial_a h h^{-1} e^b.$$

The curvature 2-form $(F_{\nabla'})^{ab}$ is given by $d(\Gamma')^{ab} + (\Gamma')^{ac} \wedge (\Gamma')^{cb}$, and thus

$$\begin{aligned}(F_{\nabla'})^{ab} = (F_{\nabla})^{ab} &+ (\nabla_c(\partial_b h h^{-1}))e^c \wedge e^a - (\nabla_c(\partial_a h h^{-1}))e^c \wedge e^b \\ &+ (\partial_c h h^{-1} e^a - \partial_a h h^{-1} e^c) \wedge (\partial_c h h^{-1} e^b - \partial_b h h^{-1} e^c).\end{aligned}$$

Write this as $\frac{1}{2}(R')_{abcd}\ (e')^c \wedge (e')^d$ to obtain the Riemann curvature tensor for g':

$$\begin{aligned}(R')_{abcd} = e^{-2u}(R_{abcd} &+ (\nabla_c \partial_b u)\delta_{ad} - (\nabla_d \partial_b u)\delta_{ca} - (\nabla_c \partial_a u)\delta_{db} \\ &+ (\nabla_d \partial_a u)\delta_{cb} + |du|^2(\delta_{ac}\delta_{bd} - \delta_{ad}\delta_{bc}) + \partial_c u \partial_b u \delta_{da} \\ &- \partial_d u \partial_b u \delta_{ca} - \partial_c u \partial_a u \delta_{db} + \partial_d u \partial_a u \delta_{cb})\end{aligned}$$

where the notation is as follows: First, $u = \ln(h)$ and $\delta_{ab} = 1$ if $a = b$ and 0 otherwise. Second, $\nabla_c u = h^{-1}\partial_c h$ are the components of du when written in terms of the basis $\{e^a\}$, and $\nabla_a \partial_c u = \partial_a \partial_c u + \Gamma^b_{ac}\partial_b u$ are the components of ∇du when written in terms of this same basis. Finally, $|du|$ is the norm of the 1-form du as defined by the metric g.

The Ricci curvature tensor $(R')_{ac}$ can then be written as

$$(R')_{ac} = e^{-2u}(R_{ac} - \Delta u \delta_{ca} - (n-2)\nabla_a \partial_b u + (n-2)(|du|^2\delta_{ac} - \nabla_a \partial_c u)),$$

where Δ is commonly used to denote $\nabla_b \partial_b$. Finally, the scalar curvatures are related by

$$R' = e^{-2u}(R - 2(n-1)\Delta u + (n-2)(n-1)|du|^2).$$

If I am looking for a metric g' conformal to g with constant scalar curvature equal to c; thus $R' = c$, then I would have to find a function u on M that solves the equation

$$R = \Delta u - \frac{n-2}{2}|du|^2 + \frac{1}{2(n-1)}ce^{2u}.$$

In the case $n = 2$, this equation reads

$$R = \Delta u + \frac{1}{2}ce^{2u}.$$

As it turns out, it is always possible to solve this equation on a compact, dimension 2 manifold.

A metric, g, on a manifold is said to be *locally conformally flat* if there exists, in a neighborhood of any given point, a positive function, f, such that f g has vanishing Riemann curvature tensor.

16.7 Sectional curvatures and universal covering spaces

Let M denote a given manifold and g a Riemannian metric for M. The Riemann curvature is viewed in what follows as a section, Riem, of $(\wedge^2 T^*M) \otimes (\wedge^2 T^*M)$. Fix a point $x \in M$, and then a 2-dimensional subspace $\Pi \subset TM|_x$. Fix a pair of orthonormal vectors u, v in Π. The latter define the element $u \wedge v \in \wedge^2 TM|_x$. Evaluate Riem on $(u \wedge v) \otimes (u \wedge v)$ to get a number. This number is called the *sectional curvature* for the subspace Π. This number does not depend on the chosen basis vectors u and v for Π.

What follows is another way to define the sectional curvature. Let $\{e^a\}$ denote an orthonormal frame for $T^*M|_x$. The section Riem of $(\wedge^2 T^*M) \otimes (\wedge^2 T^*M)$ at x is

$$\text{Riem} = \frac{1}{4} R_{abcd}(e^a \wedge e^b) \otimes (e^c \wedge e^d).$$

Granted this depiction of Riem, the sectional curvature at x for the subspace Π is

$$\text{Riem}(u, v, u, v) = R_{abcd} u^a v^b u^c v^d.$$

Here the notation is such that u^a is the pairing of u with e^a; this the component of u along the dual vector field to e^a. Meanwhile, v^a is defined analogously using v.

By way of an example, if M has dimension 2, then there is just one sectional curvature, and it is equal to $\frac{1}{2}R$.

The sectional curvatures from a metric on a given manifold M can be used in certain instances to say something about its topology. The theorem that follows is a classical result. It refers to the universal cover of a manifold. The latter is defined in Chapter 13.7.

Theorem 16.6 *Suppose that the metric is such that all sectional curvatures are nonpositive everywhere on* M. *Then the universal cover of* M *is diffeomorphic to* $\mathbb{R}^n$.

What follows sketches a proof of this: Fix a point $p \in M$ and consider the exponential map $\exp_p : TM|_p \to M$. This map was defined in Chapter 9.2; it says to move a distance 1 along the geodesic that starts at x and has initial velocity vector given by the chosen point in $TM|_p$. The assertion of the theorem follows from the claim that $\exp_p$ is a covering map. The proof of the latter claim has two parts. The first part establishes that the differential of $\exp_p$ is an isomorphism at each point of $TM|_p$. It follows as a consequence of this first part that any given vector $v \in TM|_p$ has a neighborhood that is mapped diffeomorphically by $\exp_p$ onto an open set in X. This last conclusion does not by itself imply that $\exp_p$ is a covering map. Needed is a positive constant, δ, with the following property: The inverse image of any given ball in X of radius δ is a disjoint union of open sets which are each mapped by $\exp_p$ diffeomorphically to the given ball. The second part of the argument proves that such a constant does indeed exist.

Chapter 16.8 gives Part 1 of the argument. Part 2 follows from what is said below in Chapter 16.9.

16.8 The Jacobi field equation

Part 1 of the proof of Theorem 16.6 has five steps. Steps 1–4 introduce and discuss a novel differential equation called the *Jacobi field* equation. The final step uses this equation with the assumption of nonnegative sectional curvature to complete Part 1.

Step 1

The image via $\exp_p$ of a given nonzero vector $v \in TM|_p$ is defined using the geodesic through p with tangent vector v at p. This being the case, an analysis of the differential of $\exp_p$ requires first an understanding of how the geodesics through p vary as a function of their tangent vector at p.

To investigate this issue, let $t \to \gamma(t)$ denote the geodesic, parameterized by $t \in \mathbb{R}$ with $\gamma(0) = p$ and with $\dot{\gamma}|_0 = v \in TM|_p$. A deformation of the geodesic γ is viewed in what follows as a map ϕ: $(-1, 1) \times \mathbb{R} \to M$ such that $\phi(s, \cdot)$ is a geodesic for all $s \in (-1, 1)$, and such that $\phi(0, \cdot) = \gamma$. Let ϕ denote such a deformation. There is no requirement at this point that $\phi(s, 0) = p$ for all s. Thus, the deformed geodesic is allowed to miss p.

Of interest with regards to the differential of $\exp_p$ is the section of γ^*TM given by $\eta = \phi_* \frac{\partial}{\partial s}$ at $s = 0$. This is because the section η at $t = 1$ is the image of the vector $\eta|_{t=0} \in TM|_p$ via the differential at v of $\exp_p$.

Step 2

The section $\eta = (\phi_* \frac{\partial}{\partial s})|_{s=0}$ of γ^*TM satisfies a certain differential equation by virtue of the fact that each $s \in (-1, 1)$ version of $\phi(s, \cdot)$ is a geodesic. The latter equation is called the *Jacobi field* equation. What follows is a derivation of this equation: Fix $t_* \in \mathbb{R}$ and a coordinate chart for a neighborhood in X of $\gamma(t_*)$. There exists $\varepsilon > 0$ such that if $|s| < \varepsilon$ and $|t - t_*| < \varepsilon$, then $\phi(s, t)$ will be in this coordinate chart. For such s and t, use $\{\phi^1, \ldots, \phi^{\dim(M)}\}$ to denote the coordinates of $\phi(s, t)$ with respect to the given chart. As each $\phi(s, \cdot)$ is a geodesic, each will satisfy the differential equation

$$\frac{d^2}{dt^2}\phi^i + \Gamma^i_{jk}\left(\frac{d}{dt}\phi^j\right)\left(\frac{d}{dt}\phi^k\right) = 0$$

if $|s| < \varepsilon$ and $|t - t_*| < \varepsilon$. Now differentiate this last equation with respect to s and then set $s = 0$. Doing so finds that the section $t \to \eta(t)$ of γ^*TM obeys an equation that can be written without reference to a coordinate chart as

$$\nabla_t^2\eta + (\gamma^* F_\nabla)(\eta, \dot\gamma)\,\dot\gamma = 0$$

where the notation is as follows: First ∇_t is the pull-back of the Levi-Civita covariant derivative to the bundle γ^*TM over $\mathbb{R}$. Second, F_∇ is the curvature of the Levi-Civita covariant derivative. Thus, $F_\nabla(u, v)$ for vectors u and v in TM over a given point is an endomorphism of TM at that point.

This last equation is the Jacobi field equation. It is a linear, second-order differential equation for a section of γ^*TM. A solution is a *Jacobi field* along γ.

Step 3

To see what F_∇ looks like, fix an orthonormal, covariantly constant frame $t \to \{e_a|_t\}$ for γ^*TM. This is a frame such that $\nabla_t e_a = 0$ for each index a and each $t \in \mathbb{R}$, and also such that $(\gamma^*g)(e_a, e_b)$ at any given $t \in \mathbb{R}$ is equal to 1 when $a = b$ and 0 otherwise. Recall from Chapter 11.5 that there exists frames with $\nabla_t e_a = 0$ for index a and each $t \in \mathbb{R}$. If the frame at $t = 0$ is orthonormal, then it will be orthonormal at all times by virtue of the fact that the Levi-Civita connection is metric compatible. Indeed, the latter point implies that

$$\frac{d}{dt}(\gamma^*g)(e_a, e_b) = (\gamma^*g)(\nabla_t e_a, e_b) + (\gamma^*g)(e_a, \nabla_t e_b).$$

As the right-hand side above is zero if the frame is covariantly constant, so $(\gamma^*g)(e_a, e_b)$ is constant.

Write a Jacobi field η in terms of this frame as $\eta = \eta^a e_a$ where now η^a is a function on $\mathbb{R}$. Here and in what follows, the convention is enforced that repeated indices are summed over their range of values. Meanwhile, $\dot\gamma$ defines a section of γ^*TM, and so it to can be written as $\dot\gamma = \dot\gamma^a e_a$. Since $\nabla_t e_a = 0$, the Jacobi field equation asserts that the functions $\{\eta^a\}$ obey the following system of differential equations:

$$\frac{d^2}{dt^2}\eta^a + R_{abcd}\dot{\gamma}\, b\eta^c\dot{\gamma}^d = 0, \qquad (*)$$

where R_{abcd} is the function on $\mathbb{R}$ which at time t is the indicated component of the Riemann curvature at the point $\gamma(t) \in M$ when written using the dual frame $\{e^a|_t\}$.

Step 4

The next proposition asserts what is needed concerning the question of the existence of Jacobi fields.

Proposition 16.7 *The set of solutions to the Jacobi equations is a vector space of dimension twice that of* M. *This vector space is isomorphic to* $TM|_p \otimes TM|_p$ *via the map that associates to a solution the vectors* $(\eta|_{t=0}, \nabla_t\eta|_{t=0})$.

Proof of Proposition 16.7 This is a consequence of the vector field theorem from Appendix 8.1. To elaborate, introduce the map $\mathfrak{v}\colon \mathbb{R} \times (\times_2 TM|_p) \to \mathbb{R} \times (\times_2 TM|_p)$ that sends a given element $(\tau, (u, v))$ to $(1, v, -(R_{abcd}\, \dot{\gamma}^b\, \dot{\gamma}^d)|_\tau\, u^c))$. The map $z\colon \mathbb{R} \to \mathbb{R} \times (\times_2 TM|_p)$ that solves this equation with initial value $z|_{t=0} = (\tau = 0, u = u_{,0}, v = v_0)$ has components at time t given by $z(t) = (\tau = t, u = \eta, v = \frac{d}{dt}\eta)$ where η obeys the Jacobi field equation with initial value $\eta|_{t=0} = u_0$ and $(\frac{d}{dt}\eta)|_{t=0} = v_0$. Conversely, if η obeys the Jacobi field equation with these same $t = 0$ values, then z as just described solves the corresponding vector field equation.

The following proposition restates what is said in Step 1. Note in this regard that the geodesic $\gamma_r\colon \mathbb{R} \to M$ with $\gamma_r|_{t=0} = p$ and $\dot{\gamma}_n|_{t=0} = r\, v$ is given by $\gamma_r|_t = \gamma|_{rt}$.

Proposition 16.8 *Let* $v \in TM|_p$ *denote a given vector with norm 1 and let* $\gamma\colon \mathbb{R} \to M$ *denote the geodesic with* $\gamma|_{t=0} = p$ *and with* $\dot{\gamma}|_{t=0} = v$. *Fix a vector* $u \in TM|_p$ *and let* η *denote the solution to the Jacobi field along* γ *with* $\eta|_0 = 0$ *and with* $(\frac{d}{dt}\eta)|_{t=0} = u$. *Suppose that* r *is a given positive number. Then the differential of* $\exp_p$ *at* r v *maps the vector* $u \in TM|_p$ *to* $\eta|_{t=r} \in TM|_{\gamma(r)}$.

Proof of Proposition 16.8 The assertion is a consequence of the definition of the Jacobi field equation.

Step 5

This step indicates why the signs of the sectional curvatures are relevant to the question of whether the differential of $\exp_p$ is or is not everywhere surjective. To start, suppose that γ is a geodesic with $\gamma|_{t=0} = p$ and suppose that η is a Jacobi field along γ with $\eta(0) = 0$ and $\eta^{\cdot}(0) \neq 0$. Given the second proposition in the preceding step, the surjectivity of the differential of $\exp_p$ at points on the ray in $TM|_p$ tangent to $\dot{\gamma}|_{t=0}$. follows if it is the case that $\eta|_t \neq 0$ for all such η and for all $t = 0$.

To see what to make of this, take the inner product of the expression at time t on the left-hand side of Equation (*) in Step 3 with the vector in $\mathbb{R}^{\dim(M)}$ with components $\{\eta^a|_t\}_{1\leq a\leq \dim(M)}$. The resulting equation can be manipulated so as to read

$$\frac{1}{2}\frac{d^2}{dt^2}|\eta|^2 - \left|\frac{d}{dt}\eta\right|^2 + 2R_{abcd}\eta^a\dot{\gamma}^b\eta^c\dot{\gamma}^d = 0.$$

The term with R is the sectional curvature in the plane spanned by η and $\dot{\gamma}$. If the sectional curvatures are nonpositive everywhere, then this last equation implies that

$$\frac{d^2}{dt^2}|\eta|^2 \geq \left|\frac{d}{dt}\eta\right|^2.$$

Integrating this equation finds that

$$\frac{d}{dt}|\eta|^2 > 0$$

since $\frac{d}{dt}|\eta|$ is nonzero at $t = 0$. This shows that $|\eta|^2 > 0$ for $t > 0$. Thus, the differential of $\exp_p$ on $TM|_p$ on the ray tangent to $\dot{\gamma}^b$ has trivial kernel and so the differential along this ray is at each point an isomorphism.

Note in particular that if a 2-dimensional manifold has everywhere negative scalar curvature, then it has negative sectional curvature. As a consequence, its universal cover is diffeomorphic to $\mathbb{R}^2$.

16.9 Constant sectional curvature and the Jacobi field equation

A Gaussian coordinate chart centered at a given point in M is defined in Chapter 9.3 using the exponential map. Let $p \in M$ denote the point in question. To define this chart, fix an orthonormal frame $\{e_a\}_{1\leq a\leq n}$ for $TM|_p$, where n here denotes the dimension of M. The inverse of the coordinate chart map sends a point $x = (x^1,\ldots, x^n) \in \mathbb{R}^n$ to the point $\exp_p(x^a e_a) \in M$. Here and in what follows, the summation convention for repeated indices is again used. As explained in Chapter 9.3, this map is a diffeomorphism from a ball in $\mathbb{R}^n$ about the origin onto a neighborhood of p in M. The last equation in Chapter 9.3 is a schematic formula for the pull-back of the metric via Gaussian coordinates.

A somewhat more explicit formula is derived in the first part of what follows using the Jacobi fields.

To start, suppose that γ is a geodesic with $\gamma|_{t=0} = p$ and with $\dot{\gamma}|_{t=0} = v \in TM|_p$. As done in the preceding section of this chapter, fix a covariantly constant, orthonormal frame $\{e_a\}_{1\le a\le n}$ for γ^*TM so as to write a given Jacobi field along γ as $\eta = \eta^a e_a$ where $(\eta^1, \ldots, \eta^n)$ is a map from $\mathbb{R}$ to $\mathbb{R}^n$ that obeys the Equation $(*)$ in that same section of this chapter.

The pull-back to $T^*M|_p$ of the Riemannian metric by the exponential map is obtained from the solutions to the Jacobi equations as follows: Let u and u′ denote vectors in $T^*M|_p$, and let η and η' denote the corresponding Jacobi fields along γ that vanish at $t = 0$ and are such that $\nabla_t\eta|_{t=0} = u$ and $\nabla_t\eta'|_{\tau=0} = u'$. Then the inner product at v of u and u′ as defined by $\exp_p^* g$ is given by

$$(\eta^a(\eta')^a)|_{t=1}.$$

To recover what is done at the end of Chapter 9.3, suppose that $u{\cdot}v = u^a v^a = 0$, then $\eta^a v^a = 0$ at all $t \in \mathbb{R}$ since the Jacobi equation implies that $\frac{d^2}{dt^2}(\eta{\cdot}v) = 0$. As a consequence, if v is written as $r\hat{v}$ where $r \in [0, \infty)$ and $\hat{v}$ is a unit vector, then the metric has the form

$$dr \otimes dr + h$$

where $h(u, u') = (\eta^a(\eta')^a)|_{t=1}$. Here, it is assumed that $u{\cdot}v$ and $u'{\cdot}v = 0$. Meanwhile η^a solves the equation

$$\frac{d^2}{dt^2}\eta^a + r^2 R_{abcd}\hat{v}^b\eta^c\hat{v}^d = 0$$

with initial conditions $\eta^a(0) = 0$ and $\frac{d}{dt}\eta^a|_0 u^a$, while η' solves the same equation with $(\eta')^a(0) = 0$ and $\frac{d}{dt}(\eta')^a|_0 = (u')^a$. Note that $h(u, u')$ depends both on r and $\hat{v}$ because this equation depends on them.

Suppose now that the sectional curvatures are constants, and in particular are all zero, thus $R_{abcd} = 0$. The Jacobi equation asserts in this case that $\frac{d^2}{dt^2}\eta^a = 0$. This equation integrates to give $\eta^a = tu^a$. As a consequence, the pull-back $\exp_p^* g$ has the inner product at v of u and u′ equal to $u{\cdot}u'$, which is their Euclidean inner product. This means that the metric on the universal cover of M is the Euclidean metric, and M is obtained from $\mathbb{R}^n$ as the quotient by a group of isometries of the Euclidean metric. The following proposition summarizes:

Proposition 16.9 *Suppose that* M *has a Riemannian metric with everywhere zero curvature. Then the universal cover of* M *is* $\mathbb{R}^{\dim(M)}$ *and* M *is the quotient of this Euclidean space by a group of isometries of the Euclidean metric.*

Suppose next that $R_{abcd} = c\,(\delta_{ac}\delta_{bd} - \delta_{ad}\delta_{bc})$ where c is a constant but not zero. The Jacobi equation for η with $u{\cdot}\hat{v} = 0$ reads

$$\frac{d^2}{dt^2}\eta^a + cr^2\eta^a = 0.$$

In the case where $c > 0$, the solution with $\eta|_{\tau=0} = 0$ and $\frac{d}{dt}\eta^a|_{t=0} = u^a$ is

$$\eta^a = (c^{1/2}r)^{-1}\sin(c^{1/2}rt)u^a.$$

Thus, the inner product h is given by

$$h(u', u) = c^{-1}r^{-2}\sin^2(c^{1/2}r)u{\cdot}u'.$$

This formula for $\exp_p^*g$ is identical to the formula for the metric that is obtained using the exponential map on the sphere S^n of radius $c^{-1/2}$. As a consequence, the universal cover of M is the sphere S^n; and M is the quotient of this sphere by a group of isometries of its round metric.

In the case when $R_{abcd} = c\,(\delta_{ac}\delta_{bd} - \delta_{ad}\delta_{bc})$ with $c < 0$, the solutions are

$$\eta^a = (|c|^{1/2}r)^{-1}\sinh(|c|^{1/2}rt)u^a$$

and the inner product h is given by

$$h(u, u') = |c|^{-1}r^{-2}\sinh^2(|c|^{1/2}r)u{\cdot}u'.$$

This formula for $\exp_p^*g$ is identical to the formula for metric that is obtained using the exponential map on the hyperboloid in $\mathbb{R}^{1,n}$ given by $t^2 - x^2 = |c|$. As a consequence, the universal cover of M is $\mathbb{R}^n$ with its hyperbolic metric; this is as described in Chapter 8.4.

What follows summarizes the $c > 0$ and $c < 0$ cases.

Proposition 16.10 *Suppose that* M *has a Riemannian metric with sectional curvatures given by* $R_{abcd} = c\,(\delta_{ac}\delta_{bd} - \delta_{ad}\delta_{bc})$ *with* $c > 0$. *Then the universal cover of* M *is* $S^{\dim(M)}$ *and* M *is the quotient of the sphere of radius* $c^{-1/2}$ *by a group of isometries of its round metric. Suppose, on the other hand, that* M *has a Riemannian metric with sectional curvatures given by* $R_{abcd} = c\,(\delta_{ac}\delta_{bd} - \delta_{ad}\delta_{bc})$ *with* $c < 0$. *Then the universal cover of* M *is* $\mathbb{R}^{\dim(M)}$ *and* M *is the quotient of the* $\mathbb{R}^n$ *by the group of isometries of its hyperbolic metric.*

16.10 Manifolds of dimension 3

William Thurston made a profound conjecture about 3-dimensional manifolds that implied a complete list of homeomorphism and diffeomorphism equivalence classes. (These are the same in dimension 3.) The conjecture was known as the *Geometrization conjecture*; and it was proved a few years ago by Grigory Perelman using and extending technology that was developed by Richard Hamilton. It is now known as the *Geometrization theorem*. What follows is meant to give an idea of what this theorem says.

Let M denote a compact, 3-dimensional manifold. Then M can be cut along a finite set of disjoint, embedded 2-dimensional tori and 2-dimensional spheres so that the result is a disjoint union of components. This is to say that these surfaces are deleted from M. The cutting surfaces are determined by the fundamental group of M, and by $\pi_2(M)$, the set of homotopy classes of maps from S^2 into M. In particular, the cutting surfaces are defined by algebraic data. This decomposition is known as the JSJ decomposition, named after William Jaco, Peter Shalen and Klaus Johannson.

The result of all this cutting has some finite union of components. The Geometrization theorem asserts that each component is the quotient of one, and only one, of eight possible complete Riemannian manifolds by a discrete subgroup of the diffeomorphisms that acts isometrically. Here, a Riemannian manifold is taken to be a pair (M, g) where M is the manifold and g the metric. A diffeomorphism, ϕ: M → M is said to be an isometry if $\phi^*g = g$.

All but one of these eight Riemannian manifolds can be viewed as a 3-dimensional Lie group with a left-invariant metric. In any event, the list of the eight Riemannian manifolds includes $\mathbb{R}^3$ with its flat metric, S^3 with its round metric, and the 3-dimensional hyperbolic space with its hyperbolic metric. The list also includes the product of $\mathbb{R}$ and S^2 with its round metric; and the product of $\mathbb{R}$ and 2-dimensional hyperbolic space. There are three more manifolds: The first is the universal cover of $Sl(2; \mathbb{R})$; the second other is the Heisenberg group, this the group of upper triangular 3×3 matrices with all diagonal entries equal to 1. The last is what is called *solve geometry*. The manifold is $\mathbb{R}^3$, but the group law is such that

$$(x, y, z)\cdot(x', y', z') = (e^z x' + x,\ e^{-z} y' + y, z + z').$$

A very detailed discussion of these geometries can be found in Chapters 3.8 and 4.7 of William Thurston's book *Three Dimensional Geometry and Topology* (1997).

16.11 The Riemannian curvature of a compact matrix group

Chapter 8.7 defines a matrix group to be a subgroup of some $n \geq 1$ version of SO(n) or U(n). Let G denote such connected, matrix group. It is given the

Riemannian metric that is induced from that on SO(n) or U(n) described in Chapter 8.5 or 8.6 as the case may be. This metric is such that there is an orthonormal basis of left invariant vector fields. By way of a reminder, recall from Chapter 5.4 that the tangent space to G at the identity element ι is a subvector space of the vector space of either anti-symmetric or anti-Hermitian $n \times n$ matrices. The former occurs when G is in SO(n) and the latter when G sits in U(n). This vector space is denoted by $\mathfrak{lie}(G)$. A given element $q \in \mathfrak{lie}(G)$ defines the left-invariant vector field v_q that is given at a matrix $m \in G$ by the formula

$$v_a| = -\mathrm{tr}\left(q\ m^T \frac{\partial}{\partial m}\right).$$

The inner product of v_q and $v_{q'}$ is $\mathrm{tr}(q^T q')$ when $G \subset SO(n)$ and $\mathrm{tr}(q^\dagger q)$ when $G \subset U(n)$. This inner product defines a Riemannian metric on G. What follows describes the Riemann and Ricci tensors for this metric.

For the present purposes, it proves convenient to view a subgroup of U(n) as a subgroup of the larger group SO(2n) so as to give a "one size fits all" account. By way of explanation, recall what is said in Chapter 2.5: The group U(n) is a subgroup of $Gl(n; \mathbb{C})$. Meanwhile, the group $Gl(n; \mathbb{C})$ can viewed as a subgroup of $Gl(2n; \mathbb{R})$. The depiction of $Gl(n; \mathbb{C})$ in $Gl(2n; \mathbb{R})$ makes U(n) and its subgroups appear as subgroups in SO(2n).

Use m to denote the dimension of G. Fix an orthonormal basis for $\mathfrak{lie}(G)$ and let $\{v_1, \ldots, v_m\}$ denote the corresponding orthonormal basis of left-invariant vector fields. Let $\{\omega_1, \ldots, \omega_m\}$ denote the dual basis of left invariant 1-forms. This is to say that $\omega_k(v_j)$ is zero if $j \neq k$ and one if $j = k$. The basis $\{\omega_1, \ldots, \omega_m\}$ is orthonormal. By way of a reminder from Chapter 5.4 that the left-invariant 1–forms are also in 1–1 correspondence with the elements in $\mathfrak{lie}(G)$ with the correspondence such that a given element $q \in \mathfrak{lie}(G)$ corresponds to the left-invariant 1-form that is given at m by the rule

$$\mathrm{tr}(qm^{-1}dm).$$

Keep this formula in mind for what follows.

With the basis $\{\omega^1, \ldots, \omega^m\}$ fixed, any given covariant derivative on the space of sections of TG corresponds to a set, $\{\Gamma^{ij}\}_{i \leq j,j \leq m}$, of 1-forms, and vice versa. The correspondence is such that

$$\nabla\omega^i = \sum_{1 \leq j \leq m} \omega^j \otimes \mathfrak{a}^{ij}.$$

As explained in Chapters 15.1 and 15.3, the covariant derivative ∇ is metric compatible if and only if $\{\mathfrak{a}^{ij}\}_{1 \leq i,j \leq d}$ is such that $\mathfrak{a}^{ij} = -\mathfrak{a}^{ji}$ for all index pairs (i, j). Chapter 15.2 introduces the notion of a torsion free covariant derivative. It follows from what is said in Chapter 15.3 that the Levi-Civita covariant derivative is determined completely by the preceding antisymmetry conditions $\{\mathfrak{a}^{ij} = -\mathfrak{a}^{ji}\}_{1\le i,j\le m}$ together with the $\mathfrak{m}$ additional conditions

$$d\omega^i = -\textstyle\sum_{1\le j\le m}\mathfrak{a}^{ij}\wedge\omega^j.$$

Write each 1-form $\mathfrak{a}^{ij}$ as a linear combination of the basis elements $\{\omega^1,\ldots,\omega^m\}$, thus as $\sum_{1\le k\le m}\mathfrak{a}^{ij}_k\,\omega^k$ where each $\mathfrak{a}^{ij}_k$ is a function on G. This writes the preceding equation as

$$d\omega^i = -\textstyle\sum_{1\le j,k\le m}\mathfrak{a}^{ij}_k\omega^k\wedge\omega^j. \qquad (*)$$

The fact that the basis $\{\omega^i\}_{1\le i\le m}$ is invariant under action of left translation implies that the functions $\{\mathfrak{a}^{ij}_k\}_{1\le i,j,k\le m}$ are constant functions on G. This is so because both the exterior derivative operation that appears on the left side of $(*)$ and the exterior product operation that appears on the right-hand side commute with the pull-back via a smooth map. The fact that $\{a^{ij}_k\}_{1\le i,j,k\le m}$ are constant tremendously simplifies calculations because one need only make computations at the identity element. This is done momentarily.

As explained next, the fact that $\{a^{ij}_k\}_{1\le i,j,k\le m}$ are constant also simplifies the task of computing the Riemann curvature tensor. To elaborate, recall from Chapter 15.7 that the 2-form $\{F^{ij}\}_{1\le i,j\le m}$ that determines the curvature 2-form are determined by the 1-forms $\{\Gamma^{ij}\}_{1\le i,j\le m}$ by the equation

$$(F_\nabla)^{ij} = d\mathfrak{a}^{ij} + \Sigma_{1\le k\le m}\mathfrak{a}^{ik}\wedge\mathfrak{a}^{kj}.$$

Given the defining equation $\mathfrak{a}^{ij} = \sum_{1\le k\le m}\mathfrak{a}^{ij}_k\omega^k$ and the fact that the functions $\{\mathfrak{a}^{ij}_k\}_{1\le i,j,k\le m}$ are constant, this last equation finds $(F_\nabla)^{ij}$ equal to

$$(F_\nabla)^{ij} = \textstyle\sum_{1\le k,p,q\le m}(-\mathfrak{a}^{ij}_k\mathfrak{a}^{kq}_p + \mathfrak{a}^{ik}_p\mathfrak{a}^{kj}_q)\omega^p\wedge\omega^q.$$

The components of the Riemann curvature tensor are determined by the 2-forms F^{ij} by the rule $(F_\nabla)^{ij} = 1/2\sum_{1\le p,q\le m} R_{ijpq}\,\omega^p\wedge\omega^q$. This being the case, it follows from this last formula that the Riemann tensor, Ricci tensor and scalar curvature function are

- $R_{ijpq} = \sum_{1\le k\le m}(-\mathfrak{a}^{ij}_k(\mathfrak{a}^{kq}_p + \mathfrak{a}^{ij}_k\,\mathfrak{a}^{kp}_q + \mathfrak{a}^{ik}_p\,\mathfrak{a}^{kj}_q - \mathfrak{a}^{ik}_q\,\mathfrak{a}^{kj}_p)$.
- $Ric_{ip} = \sum_{1\le k,q\le m}(-\mathfrak{a}^{iq}_k\,\mathfrak{a}^{pk}_q + \mathfrak{a}^{ik}_p\,\mathfrak{a}^{kq}_q)$.
- $R = -\sum_{1\le k\le q,p\le m}(\mathfrak{a}^{pq}_k\mathfrak{a}^{kq}_p + \mathfrak{a}^{kp}_p\,\mathfrak{a}^{kq}_q)$.

Note, by the way, that the various functions here are all constant functions on G.

It is time now to derive a formula for the constants in the set $\{\mathfrak{a}^{ij}_k\}_{1\le i,j,k\le m}$. It proves convenient for this purpose to use the exponential map to parametrize a neighborhood of the identity in G by a ball about the origin in $\mathfrak{lie}(G)$. Let $\{q_1,\ldots q_m\}$ denote the given orthonormal basis for $\mathfrak{lie}(G)$. Given $x = (x_1,\ldots, x_m)\in\mathbb{R}^m$, use $x\cdot q\in\mathfrak{lie}(G)$ to denote the matrix $\sum_{1\le i\le m} x^i\,q_i$. The exponential map associates to any given point $x = (x_1,\ldots, x_m)\in\mathbb{R}^m$ the point

$$m(x) = e^{x \cdot q} = 1 + x \cdot q + \frac{1}{2}(x \cdot q)^2 + \cdots$$

in G. As the 1-form $\omega_i = \mathrm{tr}(q_i m^{-1} dm)$, it follows from this last formula that the pull-back of ω_i to $\mathbb{R}^m$ using the exponential map is the 1-form

$$\omega^i = -dx^i - \frac{1}{2}\Sigma_{1 \le j,k \le m} \mathrm{tr}(q_i[q_j, q_k]) x^j dx^k + \cdots$$

where the unwritten terms involve product of 2 or more of the coordinate functions. Here, the terminology is such that $[q, q']$ denotes the commutator $q\, q' - q'\, q$.

This last formula tells us that ω_i and $d\omega_i$ at the identity element $\iota \in G$ are given in these coordinates by the respective forms

$$\omega^i|_\iota = -dx^i \ \textit{and} \ d\omega^i|_\iota = -\frac{1}{2}\Sigma_{1 \le j,k \le m} \mathrm{tr}(q_i[q_j, q_k]) dx^j \wedge dx^k.$$

These formulas with $(*)$ imply that the functions $\{a_k^{ij}\}_{1 \le i,j,k \le m}$ are given by

$$a_k^{ij} = \frac{1}{2} \mathrm{tr}(q_i[q_j, q_k]).$$

Note, by the way, that the expression on the right here changes sign when any two indices are interchanged. This is so because $[q, q'] = -[q', q]$ for any two matrices, and meanwhile, $\mathrm{tr}(q\,[q', q'']) = \mathrm{tr}(q''\,[q, q'])$ for any three matrices.

This antisymmetry property allows the components of the Riemann and Ricci tensors, and the scalar curvature function to be rewritten as

- $R_{ijpq} = \sum_{1 \le k \le m} (2a_k^{ij} a_k^{pq} - a_p^{ik}\, a_q^{jk} + a_q^{ik}\, a_p^{jk})$,
- $\mathrm{Ric}_{ip} = \sum_{1 \le k,q \le m} a_k^{iq}\, a_k^{pq}$,
- $R = -\sum_{1 \le k \le q,p \le m} a_k^{pq}\, a_k^{pq}$.

As can be seen from the preceding, the scalar curvature is a nonnegative constant which is zero only if all the curvature is zero. Since G is compact, it follows from what is said in Chapter 16.8 that G is the quotient of $\mathbb{R}^n$ by a group of isometries. In general, the formula above tells us that the Ricci tensor is nonnegative definite.

16.11.1 The case G = SO(n)

The tangent space is the full vector space of $n \times n$ antisymmetric matrices. An orthonormal basis is $\{\frac{1}{\sqrt{2}} q_{ab}\}_{1 \le a < b \le n}$ where q_{ab} has all entries zero except the entry in row a and column b, which is 1, and the entry in row b and column a, which is -1. The commutators are:

- $[\mathfrak{q}_{ab}, \mathfrak{q}_{cd}] = 0$ *if either* $b \neq c$ *or* $a \neq d$.
- $[\mathfrak{q}_{ab}, \mathfrak{q}_{bd}] = \mathfrak{q}_{ad}$ *and* $[\mathfrak{q}_{db}, \mathfrak{q}_{cd}] = \mathfrak{q}_{bc}$.

This implies that the only nonzero versions of $\mathfrak{a}^{abcd}_{ef}$ are $\mathfrak{a}^{abbc}_{ac} = -\frac{1}{2\sqrt{2}}$. This being the case, the only nonzero Ricci curvature components are

$$R^{ab\,ab} = \frac{n-2}{8}.$$

Note in particular that this Riemannian metric for SO(n) is an Einstein metric.

16.11.2 Simple Lie groups

Suppose that G is a given Lie group, m is a positive integer and V is a d-dimensional vector space, either $\mathbb{R}^m$ or $\mathbb{C}^m$, and ρ is a representation of G acting on V. This is to say that ρ is a map from G into $Gl(m; \mathbb{R})$ or $Gl(m; \mathbb{C})$ as the case may be such that $\rho(g_1 g_2) = \rho(g_1)\cdot\rho(g_2)$ and such that ρ maps the identity matrix $\iota \in G$ to the identity matrix in the relevant general linear group. The representation ρ is said to be *irreducible* if V cannot be written as a direct sum $V_1 \oplus V_2$ such that every $g \in G$ version of $\rho(g)$ maps V_1 to itself and also V_2 to itself. (It is assumed here that both V_1 and V_2 have dimension at least 1!) A representation that can be written as $V_1 \oplus V_2$ as above is said to be *reducible*.

A group G is said to be a *simple* Lie group when the adjoint representation on $\mathfrak{lie}(G)$ is irreducible. For example, SO(n) for $n \neq 4$ is simple, and SU(n) is simple for every n. On the other hand, no $n \geq 3$ version of U(n) is simple because its Lie algebra contains i times the identity matrix.

Proposition 16.11 *Suppose that* G *is a simple, compact matrix Lie group. Then the metric given above is an Einstein metric.*

The proof of this proposition uses a special case of Schur's lemma:

Lemma 16.12 (Schur's Lemma) *Suppose that* ρ *is a representation of a compact group* G *on either* $\mathbb{R}^m$ *or* $\mathbb{C}^m$ *that preserves the inner product. Suppose that* $\mathfrak{m}$ *is a symmetric* $m \times m$ *matrix in the first case and a Hermitian* $m \times m$ *matrix in the second; and suppose that* $\rho(g)\mathfrak{m}\rho(g^{-1}) = \mathfrak{m}$ *for all* $g \in G$. *Then* $\mathfrak{m}$ *is a multiple of the identity matrix.*

Schur's lemma is proved momentarily. What follows directly is the proof of the proposition.

Proof of Proposition 16.11 The Ricci tensor is the constant matrix $R_{ik} = \sum_{p,q} \mathfrak{a}^{ip}_{q}\mathfrak{a}^{kp}_{q}$. The metric is Einstein if and only if this matrix is a multiple of the identity matrix. Schur's lemma is used to prove this. For this purpose, suppose that $g \in G$. Let $\{\mathfrak{q}_i\}_{1 \leq i \leq m}$ denote the chosen, orthonormal basis for

$\mathfrak{lie}(G)$ and write $g\, \mathfrak{q}_i\, g^{-1}$ as $\sum_{1\le p\le m} \rho(g)_{pi}\, \mathfrak{q}_p$. The $\mathfrak{m} \times \mathfrak{m}$ matrix with entries $\{\rho(g)_{pi}\}_{1\le p,i\le m}$ is the adjoint representation matrix for g in $Gl(m; \mathbb{R})$. This matrix is orthogonal because the $-tr(\mathfrak{q}_i\mathfrak{q}_k) = -tr((g\mathfrak{q}_i g^{-1})(g\mathfrak{q}_k g^{-1})) = \sum_{1\le p\le m} \rho_{pi}\rho_{pk}$. The matrix also has the property that

$$R_{ik} = \sum_{1\le p,q\le m} \rho(g)_{pi} R_{pq} \rho(g)_{kq}.$$

To see why this is, first write any given version of $\mathfrak{a}^{ij}_k$ as $tr(\mathfrak{q}_i\, [\mathfrak{q}_j, \mathfrak{q}_k])$ and use the fact that $tr(g\mathfrak{q}_i g^{-1}\, [g\mathfrak{q}_j g^{-1}, g\mathfrak{q}_k g^{-1}]) = tr(\mathfrak{q}_i\, [\mathfrak{q}_j, \mathfrak{q}_k])$ to deduce that

$$\mathfrak{a}^{ij}_k = \sum_{1\le r,s,t\le m} \rho_{ri}(g)\rho_{sj}(g)\rho_{tk}(g)\mathfrak{a}^{rs}{}_t.$$

Then, use the formula $R_{ik} = \sum_{p,q} \mathfrak{a}^{ip}_q\, \mathfrak{a}^{kp}_q$ and the fact that the matrix $\rho(g)$ is orthogonal to deduce the desired equality. Granted that $\rho(g)$ is an orthogonal matrix, and granted that $\sum_{1\le p,q\le m} \rho(g)_{pi} R_{pq} \rho(g)_{kq} = R_{ik}$, invoke Schur's lemma to finish the proof.

Proof of Lemma 16.12 (Schur's lemma) The matrix $\mathfrak{m}$ is diagonalizable by an orthogonal or unitary transformation of the given basis. Each eigenvalue has a corresponding eigenspace. Let $V_1 \subset V$ denote one of the eigenspaces and let λ_1 denote the corresponding eigenvalue. Suppose that $v \in V_1$ and $g \in G$. Then $\rho(g)\mathfrak{m}v = \lambda\, \rho(g)v$. Meanwhile, one can write $\rho(g)\mathfrak{m}v$ as $\rho(g)\mathfrak{m}\rho(g^{-1})\, \rho(g)v$, and thus see that $\mathfrak{m}\, \rho(g)v = \lambda\, \rho(g)v$. This tells us that $\rho(g)v$ is in V_1 also. As this works for all $v \in V_1$, it follows that the representation maps V_1 to itself. If V_1 is not the whole of V or 0, then V can't be irreducible.

Additional reading

- *Riemannian Geometry*, Manfredo P. Carmo, Birkäuser, 1992.
- *Riemannian Manifolds: An Introduction to Curvature*, John M. Lee, Springer, 1997.
- *Riemannian Geometry*, Sylvestre Gallot, Dominque Hulin, and Jacques Lafontaine, Springer, 2004.
- *Comparison Theorems in Riemannian Geometry*, Jeff Cheeger and David G. Ebin, American Mathematical Society, 2008.
- *Riemannian Geometry*, Takashi Sakai, American Mathematical Society, 1996.
- *Foundations of Differential Geometry, Volume II*, Shoshichi Kobayashi and Katumi Nomizu, Wiley Interscience, 1996.
- *Ricci Flow and the Poincaré Conjecture*, John Morgan and Gang Tian, American Mathematical Society, 2007.
- *Three Dimensional Geometry and Topology*, William P. Thurston, Princeton University Press, 1997.

17 Complex manifolds

Suppose that M is a manifold of dimension 2n. An almost complex structure on TM is an endomorphism, j, such that $j^2 = -\iota$. It is traditional to denote the resulting complex, rank n bundle by $T_{1,0}M$, and the complex conjugate bundle by $T_{0,1}M$. Recall from Chapter 6.3 that the latter is the complex bundle that is obtained from TM using $-j$ for the almost complex structure. The dual bundle is denoted by $T^{1,0}M$. This is to say that the complexification of T^*M has a direct sum splitting as $T^{1,0}M \oplus T^{0,1}M$. If $\{e_a\}_{a=1,2,\ldots,2n}$ is a basis for TM over some set in M with $je_{2k} = e_{2k-1}$ and $je_{2k-1} = e_{2k}$, then the $\mathbb{C}$-valued 1-vectors $\{\upsilon_k = \frac{1}{\sqrt{2}}(e_{2k-1} - ie_{2k})\}_{1\le k\le n}$ is a basis for TM. If $\{e^a\}_{1\le a\le 2n}$ is the corresponding dual basis for T^*M, then the $\mathbb{C}$-valued 1-forms $\{\alpha^k = \frac{1}{2}(e^{2k-1} + i\, e^{2k})\}_{1\le k\le n}$ form a basis for $T^{1,0}M$, and the forms $\{\bar{\alpha}^k = \frac{1}{2}(e^{2k-1} - i\, e^{2k})\}_{1\le k\le n}$ form a basis for $T^{0,1}M$.

The following is a consequence of what is said in Chapter 6.3: An equivalent way to define an almost complex structure is to give a splitting $TM_{\mathbb{C}} = E \oplus \bar{E}$ where E is a rank n complex bundle and $\bar{E}$ is the complex conjugate bundle. One defines j on $\bar{E}$ as multiplication by i, and j on $\bar{E}$ as multiplication by $-i$. The bundle E is then $T_{1,0}$ M.

The decomposition $T^*M = T^{1,0}M \oplus T^{0,1}M$ induces an analogous decomposition of the bundle $\wedge^p T^*M_{\mathbb{C}}$ of $\mathbb{C}$-valued p-forms. In particular, the latter decomposes as direct sum

$$\wedge^p T^*M = \bigoplus_{m+q=p} (\wedge^m T^{1,0}M) \otimes (\wedge^q T^{0,1}M).$$

The summand $(\wedge^m T^{1,0}M) \otimes (\wedge^q T^{0,1}M)$ is denoted by $T^{m,q}M$. These are said to be *forms of type (p, q)*. Note that the $\wedge^{2n} T^*M_{\mathbb{C}} = T^{n,n}M$, this a complex, rank 1 line bundle that is isomorphic to $M \times \mathbb{C}$.

The standard example for an almost complex structure comes from $\mathbb{R}^{2n}$ and its canonical almost comple structure, j_0, that acts so that

$$j_0 \frac{\partial}{\partial x_{2a-1}} = \frac{\partial}{\partial x_{2a}}.$$

In this case, $T^{1,0}$ M is spanned by $\{dz_k = dx_{2k-1} + i\, dx_{2k}\}_{1\le k\le n}$.

Other examples come from complex manifolds.

17.1 Some basics concerning holomorphic functions on $\mathbb{C}^n$

The definition of a complex manifold requires the brief digression that follows to introduce some definitions from complex function theory. To start the digression, fix an integer $n \geq 1$ and use j_0 to view $\mathbb{R}^{2n}$ as $\mathbb{C}^n$. Keep in mind that the complex coordinate functions, $(z_1, \ldots, z_n)$, on the $\mathbb{C}^n$ are defined in terms of the real coordinates $(x_1, \ldots, x_{2n})$ on $\mathbb{R}^{2n}$ by the rule $z_k = x_{2k-1} + ix_{2k}$. A smooth map, $f = (f_1, f_2)$, from a domain in $\mathbb{R}^{2n}$ to $\mathbb{R}^2 \to \mathbb{R}^2$ can be viewed as a complex valued function on a domain in $\mathbb{C}^n$ when $\mathbb{R}^{2n}$ is viewed as $\mathbb{C}^n$ and $\mathbb{R}^2$ is viewed as $\mathbb{C}$. Such a function is said to be *holomorphic* when $\frac{\partial}{\partial \bar{z}_k} f = 0$ for each $k \in \{1, \ldots, n\}$. Here, $\frac{\partial}{\partial \bar{z}_k} = \frac{1}{2}(\frac{\partial}{\partial x_{2k-1}} + i\frac{\partial}{\partial x_{2k}})$. This is to say that the $\mathbb{R}^2$ components (f_1, f_2) of f obey the *Cauchy–Riemann* equations

$$\frac{\partial f_1}{\partial x_{2k-1}} - \frac{\partial f_2}{\partial x_{2k}} = 0 \quad \textit{and} \quad \frac{\partial f_2}{\partial x_{2k-1}} + \frac{\partial f_1}{\partial x_{2k}} = 0$$

for each $k \in \{1, \ldots, n\}$. For example, a function f that is given as a polynomial or convergent power series in the complex coordinates $(z_1, \ldots, z_n)$ is holomorphic. Note that the Cauchy–Riemann equations say the following: View f again as a map from $\mathbb{R}^{2n}$ to $\mathbb{R}^2$. View the differential of f at each point of its domain as a linear transformation, $f_*: \mathbb{R}^{2n} \to \mathbb{R}^2$, thus as a matrix with 2 rows and 2n columns. Then

$$f_* j_0 = j_0 f_*.$$

This is to say that f_* intertwines the action of the 2n-dimensional version of j_0 with that of the 2-dimensional version. Indeed, this follows by virtue of the fact that columns $2k-1$ and $2k$ of the differential f_* are

$$\begin{pmatrix} \partial_{2k-1} f_1 & \partial_{2k} f_1 \\ \partial_{2k-1} f_2 & \partial_{2k} f_2 \end{pmatrix};$$

where ∂_i is shorthand here (and in what follows) for $\frac{\partial}{\partial x_i}$. In particular, the latter has the form

$$\begin{pmatrix} a_k & -b_k \\ b_k & a_k \end{pmatrix} = a_k \iota + b_k j_0$$

if and only if f is holomorphic. Note, by the way that $a_k + ib_k$ here is $\frac{\partial f}{\partial z_k}$ where $\frac{\partial}{\partial z_k} = \frac{1}{2}(\frac{\partial}{\partial x_{2k-1}} - i\frac{\partial}{\partial x_{2k}})$.

If $m \geq 1$ is another given integer, then a map f from a domain in $\mathbb{R}^{2n}$ to $\mathbb{R}^{2m}$ can be viewed as a map from a domain in $\mathbb{C}^n$ to $\mathbb{C}^m$. Such a map is said to be holomorphic if its entries, $(f_1, \ldots, f_m) \in \mathbb{C}^m = \times_m \mathbb{C}$, are each holomorphic functions on its domain in $\mathbb{C}^n$. Here is an equivalent definition: View f in its original guise as a map from $\mathbb{R}^{2n}$ to $\mathbb{R}^{2m}$. Its differential at each point is a matrix, f_*, with 2m rows and 2n columns; and as in the $m = 1$ case, f is holomorphic if and only if $f_* j_0 = j_0 f_*$. To see this, note that the j'th row of the matrix f_* has k'th entry $\partial f_j/\partial x_k$ and so f_* can be written in blocks of size 2×2 with the block corresponding to rows $2j - 1$ and $2j$, columns $2k - 1$ and $2k$ given by

$$\begin{pmatrix} \partial_{2k-1}f_{2j-1} & \partial_{2k}f_{2j-1} \\ \partial_{2k-1}f_{2j} & \partial_{2k}f_{2j} \end{pmatrix} = \begin{pmatrix} a_{j,k} & -b_{j,k} \\ b_{j,k} & a_{j,k} \end{pmatrix}$$

where the equality holds if and only if f is holomorphic. Note that the condition $f_* j_0 = j_0 f_*$ means that f_* can be viewed as a $\mathbb{C}$-linear transformation from $\mathbb{C}^n$ to $\mathbb{C}^m$; which is to say a matrix with m columns and n rows whose entries are complex numbers. This matrix is denoted by ∂f. The $\mathbb{C}$-valued entry for the j'th row and k'th column of ∂f is $\frac{\partial f_j}{\partial z_k} = a_{j,k} + i b_{j,k}$ with the numbers $a_{j,k}$ and $b_{j,k}$ as above. In the particular case where $m = n$, the condition $f_* j_0 = j_0 f_*$ means that ∂f at each point defines an element in $\mathbb{M}(n; \mathbb{C})$, this the space of $n \times n$ matrices with complex number entries.

As a parenthetical remark, note that the matrix ∂f in $\mathbb{M}(n; \mathbb{C})$ is invertible if and only if the $2n \times 2n$ real matrix f_* is invertible. This follows from the fact that $\det(f_*)$ is equal to $|\det(\partial f)|^2$ where $\det(\partial f)$ signifies the determinant of the $\mathbb{C}$-valued matrix ∂f. This means that when checking the invertibility of a holomorphic map from $\mathbb{C}^n$ to $\mathbb{C}^n$, one can work with the holomorphic coordinates instead of the real coordinates.

17.2 The definition of a complex manifold

Turn now to the definition of a complex manifold. To start, fix an integer $n \geq 1$ and let M denote a smooth manifold of dimension 2n. A *complex manifold structure* on M is defined by a coordinate cover $\mathcal{U}$ with a certain special property. To elaborate, keep in mind that any given element from $\mathcal{U}$ is a pair (U, φ_U) where $U \subset M$ is an open set and $\varphi_U\colon U \to \mathbb{R}^{2n}$ is a diffeomorphism. These pairs from $\mathcal{U}$ must be such that for each $(U, \varphi_U) \in \mathcal{U}$, there is a diffeomorphism, ψ_U, from $\mathbb{R}^{2n}$ onto an open set in $\mathbb{R}^{2n}$ such that the following condition is satisfied:

Let (U, φ_U) and (V, φ_V) denote any two pairs from $\mathcal{U}$. Use j_0 to identify $\mathbb{R}^{2n}$ with $\mathbb{C}^n$. Let $\phi_U = \psi_U \circ \varphi_U$ and likewise define ϕ_V. View the transition function $\varphi_V \circ \phi_U^{-1}$ as a map from one domain in $\mathbb{C}^n$ another. This map must define a holomorphic function on its domain.

At the risk of being overly pedantic, this means the following: For each $k \in \{1, \ldots, n\}$, define the complex coordinate z_k from the real coordinates $(x_1, \ldots, x_{2n})$ by $z_k = x_{2k-1} + ix_{2k}$. To say that $\phi_V \circ \phi_U^{-1}$ is holomorphic means that $\frac{\partial}{\partial \bar{z}_k}(\phi_V \circ \phi_U^{-1})$ must equal zero at each point of the domain of $\phi_V \circ \phi_U^{-1}$ and for each $k \in \{1, \ldots, n\}$. With what is said in Chapter 17.1 in mind, what follows is another way to present the constraint on U:

Let (U, φ_U) *and* (V, φ_V) *denote any two pairs from* $\mathcal{U}$*. Then the differential of the corresponding* $\phi_V \circ \phi_U^{-1}$ *must commute with* j_0 *at each point of the domain of* $\phi_V \circ \phi_U^{-1}$.

Two complex structures on M are said to be equivalent when there is a diffeomorphism, ψ: M $\to$ M with the following property: Let (U, φ_U) denote a chart for the first complex structure, and let $(U', \varphi_{U'})$ denote a chart for the second if $\psi(U)$ intersects U', then $\phi_{U'} \circ \psi \circ \phi_U^{-1}$ defines a holomorphic map between one domain in $\mathbb{C}^n$ and another. It is customary to view a complex manifold as a manifold with an equivalence class of complex structures.

To make contact with what was said at the end of this chapter's introduction, keep in mind that the differential of the coordinate transition functions give the vector bundle transition functions for TM. The fact that these commute with j_0 implies that TM has the structure of a complex, rank n vector bundle. Said differently, the tangent bundle of a complex manifold has a canonical almost complex structure, this given over the open set from an element of $\mathcal{U}$ by $j = (\phi_U^{-1})_* j_0 (\phi_U)_*$. Here, $(\phi_U)_*$: $TM|_U \to T\mathbb{R}^n$ denotes the differential of the coordinate chart ϕ_U.

17.3 First examples of complex manifolds

Of course, $\mathbb{C}^n$, and any open set in $\mathbb{C}^n$ is a complex manifold. There are others.

Example 17.1 Introduce from Chapter 6.8 the space $\mathbb{CP}^n$ of complex, 1-dimensional subspaces in $\mathbb{C}^{n+1}$. As noted in Chapter 10.5, it is also the quotient space S^{2n+1}/S^1. It is also the quotient space $(\mathbb{C}^{n+1}-\{0\})/\mathbb{C}^*$ where $\mathbb{C}^* = \mathbb{C}-\{0\}$ is the multiplicative group of nonzero complex numbers. This group acts on $\mathbb{C}^{n+1}$ by the rule $(u, z = (z_1, \ldots, z_n)) \to uz = (uz_1, \ldots, uz_n)$. This view of $\mathbb{CP}^n$ is equivalent to the other because $\mathbb{C}^* = \mathbb{R}_+ \times S^1$ where $\mathbb{R}_+ = (0, \infty)$ and $(\mathbb{C}^{n+1}-\{0\})/\mathbb{R}_+ = S^{2n+1}$. As noted in Example 10.4, the manifold $\mathbb{CP}^1$ is also S^2. In general, the manifold $\mathbb{CP}^n$ is an example of a complex manifold. To see this complex structure, recall from Chapter 6.8 that it can be covered by n+1 coordinate charts. The k'th chart consists of the 1-dimensional subspaces of the form $\mathbb{C}\, z \subset \mathbb{C}^{n+1}$ where $z = (z_1, \ldots, z_{n+1})$ is such that $z_k \neq 0$. Let $\mathcal{O}_k$ denote this chart. As explained in Chapter 6.8, a diffeomorphism $\varphi_k\colon \mathcal{O}_k \to \mathbb{C}^n$ sends z to the n-tuple whose j'th component for $j < k$ is z_j/z_k, and whose j'th component for $j \geq k$ is z_{j+1}/z_k. The transition function $\varphi_k \circ \varphi_m^{-1}$ is defined as follows: Assume that $m < k$. The transition function sends $(u_1, \ldots, u_k)$ to the point whose j'th component is u_j/u_{k-1} for $j < m$, whose m'th component is $1/u_{k-1}$, any given $j \in \{m+1, \ldots, k-1\}$ component is u_{j-1}/u_{k-1}, and any $j \geq k$ component is u_j/u_{k-1}. This transition function is holomorphic on its domain of definition. Note that in this case, the $U = \mathcal{O}_k$ version of the map ψ_U can be taken to be the identity map.

Example 17.2 Fix integers $m > n \geq 1$ and introduce from Chapter 6.8 the complex Grassmannian $Gr_{\mathbb{C}}(m; n)$ whose points are the n-dimensional complex subspaces in $\mathbb{C}^m$. It follows from the description given in Chapter 6.8 that this Grassmannian is a complex manifold. To elaborate, Chapter 6.8 gives a coordinate chart cover of $Gr_{\mathbb{C}}(m; n)$ with the coordinate chart map sending a point in the coordinate chart to the version of $\mathbb{C}^{n(m-n)}$ that is obtained by taking the entries of a complex $n \times (m - n)$ matrix. Moreover, the transition functions given in Chapter 6.8 between any two such charts are holomorphic functions of the entries of the matrix. This implies that $Gr_{\mathbb{C}}(m; n)$ is a complex manifold. Here, again, the relevant version of the map ψ_U is the identity map.

Example 17.3 An even dimensional torus, $\mathbb{T}^{2n} = \times_{2n} S^1$, has the structure of a complex manifold. This can be seen as follows: One can view S^1 as the quotient space $\mathbb{R}/(2\pi\mathbb{Z})$ with the quotient map sending any given $t \in \mathbb{R}$ to the unit length complex number e^{it}. The product of such maps identifies $\times_{2n} S^1$ with $\mathbb{R}^{2n}/(2\pi\mathbb{Z}^{2n})$. View $\mathbb{R}^{2n}$ in the standard way as $\mathbb{C}^n$. This done, the quotient map identifies points z and z′ from $\mathbb{C}^n$ if and only if $z' = z + 2\pi\, n$ where $n \in \mathbb{C}^n$ is a vector whose real and imaginary parts have integer entries. It follows from this that the torus $\mathbb{T}^{2n}$ has a coordinate atlas $\mathfrak{U}$ such that each $(U, \varphi_U) \in \mathfrak{U}$ is such that U is the image via the projection from $\mathbb{C}^n$ of a ball in $\mathbb{C}^n$ centered with radius less than 2π. In this case, the map ϕ_U can be taken to be the identity map on this ball. This being the case, then the transition function $\phi_V \circ \phi_U^{-1}$ has the form $z \to z + 2\pi\, n$ where $n \in \mathbb{C}^n$ is a fixed vector whose real and imaginary parts have integer entries. I hope it is evident that such a translation is a holomorphic map.

Example 17.4 A surface of genus greater than one has the structure of a complex manifold. The genus zero case is S^2 which is $\mathbb{CP}^1$ and so has a complex structure. The genus one case is the torus $\mathbb{T}^2$ and so a surface of genus one also has a complex structure. What follows describes how to construct a complex manifold structure for a surface of genus $g > 1$. The construction does not speak of the surface as embedded in $\mathbb{R}^3$ or, for that matter, in any other ambient manifold. To set the stage however, picture the following construction of a surface of genus g in $\mathbb{R}^3$: Start with a round ball of modeling clay, so the boundary is S^2. Now, make g solid, reasonably long cylinders of clay, bend each one into a U shape, and then attach it to the ball by sticking its ends to the boundary of the ball using 2g separate disks on the boundary for the attaching regions—one disk for each end of each cylinder. Smooth the creases where the attachments are made, and—Voilà—the boundary of what you have is a surface of genus g.

I am going to make this same handle attaching construction without reference to $\mathbb{R}^3$ and without reference to the interiors of either the balls or the solid cylinders. To elaborate, I describe in what follows a surface of genus g as a union of g + 1 open sets, $\{U_0, U_1, \ldots, U_g\}$. The set U_0 will be diffeomorphic to the complement in S^2 of 2g disjoint, small radius disks. Each of the other sets is diffeomorphic to the annulus in $\mathbb{C}$ where the complex coordinate z obeys $e^{-2} < |z| < e^2$, where e here denotes Euler's constant, whose natural log is 1.

To put this in perspective with regards to the modeling clay construction, the set U_0 will represent the boundary of the original ball of clay that remains visible after the solid clay cylinders are attached. To identify the role of the annuli, I point out that a solid cylinder of the sort I described in the modeling clay discussion can be viewed (before being bent into a U shape) as the subset of $\mathbb{R}^3$ where the Euclidean coordinates (x_1, x_2, x_3) obeys

$$x_1^2 + x_2^2 \leq 1 \ \textit{and} \ -1 \leq x_3 \leq 1.$$

The boundary of this solid cylinder can be decomposed into three parts. The first two are the disks where $x_3 = 1$ and $x_3 = -1$. These are the regions that are glued to the boundary of S^2 in the modeling clay construction. The remainder of the boundary is the set of points where $x_1^2 + x_2^2 = 1$ and $-1 \leq x_3 \leq 1$. The part of this boundary where $-1 < x_3 < 1$ is a smooth manifold that is diffeomorphic to the annulus in $\mathbb{C}$ where $e^{-1} < |z| < e$; the diffeomorphism sends $z \in \mathbb{C}$ to

$$(x_1 = \cos(z/|z|), x_2 = \sin(z/|z|), x_3 = \ln|z|).$$

I have described g+1 open sets $\{U_0, U_1, \ldots, U_g\}$. I describe next the identifications that are used to construct a surface. To begin, I must say more about U_0. View S^2 as $\mathbb{CP}^1$, and as such it has an open cover with two sets, both diffeomorphic to $\mathbb{C}$. The first set I denote as $\mathcal{O}_0$. It consists of the complex lines through 0 in $\mathbb{C}^2$ that are the linear span of a vector that can be written as (1, u) with $u \in \mathbb{C}$. The value of u gives the coordinate in $\mathbb{C}$ for this open set. The other open set consists of the linear span of the vectors through the origin in $\mathbb{C}^2$ that can be written as (w, 1) where $w \in \mathbb{C}$ gives the complex coordinate on this open set. The transition function identifies $w = 1/u$ where both u and w are nonzero. Fix 2g points in $\mathbb{C}$ that are pair-wise separated by distance at least e^{10}. The set U_0 is the complement in the $\mathbb{CP}^1$ of the subset of points in the coordinate chart $\mathcal{O}_0$ where the coordinate u has distance e^{-2} or less from any one of these 2g points. Group these 2g points into g sets of two points, $\{(p_k, q_k)\}_{1 \le k \le g}$.

Now consider the identifications between the various open sets from the collection $\{U_0, U_1, \ldots, U_g\}$. The identification will pair points in any given $k > 0$ version of U_k with points in U_0. If $k \neq k'$ are both greater than zero, no point in U_k will be paired with a point in $U_{k'}$. This understood, I will describe only the identifications that concern the set U_1 as the description for the other sets is the same but for the notation. I use (p, q) in what follows to denote the pair (p_1, q_1). The identification has two steps, one for near p and the other for near q.

Step 1: The part of the disk of radius e^{-1} centered at p that lies in U_0 is the annulus centered at p where $e^{-2} < |u - p| < e^{-1}$. This annulus is identified with the part of the cylinder U_1 where $e^{-2} < |z| < e^{-1}$ by the rule

$$z = e^{-3}/(u - p).$$

Step 2: The part of the disk of radius e^{-1} at q that lies in U_0 is the annulus centered at p where $e^{-2} < |u - q| < e^{-1}$. This annulus is identified with the part of the cylinder U_1 where $e < |z| < e^2$ by the rule

$$z = 1/(u - q).$$

Let Σ denote the space obtained by making all of these equivalences. It follows from what is said in Chapter 1.5 that the Σ has the structure of a smooth manifold, one where the quotient map from each set $\{U_0, U_1, \ldots, U_g\}$ to Σ defines an embedding onto an open set in Σ. This understood, I now view each of these sets as sitting in Σ. The complex manifold structure for Σ comes from that on each of these sets. Indeed, I presented each as an open subset of a complex manifold, U_0 in $\mathbb{CP}^1$ and each $k > 0$ version of U_k in $\mathbb{C}$. The transition functions between these open sets are relevant only for $U_k \cap U_0$, and the latter are the identification maps in Steps 1 and 2 above. In particular, they are holomorphic maps on their domains of definition. I leave it to you to verify that Σ is a surface of genus g. (You might try mimicking the construction in Example 14.9 to construct a connection on the complex line bundle $T_{1,0}\Sigma$, and then compute the resulting first Chern class.)

17.4 The Newlander–Nirenberg theorem

Suppose that M is a 2n-dimensional manifold and that j is a given almost complex structure on TM. The question arises as to whether j comes from an underlying complex structure on M. The Newlander–Nirenberg theorem gives necessary and sufficient conditions for this to occur.

To set the stage for this theorem, consider that Chapter 12.1's exterior derivative, d, acts on sections of $T^{1,0}M$ since such a section is a $\mathbb{C}$-valued 1-form; albeit one with some special properties with regards to j. In any event, if v is a section of $T^{1,0}M$, then $d\alpha$ will be a $\mathbb{C}$-valued 2-form on M which is to say a section of $\wedge^2 T^*M_{\mathbb{C}}$. As such, it decomposes into forms of type (2, 0), (1, 1) and (0, 2). Introduce

$$\mathfrak{N}_j v = (dv)_{(0,2)}$$

to denote the (0, 2) part of dv. Note that $\mathfrak{N}_j(f v) = f\, \mathfrak{N}_j v$ for any function f. This is because $d(f v) = df \wedge v + f\, dv$. In particular, as v is of type (1, 0), so $df \wedge v$ has no components of type (0, 2). The fact that $\mathfrak{N}_j(fv) = f\, \mathfrak{N}_j v$ for all functions f implies that $\mathfrak{N}_j$ defines a section of $\mathrm{Hom}(T^{1,0}M; T^{0,2}M)$. This follows from the Lemma 11.1 in Chapter 11.2 as applied to the case where $E = T^{1,0}M$, $E' = T^{0,2}M$ and $\mathcal{L} = \mathfrak{N}_j$. This section $\mathfrak{N}_j$ of $\mathrm{Hom}(T^{1,0}M; T^{0,2}M)$ is called the *Nijenhuis* tensor.

Note that $\mathfrak{N}_j = 0$ if j comes from a complex manifold structure on M. Here is how to see this: Let Λ denote a set of charts that define the complex manifold structure, let $U \subset \Lambda$ and let φ_U denote the map to $\mathbb{R}^{2n}$. Since $\mathfrak{N}_j$ is a tensor, it is enough to check that it acts as zero on a local basis for $T^{1,0}M$. Such a local basis is $\{\alpha^k = \varphi_U^* dz^k\}_{1\leq k\leq n}$. Then $d\alpha^k = 0$ so $(d\alpha^k)_{0,2} = 0$.

Theorem 17.5 (Newlander and Nirenberg) *An almost complex structure on* M *comes from a complex structure if and only if its Nijenhuis tensor is zero.*

To state an equivalent condition, remember that a section of $T_{1,0}M$ is a $\mathbb{C}$-valued vector field on M on which j acts as multiplication by i. The commutator of two vector fields is also a vector field, and so one can ask if the commutator of two sections of $T_{1,0}M$ is a section of $T_{1,0}M$ or not. Does the commutator have any $T_{0,1}M$ part?

With the preceding understood, here is the equivalent condition: The Nijenhuis tensor $\mathfrak{N}_j = 0$ if and only if the commutator of two sections of $T_{1,0}M$ is also a section of $T_{1,0}M$. To prove this, use the relation between commutator and exterior derivative given by the formula $(d\alpha)(\mathfrak{v}, \mathfrak{v}') = -\alpha([\mathfrak{v}, \mathfrak{v}']) + \mathfrak{v}\alpha(\mathfrak{v}') - \mathfrak{v}'\alpha(\mathfrak{v})$. Here, $\alpha(\mathfrak{v})$ is viewed as the pairing between a 1-form and a vector field, thus a function on M. Meanwhile, $\mathfrak{v}'\alpha(\mathfrak{v})$ denotes the directional derivative of the function $\alpha(\mathfrak{v})$ in the direction of $\mathfrak{v}'$.

The proof of the Newlander–Nirenberg theorem would take us some way into complex analysis, and so I won't say more about it. See A. Newlander and L. Nirenberg, "Complex analytic coordinates in almost complex manifolds," *Annals of Mathematics, Second Series* 65 (1957), 391–404.

What follows is an immediate corollary:

Theorem 17.6 *Let* M *denote an oriented manifold of dimension 2. Then every almost complex structure on* TM *comes from a complex manifold structure.*

The reason is that $\wedge^2 TM_{\mathbb{C}} = \wedge^{1,1} TM_{\mathbb{C}}$ in dimension 2. There are neither (2, 0) or (0, 2) forms. By the way, every oriented, 2-dimensional manifold has a plethora of almost complex structures for its tangent bundle as each Riemannian metric determines one. This almost complex structure can be seen as follows: Take an oriented, orthonormal frame for TM near any given point. Let $\{e_1, e_2\}$ denote this frame. Then $je_1 = e_2$ and $je_2 = -e_1$. This doesn't depend on the choice of frame because j as just defined commutes with any the action of a matrix in SO(2) on the set of oriented, orthonormal frames. Indeed, a new frame $\{e_1', e_2'\}$ is related to the old by $e_1' = \cos\theta\, e_1 + \sin\theta\, e_2$ and $e_2' = -\sin\theta\, e_1 + \cos\theta\, e_2$. Hence, $je_1' = \cos\theta\, e_2 - \sin\theta\, e_1 = e_2'$. Note as well that j depends only on the conformal class of the given metric. To see this, suppose that $h > 0$ and $g' = h^2 g$. Then if $\{e_1, e_2\}$ are orthonormal for g, then the pair $\{he_1, he_2\}$ are orthonormal for g'. This understood, the g and g' versions of the almost complex structure both map e_1 to e_2 and e_2 to $-e_1$.

To indicate what a proof of the Newlander–Nirenberg theorem involves in dimension 2, consider the following situation: Suppose that j is a given almost complex structure on a surface. Fix a point p and a coordinate chart centered at p so as to identify a neighborhood of p with a disk centered at the origin in $\mathbb{R}^2$. View $\mathbb{R}^2$ as $\mathbb{C}$ with the usual complex coordinate $z = x + iy$ coordinate on $\mathbb{C}$. Let $dz = dx + idy$ and $d\bar{z} = dx - idy$ denote the standard basis for the respective j_0 versions of $T^{1,0}\mathbb{C}$ and $T^{0,1}\mathbb{C}$. The given almost complex structure, j, appears in this coordinate chart as an almost complex structure on the disk in $\mathbb{C}$. The latter's version of $T^{1,0}\mathbb{C}$ will be spanned by a complex number valued 1-form that can be written in terms of the basis dz and $d\bar{z}$ as

$$e = a dz + b d\bar{z}$$

where a and b are $\mathbb{C}$-valued functions that are constrained so that $|a|$ and $|b|$ are nowhere equal. Here is why they can't be equal: The $\mathbb{C}$-valued 1-form e and its complex conjugate $\bar{e}$ are supposed to be linearly independent over $\mathbb{C}$ because they must span at each point the fiber of the 2-complex dimensional vector bundle $T\mathbb{C}|_{\mathbb{C}}$. However, these to $\mathbb{C}$-valued 1-forms are linearly independent over $\mathbb{C}$ at a given point if and only if $|a| \neq |b|$ at this point. Note in this regard that $e = (a/\bar{b})\, \bar{e}$ if e and $\bar{e}$ are linearly dependent; and this equation can hold if and only if $|a| = |b|$.

With the preceding understood, suppose that $|a| > |b|$. The discussion in the alternative case is the same once z and $\bar{z}$ are switched. If j comes from an underlying complex structure, then there is a coordinate chart map that gives a diffeomorphism from a neighborhood of the origin in $\mathbb{C}$ to $\mathbb{R}^2$ sending 0 to 0 and with the following property: Write the map as $z \to w(z) = w_1(z) + i\, w_2(z)$, thus as a $\mathbb{C}$-valued function. Use dw to denote the $\mathbb{C}$-valued 1-form $dw_1 + idw_2$. Then j's version of $T^{1,0}\mathbb{C}$ is spanned by dw. This is to say that e can be written as α dw with α a nowhere zero, $\mathbb{C}$-valued function. To see what this means for w, use the Chain rule to write

$$dw = \frac{\partial w}{\partial z} dz + \frac{\partial w}{\partial \bar{z}} d\bar{z}.$$

Note in this regard that this version of the Chain rule amounts to no more than using the standard 2-variable, real version to first write $dw_1 + idw_2$ as

$$\left(\frac{\partial}{\partial x} w_1 + i \frac{\partial}{\partial x} w_2\right) dx + \left(\frac{\partial}{\partial y} w_1 + i \frac{\partial}{\partial y} w_2\right) dy,$$

then write dx as $\frac{1}{2}(dz + d\bar{z})$ and dy as $\frac{1}{2}(d\bar{z} - dz)$; and finally write $\frac{\partial}{\partial x}$ as $\frac{\partial}{\partial z} + \frac{\partial}{\partial \bar{z}}$ and $\frac{\partial}{\partial y}$ as $(\frac{\partial}{\partial z} - \frac{\partial}{\partial \bar{z}})$. In any event, with dw written as above, and with $e = adz + bd\bar{z}$, you can see by directly comparing respective dz and $d\bar{z}$ components that e can be written as α dw if and only if the $\mathbb{C}$-valued function w obeys the equation

$$\frac{\partial w}{\partial \bar{z}} = \left(\frac{a}{b}\right) \frac{\partial w}{\partial z}.$$

This said, the proof of the Newlander–Nirenberg theorem in this context amounts to proving that there does indeed exist a diffeomorphism from a neighborhood of the origin to a neighborhood of the origin in $\mathbb{C}$ that satisfies the preceding equation. (This existence theorem was known in the 2-dimensional case before the work of Newlander and Nirenberg: It can be proved using a version of the contraction mapping theorem (Theorem 1.6) with the help of some basic facts about the Green's function for the Laplacian on $\mathbb{R}^2$.)

In dimension 4, there are manifolds that admit almost complex structures but no complex structures. For example, the connect sum of 3 copies of $\mathbb{CP}^2$. (The connect sum of two copies of $\mathbb{CP}^2$ has no almost complex structures: There is a topological obstruction to the existence of a square -1 endomorphism of its tangent bundle.)

A famous open question concerns S^6: This manifold has almost complex structures, but it is not known whether S^6 can be given the structure of a complex manifold. What follows describes an almost complex structure for TS^6 that comes from the *octonion* algebra.

The vector space, $\mathbb{O}$, of octonions is 8-dimensional; it can be viewed as the space of pairs (a, b) where a and b are quaternions. Multiplication is defined by the rule:

$$(a, b)\cdot(c, d) = (ac - d^{\dagger}b, da + bc^{\dagger}).$$

Here, the notation is as follows: Write a quaterion q as $q_0 + q_i\tau^i$ where (q_0, q_1, q_2, q_3) are real and τ^1, τ^2 and τ^3 are the Pauli matrices from Chapter 6.4. Then $q^{\dagger} = q_0 - q_i\tau^i$. Thus, $qq^{\dagger} = |q|^2$. This algebra is nonassociative.

Terminology is such that q_0 is called the real part of q and $q_j\tau^j$ is the imaginary part. Identify $\mathbb{R}^7$ as the subspace of pairs (a, b) $\in \mathbb{O}$ where a has zero real part. Identify S^6 as the unit sphere in S^7, thus the set of pairs (a, b) $\in \mathbb{O}$ with real(a) = 0 and $|a|^2 + |b|^2 = 1$. This done, then the tangent space to S^6 at a given point (a, b) consists of the vector subspace in $\mathbb{O}$ whose elements are pairs (c, d) where both c and $(a^\dagger c + b^\dagger d)$ have vanishing real part. Granted this identification of $TS^6|_{(a,b)}$, a linear map $j_{(a,b)}$: $TS^6|_{(a,b)} \to \mathbb{O}$ is defined by the rule

$$j|_{(a,b)}(c, d) = (a, b)\cdot(c, d) = (ac - d^\dagger b, da + bc^\dagger).$$

I leave it as an exercise to verify that j maps TS^6 to itself and has square -1.

17.5 Metrics and almost complex structures on TM

Suppose that M is a 2n-dimensional manifold with an almost complex structure, j, and metric g. If the metric g is such that j acts as an orthogonal transformation, then j and g are said to be compatible. Note that j acts orthogonally if and only if

$$g(u, jv) = -g(ju, v).$$

The metric g defines a Hermitian metric on the complex vector bundle $T_{1,0}M$ if and only if the preceding condition is obeyed. To see why this is, remember that vectors in $T_{1,0}M$ are of the form $\mathfrak{u} = u - ij\cdot u$ with u a vector in TM. This understood, the Hermitian metric on $T_{1,0}M$ is defined so that the inner product of $\mathfrak{u}$ and $\mathfrak{v} = v - i\,jv$ is

$$\langle \mathfrak{u}, \mathfrak{v}\rangle = g(u, v) + g(ju, jv) + i\big(g(ju, v) - g(u, jv)\big).$$

The condition for defining a Hermitian form is that the complex conjugate of $\langle \mathfrak{u}, \mathfrak{v}\rangle$ must equal $\langle \mathfrak{u}, \mathfrak{v}\rangle$. As you can see, this is the case for all $\mathfrak{u}$ and $\mathfrak{v}$ if and only if g(u, jv) and g(ju, v) have opposite signs.

What follows is a converse to the preceding observation: A Hermitian metric $\langle\ ,\ \rangle$ on $T_{1,0}M$ defines a Riemannian metric on TM by setting g(u, v) to equal $\frac{1}{2}$ of the real part of $\langle \mathfrak{u}, \mathfrak{v}\rangle$. The almost complex structure j defines an orthogonal transformation with respect to this metric.

17.6 The almost Kähler 2-form

Given an almost complex structure on TM, the standard default option is to choose the metric g so that j acts as an orthogonal transformation. Assuming that such is the case, then the bilinear form that assigns g(u, jv) to vectors u and v is antisymmetric. Thus, it defines a 2-form on M, which is to say a section of $\wedge^2 T^*M$. This 2-form is denoted in what follows by ω. It has the nice property that its n'th power, ω^n, is n! times the volume form that is defined by the metric. Remember in this regard that this volume form is given at any given point in terms of an orthonormal basis $\{e^1, \ldots, e^{2n}\}$ by $\Omega = e^1 \wedge \cdots \wedge e^{2n}$. To see why ω has this property, choose orthonormal basis for T^*M at a given point so that $je^{2k-1} = e^{2k}$ and $je^{2k} = -e^{2k-1}$ for each $k \in \{1, \ldots, n\}$. Written with respect to this basis finds that

$$\omega = \sum_{1 \leq k \leq n} e^{2k-1} \wedge e^{2k};$$

and so taking the wedge product of n copies of ω finds $\omega^n = n!\, e^1 \wedge \cdots \wedge e^n$. Note that this implies that a manifold with an almost complex structure has a canonical orientation, this as defined by ω^n. The form ω is called the *almost Kähler* form unless j comes from a complex structure on M and ω is a closed 2-form. In this case, ω is called the *Kähler form*, and M is called a *Kähler* manifold.

Suppose that j is an almost complex structure on M. If ω is a given 2-form on M, then j is said to be *ω-compatible* when the bilinear form ω(·, j(·)) on TM defines a Riemannian metric. The complex structure j acts as an orthogonal transformation for this metric, since $g(u, jv) = \omega(u, j^2 v) = -\omega(u, v)$. As ω is antisymmetric, $\omega(u, v) = -\omega(v, u)$, which is $-\omega(v, j^2(u))$, and thus $-g(v, ju)$. Since g is a metric, it is symmetric, so we see that $g(u, jv) = -g(ju, v)$, thus proving that the transpose of j is −j. Since j(−j) is the identity, j's transpose is its inverse and so j is orthogonal. It then follows that this metric defines a Hermitian metric on $T_{1,0}M$ and ω is its almost Kähler form. Thus, one can go from 2-forms and almost complex structures to metrics, or from metrics and almost complex structures to 2-forms.

By the way, if one is given only a 2-form ω whose top exterior power is nowhere zero, there is in all cases a complex structure, j, on M such that the bilinear form given by ω(·, j(·)) is a Riemannian metric. The proof that ω-compatible almost complex structures exist is given in Appendix 17.1.

17.7 Symplectic forms

A *closed* 2-form, ω, on an even dimensional manifold is said to be *symplectic* if its top exterior power is nowhere zero. There is a cohomology obstruction that must vanish if ω is to be closed. To see this, note that the integral of ω^n over

M is nonzero. This has the following consequence: The class in $H^2_{\text{De Rham}}(M)$ that is defined by ω must be nonzero, as must all of its powers. In particular, if the De Rham cohomology of M has no 2-dimensional class with nonzero n'th power, then there can be no symplectic form on M. For example, there are no closed, almost Kähler forms on S^6.

There may be other obstructions to the existence of a symplectic form.

17.8 Kähler manifolds

Suppose that M is a 2n-dimensional manifold with an almost complex structure j. As noted above, it is natural to choose a metric, g, such that j acts in an orthogonal transformation. The question then arises as to whether g can be chosen so that j is also covariantly constant with respect to the g's Levi-Civita connection. These sorts of metrics can be viewed as being maximally compatible.

As it turns out, a metric g that is maximally compatible in this sense may not exist:

Proposition 17.7 *If* j *and a metric* g *are compatible and* j *is covariantly constant with respect to the Levi-Civita covariant derivative, then* j*'s Nijenhuis tensor is zero. As a consequence*, j, *comes from a complex manifold structure.*

Proof of Proposition 17.7 To see why, remark that if j is covariantly constant with respect to a covariant derivative, ∇, on TM, then ∇ maps sections of $T^{1,0}M$ to sections of $T^{1,0}M \otimes T^*M$. To elaborate, suppose that $\{\upsilon^k\}_{k=1,2,\ldots,n}$ is a basis of sections for $T^{1,0}M$ on some open set in M. Then a randomly chosen covariant derivative will act so that

$$\nabla\upsilon^k = \upsilon^m \otimes B^{mk} + \bar{\upsilon}^m \otimes C^{mk}$$

where each B^{mk} and C^{mk} are 1-forms. Here and in what follows, the summation convention is employed whereby repeated indices are implicitly to be summed over their range of values, $\{1,\ldots, n\}$. The condition that j is covariantly constant means the condition that $\nabla(j\upsilon^k) = j\nabla\upsilon^k$. Since $j\upsilon^k = i\upsilon^k$, and since

$$j\nabla\upsilon^k = j\upsilon^m \otimes B^{mk} + j\bar{\upsilon}^m \otimes C^{mk} = i\upsilon^m \otimes B^{mk} - i\bar{\upsilon}^m \otimes C^{mk},$$

the condition that j commute with ∇ requires that $C^{mk} = 0$ for all indices $1 \le a, b \le n$.

On the other hand, suppose that ∇ is a torsion free connection. This means that $d\upsilon^a = d_\nabla\upsilon^a$ and so

$$d\upsilon^k = -B^{mk} \wedge \upsilon^m - C^{mk} \wedge \bar{\upsilon}^m.$$

Thus, the Nijenhuis tensor can be obtained from the (0, 1) parts of the 1-forms $\{C^{km}\}$. Indeed, writing $C^{mk} = c^{mk}_{+i}\, \upsilon^i + c^{mk}_{-i}\, \bar{\upsilon}^i$, one sees that the Nijenhuis tensor is given in this basis by $c^{mk}_{-i} \bar{\upsilon}^m \wedge \bar{\upsilon}^i$. In particular, if $C = 0$, then the Nijenhuis tensor is zero.

If j is covariantly constant with respect to the metric's Levi-Civita covariant, then so is the 2-form $\omega(\cdot, \cdot) = g(\cdot, j(\cdot))$. (Remember that the metric is also covariantly constant.) Since the covariant derivative is torsion free, this implies that ω is also closed, and so j and the metric g give M the structure of a Kähler manifold.

17.9 Complex manifolds with closed almost Kähler form

As explained in the previous part of this chapter, if M has an complex structure, j, with compatible metric g such that j is covariantly constant with respect to the Levi-Civita covariant derivative, then the Nijenhuis tensor of j vanishes and so j comes from a complex manifold structure. The previous part of the chapter also explained why the almost Kähler 2-form ω is closed in this case. A converse of sorts to this conclusion is also true:

Proposition 17.8 *Suppose that* j *comes from a complex manifold structure on* M *and that* g *is a compatible metric such that the associated almost Kähler form* ω *is closed. Then* j *and* ω *are both covariantly constant.*

Proof of Proposition 17.8 Fix an orthonormal frame $\{e^1, \ldots, e^{2n}\}$ for T^*M on a neighborhood of a given point, one such that $je^{2k-1} = e^{2k}$. Write $\nabla e^b = e^a \otimes W^{ab}$ where each W^{ab} is a 1-form, and where $W^{ba} = -W^{ab}$. Here, ∇ denotes the Levi-Civita connection. Introduce the complex basis $\{\upsilon^k = e^{2k-1} + ie^{2k}\}$ for the dual space, $T^{1,0}M$ of $T_{1,0}M$. A computation finds

$$\nabla \upsilon^k = \upsilon^m \otimes B^{mk} + \bar{\upsilon}^m \otimes C^{mk}$$

where

- $B^{mk} = \frac{1}{2}\left(W^{(2m-1)(2k-1)} + W^{(2m)(2k)} + i\left(W^{(2m-1)(2k)} - W^{(2m)(2k-1)}\right)\right)$
- $C^{mk} = \frac{1}{2}\left(W^{(2m-1)(2k-1)} - W^{(2m)(2k)} - i\left(W^{(2m-1)(2k)} + W^{(2m)(2k-1)}\right)\right)$.

Note that $\overline{B^{km}} = -B^{mk}$, so the matrix of 1-forms, B, with entries $\{B^{mk}\}$ is anti-Hermitian. Meanwhile, $C^{km} = -C^{mk}$. Write the component of C^{mk} in $T^{0,1}M$ as $c^{mk}_{-i}\bar{\upsilon}^i$. The condition that j have zero Nijenhuis tensor requires that $c^{mk}_{-i} = c^{ik}_{m}$. Given that $c^{mk}_{-i} = -c^{km}_{-i}$, this requires that $c^{mk}_{-i} = 0$ for all indices k, m and i. Essentially the same calculation was done in Chapter 15.3 to prove that there is but a single torsion free, metric compatible connection.

Meanwhile, the form ω can written as

$$\omega = \frac{i}{2\pi} \upsilon^k \wedge \bar{\upsilon}^k.$$

Thus,

$$d\omega = \frac{i}{2\pi}\left(-B^{mk} \wedge \upsilon^m \wedge \bar{\upsilon}^k - C^{mk} \wedge \bar{\upsilon}^m \wedge \bar{\upsilon}^k - \overline{B^{km}} \wedge \upsilon^k \wedge \bar{\upsilon}^m - \overline{C^{km}}\, \upsilon^k \wedge \upsilon^m\right).$$

This uses the fact that ∇ is torsion free. The terms involving the matrix B and its conjugate vanish because B is an anti-Hermitian matrix of 1-forms. The term with C cannot cancel the term with its conjugate as the C-term is a section of $T^{1,2}M$ and its conjugate is a section of $T^{2,1}M$. Given that $C^{km} = -C^{mk}$, the C term is zero if and only if $C = 0$ identically.

These last conclusions tell us that

$$\nabla \upsilon^k = \upsilon^m \otimes B^{mk}$$

and so the Levi-Civita covariant derivative commutes with j. This is because the preceding equation asserts that if υ is a section of $T^{1,0}M$, then ∇υ is a 1-form on M with values in $T^{1,0}M$. In particular, j is covariantly constant. This means that ω too is covariantly constant because the metric g is covariantly constant. One can see this directly because with the C-term absent in the covariant derivative of υ, one has

$$\nabla \omega = \frac{i}{2\pi}\left(\upsilon^m \otimes \bar{\upsilon}^k \otimes B^{mk} + \upsilon^k \otimes \bar{\upsilon}^m \otimes \overline{B^{km}}\right),$$

and this is zero because B is an anti-Hermitian matrix of 1-forms.

17.10 Examples of Kähler manifolds

What follows are some examples of Kähler and non-Kähler complex manifolds.

Example 17.9 The manifold $S^1 \times S^3$ is a nonexample. It is an example of a manifold with a complex structure but no Kähler structure. (There is no 2-dimensional cohomology.) To see the complex structure, note that $\mathbb{C}^2 - \{0\}$ is diffeomorphic to $\mathbb{R} \times S^3$ with the diffeomorphism given as follows: Identify S^3 with the unit sphere in $\mathbb{C}^2$. Then the diffeomorphism sends a pair $(t, (a, b)) \in \mathbb{R} \times S^3$ to $(e^t a, e^t b) \in \mathbb{C}^2 - \{0\}$. View the manifold $S^1 \times S^3$ as $(\mathbb{R}/\mathbb{Z}) \times S^3$

where the $\mathbb{Z}$-action on $\mathbb{R}$ has the integer n acting to send a given $t \in \mathbb{R}$ to $t + n$. An equivalent way to view $S^1 \times S^3$ is as the quotient of $\mathbb{C}^2-\{0\}$ by the $\mathbb{Z}$-action that has $n \in \mathbb{Z}$ acting to send a pair $z = (z_1, z_2)$ to the pair $\phi_n(z) = (e^n z_1, e^n z_2)$. The pull-back of the basis $\{dz_1, dz_2\}$ of $T^{1,0}\mathbb{C}^2$ by φ_n is $\{e^n dz_1, e^n dz_2\}$, and so the φ_n pull-back of $T^{1,0}\mathbb{C}^2$ is $T^{1,0}\mathbb{C}^2$. This is to say that φ_n preserves the standard complex structure, j_0, on $\mathbb{C}^2-\{0\}$. As a consequence, this complex structure descends to the quotient to define one on $S^1 \times S^3$.

Note that $S^1 \times S^{2n-1}$ for any $n > 2$ gives a very analogous example. The latter can be viewed as the quotient of $\mathbb{C}^n-\{0\}$ by the action of $\mathbb{Z}$ that has $n \in \mathbb{Z}$ acting on the complex coordinates $(z_1, \ldots, z_n)$ to give $(e^n z_1, \ldots, e^n z_n)$.

As it turns out, the product $S^{2p-1} \times S^{2q-1}$ for any $p, q \geq 1$ has a whole family of complex structures, no two of which are equivalent. The construction is due to Calabi and Eckmann: Let $u \in \mathbb{C}$ denote a complex number with nonzero imaginary part. View $\mathbb{C}$ as an additive group, and let $\mathbb{C}$ act on $(\mathbb{C}^p-\{0\}) \times (\mathbb{C}^q-\{0\})$ as follows: An element $z \in \mathbb{C}$ sends a pair $(w_1, w_2) \in (\mathbb{C}^p-\{0\}) \times (\mathbb{C}^q-\{0\})$ to $(e^z w_1, e^{uz} w_2)$. Let X_U denote the quotient space by this action. The result is a complex manifold by virtue of the fact that the action of $\mathbb{C}$ defined here is complex linear with respect to the coordinates $(w_1, w_2) \in \mathbb{C}^p \times \mathbb{C}^q$. This is to say that it commutes with the standard complex structure. It is easiest to see that the resulting space X_U is diffeomorphic to $S^{2p-1} \times S^{2q-1}$ when $u = i$. In this case, there is a unique point in $S^{2p-1} \times S^{2q-1}$ on the orbit of any given pair (w_1, w_2), this the point (w_1', w_2') where $w_1' = w_1|w_1|^{-1}|w_2|^{i}$ and $w_2' = w_2\, |w_2|^{-1}|w_1|^{-i}$.

Example 17.10 An oriented manifold of dimension 2. Let Σ denote an oriented surface with metric g. As noted in Chapter 17.4, there is an associated, compatible almost complex structure. This is defined with respect to an oriented, orthonormal basis $\{e^1, e^2\}$ by the rule $je^1 = e^2$ and $je^2 = -e^1$. The volume form $\omega = e^1 \wedge e^2$ is the Kähler form.

Example 17.11 The manifold $\mathbb{CP}^n$. I have defined $\mathbb{CP}^n$ as S^{2n+1}/S^1 where I am viewing S^{2n+1} as the unit sphere in $\mathbb{C}^{n+1}$ and where S^1 is viewed as the unit circle in $\mathbb{C}$. A given element $\lambda \in S^1$ sends a given $(z_1, \ldots, z_{n+1}) \in S^{2n+1}$ to $(\lambda z_1, \ldots, \lambda z_{n+1})$. Let $\mathbb{C}^* \subset \mathbb{C}$ denote the multiplicative group of nonzero complex numbers, this, $\mathbb{C}-\{0\}$. As explained in Example 17.1, $\mathbb{CP}^n$ is also $(\mathbb{C}^{n+1}-\{0\})/\mathbb{C}^*$. The view of $\mathbb{CP}^n$ as $(\mathbb{C}^{n+1}-\{0\})/\mathbb{C}^*$ is taken here as it allows for a direct view of the complex manifold structure: Recall from Example 17.1, an open cover of $\mathbb{CP}^n$ is given by sets $\mathcal{U} = \{\mathcal{O}_k\}_{1 \leq k \leq n+1}$ where $\mathcal{O}_k$ is the image in $\mathbb{CP}^n$ of the set of points $z = (z_1, \ldots, z_{n+1}) \in \mathbb{C}$ where $z_k \neq 0$. The coordinate chart diffeomorphism from $\mathcal{O}_1$ to $\mathbb{C}^n$ sends $(z_1, \ldots, z_{n+1})$ to the point $(z_2/z_1, z_3/z_1 \ldots, z_{n+1}/z_1) \in \mathbb{C}^n$. That from $\mathcal{O}_2$ to $\mathbb{C}^n$ sends $(z_1, \ldots, z_{n+1})$ to the point $(z_1/z_2, z_3/z_2, \ldots, z_{n+1}/z_2) \in \mathbb{C}^n$, etc. The transition functions are holomorphic; for example, if $\varphi_1: \mathcal{O}_1 \to \mathbb{C}^n$ denotes the coordinate chart map just described, and likewise $\varphi_2: \mathcal{O}_2 \to \mathbb{C}^n$, then transition function $\varphi_2 \circ \varphi_1^{-1}$ is defined on the set $(w_1, w_2, \ldots, w_n)$ in $\mathbb{C}^n$ where w_1 is nonzero, and it sends such a point $(w_1, \ldots, w_n)$ to $(1/w_1, w_2/w_1, \ldots, w_n/w_1)$.

There is a special Kähler metric for $\mathbb{CP}^n$. This metric is called the *Fubini–Study* metric, and it is denoted here by g_{FS}. It is customary to define g_{FS} by giving its pull-back to $\mathbb{C}^{n+1}-\{0\}$ via the projection map to $\mathbb{CP}^n$. This pull-back is

$$|z|^{-2}dz_a \otimes d\bar{z}_a - |z|^{-2}\bar{z}_a dz_a \otimes z_b d\bar{z}_b.$$

The metric on any given $k \in \{1, \ldots, n+1\}$ version of $\mathcal{O}_k$ pulls back to $\mathbb{C}^n$ via the inverse of the corresponding coordinate chart map φ_k as

$$({\varphi_k}^{-1})^*(g_{FS}) = (1+|w|^2)^{-1}dw_j \otimes d\bar{w}_j - (1+|w|^2)^{-2}\bar{w}_j dw_j \otimes w_m d\bar{w}_m.$$

Here, I have written it so that it is evident that g_{FS} defines a Hermitian, positive definite form on $T_{1,0}\mathcal{O}_k$. The metric on the underlying real vector space $T\mathbb{CP}^n$ is obtained from this by writing each w_k in terms of its real and imaginary part, thus as $w_k = x_{2k-1} + ix_{2k}$. This done, then write $dw_k = dx_{2k-1} + idx_{2k}$ and insert into the expression above.

I leave it as an exercise to verify using the coordinate chart transition functions that the expression written above for a given $k \in \{1, \ldots, n+1\}$ is consistent with that defined by a different value of k on the overlap of the corresponding coordinate patches.

The associated almost Kähler form ω pulls up to $\mathbb{C}^{n+1}-\{0\}$ as

$$\frac{i}{2}|z|^{-2}\, dz_a \wedge dz_a - \frac{i}{2}|z|^{-2}\bar{z}_a\, dz_a \wedge z_b d\bar{z}_b.$$

On any given $\mathcal{O}_k$, this form pulls back via φ_k^{-1} as

$$(\varphi_k^{-1}) * \omega_{FS} = \frac{i}{2}(1+|w|^2)^{-1}dw_j \wedge d\bar{w}_j - \frac{i}{2}(1+|w|^2)^{-2}\bar{w}_j dw_j \wedge w_m d\bar{w}_m.$$

To see that it is closed, it is sufficient to note that

$$(\varphi_k^{-1}) * \omega_{FS} = \frac{i}{4}d((1+|w|^2)^{-1}(w_j d\bar{w}_j - \bar{w}_j dw_j)).$$

Appendix 17.1 Compatible almost complex structures

Let ω denote a 2-form on a 2n-dimensional manifold M whose n'th exterior power is nowhere zero. Suppose that j is an almost complex structure on TM. This almost complex structure is said to be ω-compatible when the section $\omega(\cdot, j(\cdot))$ of $\bigotimes^2 T^*M$ defines a Riemannian metric. This is to say that it defines a symmetric, positive definite inner product on TM. As noted in Chapter 17.6, every almost Kähler form has compatible almost complex structures. What follows here is a proof of this assertion in the case when M is compact. The argument in general is almost the same but for the use of locally finite covers. The proof for compact M has five parts.

Part 1

Use ω^n to define an orientation for M. Fix $p \in M$ and fix an oriented basis, $\mathfrak{v} = (v_1, \ldots, v_{2n})$ for TM. Define from $\mathfrak{v}$ and ω the $2n \times 2n$, antisymmetric matrix $A^{\mathfrak{v}}$ whose $(i, j) \in \{1, \ldots, 2n\}$ component is $\omega(v_i, v_j)$. This matrix at p is nondegenerate, and so it has 2n eigenvalues, of which n are of the form $i\,\lambda$ with λ a positive real number. The other n eigenvalues are the complex conjugates of the first n. It proves useful to require that these eigenvalues are distinct. If they are not for the given basis, then there is, in all cases, a change of basis that makes them distinct. I leave the proof of this to you as an exercise in linear algebra.

With it understood now that the eigenvalues of $A^{\mathfrak{v}}$ are distinct, suppose that $i\lambda$ is any given eigenvalue of $A^{\mathfrak{v}}$ with $\lambda > 0$. Let $e \in \mathbb{C}^{2n}$ denote the corresponding eigenvector. Both the real and imaginary parts of e are nonzero as can be seen from the fact that the complex conjugate vector is an eigenvector with eigenvalue $-i\lambda$. The point being that $-i\lambda \neq i\lambda$ and so e cannot be a complex multiple of its complex conjugate. Write e as $e_1 + ie_2$ where e_1 and e_2 are in $\mathbb{R}^{2n}$. Write their respective components as $\{e_1^i\}_{1\leq i\leq 2n}$ and $\{e_2^i\}_{1\leq i\leq 2n}$. Use $\mathfrak{e}_1 \in TM|_p$ to denote the point $\sum_{1\leq i\leq 2n} e_1^i v_i$ and use $\mathfrak{e}_2 \in TM|_p$ to denote $\sum_{1\leq i\leq 2n} e_2^i v_i$. Define an almost complex structure, j, on the span of $\{\mathfrak{e}_1, \mathfrak{e}_2\}$ by the rule $j\mathfrak{e}_1 = \mathfrak{e}_2$ and $j\mathfrak{e}_2 = -\mathfrak{e}_1$. Making this definition of j for each eigenspace defines j on the whole of $TM|_p$ as an almost complex structure.

Now suppose that λ, $\mathfrak{e}_1$ and $\mathfrak{e}_2$ are as just described. Given that e_1 and e_2 come from the $i\lambda$ eigenvector of $A^{\mathfrak{v}}$, it follows that $\omega(\mathfrak{e}_1, j\mathfrak{e}_1) = \lambda = \omega(\mathfrak{e}_2, j\mathfrak{e}_2)$ and $\omega(\mathfrak{e}_1, j\mathfrak{e}_2) = 0 = \omega(\mathfrak{e}_2, j\mathfrak{e}_1)$. Thus, the bilinear form given by $\omega(\cdot, j(\cdot))$ defines a Riemannian metric on the span of $\{\mathfrak{e}_1, \mathfrak{e}_2\}$. Suppose that $\lambda' \neq \lambda$, λ' is positive and $i\lambda'$ is an eigenvalue of $A^{\mathfrak{v}}$. Let $\mathfrak{e}'$ denote the correponding eigenvector. Write it in terms of real vectors in $\mathbb{R}^{2n}$ as $\mathfrak{e}_1' + i\mathfrak{e}_2'$. As $\lambda' \neq \lambda$, one has $\omega(\mathfrak{e}_1, j\mathfrak{e}_1') = 0 = \omega(\mathfrak{e}_1, j\mathfrak{e}_2')$ and likewise $\omega(\mathfrak{e}_2, j\mathfrak{e}_1') = \omega(\mathfrak{e}_2, j\mathfrak{e}_1{}')$. Reversing the roles of the prime and unprimed components shows that the span of $\{\mathfrak{e}_1, \mathfrak{e}_2\}$ is orthogonal with respect to $\omega(\cdot, j(\cdot))$ to the span of $\{\mathfrak{e}_1', \mathfrak{e}_2'\}$. Granted this, it follows that j as just defined is an ω-compatible almost complex structure at p.

Part 2

The task for this part of the argument is to extend this almost complex structure to an open neighborhood of p. This is done as follows: First extend the basis v to a neighborhood of p. This is a straightforward procedure that can be done, for example, by fixing an open set containing p which comes with an isomorphism from TM to the product bundle. With this isomorphism to the product bundle understood, extend the basis as the constant basis.

Given such an extension, any given point p′ in a neighborhood of p has a corresponding antisymmetric matrix, $A^{\upsilon}|_{p'}$. If the eigenvalues are distinct at p, they will be distinct at each such point in some small neighborhood of p. This said, the corresponding eigenspaces vary near p so as to define a set of n complex line bundles. To elaborate, let $U \subset M$ denote such a neighborhood. Then the eigenspaces define a decomposition $TM_{\mathbb{C}}|_U = (\bigoplus_{1 \le a \le n} E_a) \oplus (\bigoplus_{1 \le a \le n} \bar{E}_a)$ where each E_a is a complex line bundle. Define j on $TM|_U$ so that its (1, 0) tangent space in $TM_{\mathbb{C}}|_U$ is the summand $(\bigoplus_{1 \le a \le n} E_a)$. See Chapter 6.3 if your memory fails you with regards to this definition of an almost complex structure. In any event, it follows from the observations from Part 1 that $\omega(\cdot, j(\cdot))$ is a Riemannian metric on $TM|_U$.

Part 3

Part 5 explains how to obtain an ω-compatible almost complex structure for the whole of TM from those constructed in Part 3 on small open sets. As you might guess from the discussion at the outset of Chapter 7, the construction involves partitions of unity for an open cover. However, the story here is not as simple because the sum of two almost complex structures gives an endomorphism of TM whose square is not, in general, equal to minus the identity. This part and Part 4 set up the machinery to deal with this complication.

To start, consider an open set $U \subset M$ where TM has a metric compatible, almost complex structure, j. Take a smaller open set if necessary so as to guarantee that there is an orthonormal basis of sections of $TM|_U$ for the metric $g(\cdot, \cdot) = \omega(\cdot, j(\cdot))$. Use the metric to identify $TM|_U$ with $T^*M|_U$ and use the basis to identify both isometrically with $U \times \mathbb{R}^n$. With this identification understood, both ω and j appear as maps from U to the space $\mathbb{M}(2n; \mathbb{R})$ of $2n \times 2n$ matrices. Use W to denote the former and J to denote the latter. This is to say that the (i, j) component of W is $\omega(e_i, e_j)$ where e_i and e_j denote the i'th and j'th elements of the chosen orthonormal basis. Meanwhile, the (i, j) component of J is $e^i(Je_j)$ where here e^i denotes the i'th component of the dual basis for T*M. Since the identification between $TM|_U$ and $T^*M|_U$ is isometric, the identification sends the metric to the identity matrix ι. This implies that $W = J^{-1}$ at each point of U.

Meanwhile, $J^2 = -\iota$, and so $J^{-1} = -J$ at each point of U. Moreover, $J^T = -J$ as j and hence J acts orthogonally.

Now suppose that j′ is a second, ω-compatible almost complex structure on $TM|_U$. Let J′ denote the corresponding map from U to the space 2n × 2n matrix. This map is also defined using the orthonormal frames as defined by the metric g for $TM|_U$ and $T^*M|_U$. To say that j′ is ω-orthogonal means neither more nor less than the following: The matrix WJ′ is a symmetric, positive definite matrix at each point of U. Let H denote the latter. Since $W = J^{-1}$, this means that $J' = JH$. If $(J')^2$ is to equal $-\iota$, then $JH = -(JH)^{-1} = -H^{-1}J^{-1}$. Multiplying both sides of this equation by J finds that $H = JH^{-1}J^{-1}$. This understood, the task for what follows is to identify the set of maps from U to the space of symmetric, positive definite, 2n × 2n matrices that obey $H = JH^{-1}J^{-1}$ at each point. This task is accomplished in the next part of the Appendix.

Part 4

This part of the appendix identifies the space of maps from U to the positive definite, 2n × 2n symmetric matrices that obey $H = JH^{-1}J^{-1}$ with a space of maps from U to a certain *linear* subspace in the space of 2n × 2n matrices. The key point for use later is that the subspace in question is linear. The identification is given in three steps.

Step 1: Fix a positive integer m, and let $\mathrm{Sym}^+(m; \mathbb{R}) \subset \mathbb{M}(m; \mathbb{R})$ denote the open subspace of positive definite, symmetric m × m matrices. In the application to the problem at hand, m will be equal to 2n. Let Sym(m; ℝ) denote the space of symmetric, m × m matrices. The lemma below refers to the exponential map that sends 𝕄(n; ℝ) to Gl(n; ℝ). The latter was introduced in Chapter 5.5; it sends a matrix $\mathfrak{m}$ to $e^{\mathfrak{m}}$ with $e^{\mathfrak{m}}$ defined by its power series: $e^{\mathfrak{m}} = \iota + \mathfrak{m} + \frac{1}{2}\mathfrak{m}^2 + \frac{1}{3!}\mathfrak{m}^3 + \cdots$.

Lemma 17.1 *The exponential map* $\mathfrak{m} \to e^{\mathfrak{m}}$ *restricts to* Sym(m; ℝ) *as a diffeomorphism onto* $\mathrm{Sym}^+(m; \mathbb{R})$.

Proof of Lemma 17.1 Look at the power series definition of $e^{\mathfrak{m}}$ to see that each term is symmetric when $\mathfrak{m}$ is symmetric. To see that $e^{\mathfrak{m}}$ is positive definite, it is enough to note that $e^{\mathfrak{m}}$ can be written as $e^{\mathfrak{m}/2}e^{\mathfrak{m}/2}$ which is the square of a symmetric matrix. Hence, all the eigenvalues of $e^{\mathfrak{m}}$ are squares and so all are positive. None are zero because the exponential map lands in Gl(m; ℝ). To see that this map is onto $\mathrm{Sym}^+(m; \mathbb{R})$, remark that any given positive definite, symmetric matrix $\mathfrak{h}$ can be written as $\mathfrak{o}\mathfrak{d}\mathfrak{o}^{-1}$ where $\mathfrak{o}$ is orthogonal and $\mathfrak{d}$ is a diagonal matrix with positive entries. This said, let $\ln(\mathfrak{d})$ denote the diagonal matrix whose entry on any given diagonal is the natural log of the corresponding entry of $\mathfrak{d}$. Set $\mathfrak{m} = \mathfrak{o}\ln(\mathfrak{d})\mathfrak{o}^{-1}$. A look at the power series expansion shows that $\mathfrak{h} = e^{\mathfrak{m}}$.

The map $\mathfrak{m} \to e^{\mathfrak{m}}$ is also 1–1 as can be seen by the following argument: Suppose that $e^{\mathfrak{m}} = e^{\mathfrak{m}'}$. Since $\mathfrak{m}$ is symmetric, it too can be written as $\mathfrak{o}\mathfrak{d}\mathfrak{o}^{-1}$ with $\mathfrak{d}$ a diagonal matrix. This implies that $e^{\mathfrak{m}} = \mathfrak{o}e^{\mathfrak{d}}\mathfrak{o}^{-1}$ and so the eigenvalues of $e^{\mathfrak{m}}$ are the exponentials of those of $\mathfrak{m}$. It follows from this that $\mathfrak{m}$ and $\mathfrak{m}'$ have the same set of eigenvalues and corresponding eigenvalues have the same multiplicity. This implies that $\mathfrak{m}'$ can be written as $\mathfrak{u}\mathfrak{m}\mathfrak{u}^{-1}$ with $\mathfrak{u}$ a diagonal matrix. But if the latter is true, then $e^{\mathfrak{m}'} = \mathfrak{u}e^{\mathfrak{m}'}\mathfrak{u}^{-1}$, and this can happen only if $\mathfrak{u}$ preserves the eigenspaces of $\mathfrak{m}$. This is to say that $\mathfrak{u}\mathfrak{m}\mathfrak{u} = \mathfrak{m}$ and so $\mathfrak{m}' = \mathfrak{m}$.

The next point to make is that the differential of the map $\mathfrak{m} \to e^{\mathfrak{m}}$ on Sym(m; $\mathbb{R}$) defines an isomorphism at each point. To see why this is, fix $\mathfrak{m}$, $\mathfrak{q} \in \text{Sym}(m; \mathbb{R})$ and consider for $t \in (0, 1)$ but small the map $t \to e^{\mathfrak{m}+t\mathfrak{q}}$. A look at the derivative of this map at $t = 0$ finds the latter given by

$$\int_0^1 e^{(1-s)\mathfrak{m}}\mathfrak{q}e^{s\mathfrak{m}}\, ds.$$

This is $\mathfrak{m}$'s image via the differential at $\mathfrak{m}$ of the exponential map. To see when this is zero, write $\mathfrak{m} = \mathfrak{o}\,\mathfrak{d}\,\mathfrak{o}^{-1}$ with $\mathfrak{d}$ a diagonal matrix and $\mathfrak{d}$ an orthogonal matrix. What is written above is then

$$\mathfrak{o}\left(\int_0^1 e^{(1-s)\mathfrak{d}}\left(\mathfrak{o}^{-1}\mathfrak{o}\mathfrak{m}\right)e^{s\mathfrak{d}}ds\right)\mathfrak{o}^{-1}.$$

This can vanish only if the integral is zero. To see when such is the case, introduce z_{ij} denote the (i, j) entry of the matrix $\mathfrak{o}^{-1}\mathfrak{q}\,\mathfrak{o}$. Let d_i denote the i'th diagonal entry of $\mathfrak{d}$. Then the (i, j) entry of the matrix that appears in the integrand above is equal to

$$e^{(1-s)d_i+sd_j}\, Z_{ij}.$$

In particular, the integral of this with respect to s is zero if and only if $z_{ij} = 0$.

As just argued, the map $\mathfrak{m} \to e^{\mathfrak{m}}$ is 1–1 and onto; and its differential is everywhere an isomorphism. This implies that it is a diffeomorphism.

Step 2: Suppose that H is a map from U to the space $\text{Sym}^+(2n; \mathbb{R})$. It follows from the lemma in Step 1 that this map can be written in a unique way as $H = e^Q$ where Q is a smooth map from U to Sym(2n; $\mathbb{R}$). The inverse matrix H^{-1} is e^{-Q}. This being the case, then the condition $H = JH^{-1}J^{-1}$ requires that $JQJ^{-1} = -Q$. Such is the case if and only if Q can be written as $Q = V - J^{-1}VJ$ with V a smooth map from U to Sym(2n; $\mathbb{R}$).

To put this in perspective, remark that the map $Q \to JQJ^{-1}$ defines a linear map from Sym(2n; $\mathbb{R}$). This is because $J^{-1} = -J$ and $J = -J^T$, so this is the same as the map sending Q to JQJ^T. Let ϑ: Sym(2n; $\mathbb{R}$) $\to$ Sym(2n; $\mathbb{R}$) denote this map

$Q \to JQJ^T$. Since $J^2 = -1$, this map ϑ obeys $\vartheta^2 = 1$, so its eigenvalues are ± 1. Of interest here is the -1 eigenspace, because if $\vartheta(Q) = -Q$, then $Q = -JQJ^{-1}$, and conversely if $Q = -JQJ^{-1}$ then Q is an eigenvector of ϑ with eigenvalue -1. The dimension of this eigenspace is the same as that of the vector space (over $\mathbb{R}$) of Hermitian, $n \times n$ matrices. This follows from what is said in the final paragraph of the proof of Lemma 6.4. Thus, this -1 eigenspace has dimension $n(n + 1)$. Denote this -1 eigenspace by $S_-(2n; \mathbb{R})$.

Step 3: To summarize: A map H from U to $Sym^+(2n; \mathbb{R})$ that obeys $H = JH^{-1}J^{-1}$ can be written in a unique fashion as e^Q where Q is a smooth map from U to $S_-(2n; \mathbb{R})$. Conversely, if Q is such a smooth map, then $H = e^Q$ is a map from U to $Sym^+(2n; \mathbb{R})$ that obeys $H = JH^{-1}J^{-1}$.

Part 5

It follows from what was said in Part 3 that M has a finite cover $\mathfrak{U}$ such that each $U \in \mathfrak{U}$ comes with an ω-compatible almost complex structure on TM. This cover should be chosen so as to have the following two additional features: First, each $U \subset \mathfrak{U}$ comes with a diffeomorphism, φ_U, to the interior of the ball of radius 3 centered at the origin in $\mathbb{R}^{2n}$. Note that this implies that $TM|_U$ is isomorphic to the product bundle $U \times \mathbb{R}^{2n}$. To state the second, introduce $B_1 \subset \mathbb{R}^{2n}$ to denote the ball of radius 1 centered at the origin. Here is the second requirement: The collection $\{\varphi_U^{-1}(B_1)\}_{U \in \mathfrak{U}}$ is also an open cover of M. Let N denote the number of sets in this cover. Label these sets as $\{U_k\}_{1 \leq k \leq N}$. For each $k \in \{1, \ldots, N\}$, let φ_k to denote the $U = U_k$ version of φ_U. Likewise, let j_k denote the ω-compatible, almost complex structure given on U_k.

For each $k \in \{1, \ldots, N\}$, use $B_{2-k/N} \subset \mathbb{R}^{2n}$ to denote the ball of radius $2 - k/N$ centered at the origin. Introduce $M^{(k)} \subset M$ to denote $\cap_{1 \leq m \leq k} \varphi_k^{-1}(B_{2-k/N})$. Note that $M^{(N)} = M$ because the collection $\{\varphi_k{}^{-1}(B_1)\}_{1 \leq k \leq N}$ covers M.

The construction of the desired ω compatible almost complex structure will be done in a step-by-step fashion. The k'th step will result in an ω-compatible almost complex structure over $M^{(k)}$. The latter will be denoted in what follows by $j^{(k)}$. To start, consider $M^{(1)}$. This is a subset of U_1. The desired ω-compatible almost complex structure on $M^{(1)}$ is $j^{(1)} = j_1$. Now to proceed by induction, suppose that $k \in \{1, 2, \ldots, N - 1\}$ and that $j^{(k)}$ has been defined on $M^{(k)}$. What follows defines $j^{(k+1)}$ on $M^{(k+1)}$.

There are two cases to consider. In the first case, U_{k+1} is disjoint from $M^{(k)}$. In this case, $M^{(k+1)}$ is the disjoint union of $\varphi_{k+1}^{-1}(B_{2-(k+1)/N})$ with the subset $\cup_{1 \leq m \leq k} \varphi_k^{-1}(B_{2-(k+1)/N})$ of $M^{(k)}$. This understood, set $j^{(k+1)}$ to equal j_{k+1} in $\varphi_{k+1}^{-1}(B_{2-(k+1)/N})$ and equal to $j^{(k)}$ on the component of $M^{(k+1)}$ in $M^{(k)}$. Now suppose that $U_{k+1} \cap M^{(k)}$ is not empty. Denote this subset of U_{k+1} by V_k. Let $V_{k+1} \subset U_{k+1}$ denote $U_{k+1} \cap M^{(k+1)}$. Note that $V_{k+1} \subset V_k$ and that the closure of V_{k+1} in M is disjoint from that of $U_{k+1} - V_k$ in M. This is an important observation so keep it in mind.

Define $j^{(k+1)}$ on the complement in $M^{(k+1)}$ of $V_k - V_{k+1}$ to equal $j^{(k)}$. To define $j^{(k+1)}$ on $U_{k+1} \cap M^{(k+1)}$, first introduce the metric $g_{k+1} = \omega(\cdot, j_{k+1}(\cdot))$ for TU_{k+1}. As TU_{k+1} is isomorphic to the product bundle, there exists an orthonormal basis of sections for this metric. Use this basis as in Part 3 to identify j_{k+1} with a map, J, from U_{k+1} to the space of $2n \times 2n$ matrices. The desired almost complex structure $j^{(k+1)}$ will also be given over the part of $M^{(k+1)}$ in U_{k+1} as a map, J^*, to this same space of matrices. To define J^*, use the chosen g_{k+1}-orthonormal frame for TU_{k+1} to identify $j^{(k)}$ on V_k with a map, J', from V_k to the space of $2n \times 2n$ matrices. As seen in Part 4, the map J' is such that $H = J^{-1}J'$ has the form e^Q where Q is a map from V_k to $S_-(2n; \mathbb{R})$. This is to say that $J' = Je^Q$.

Given that the closure of V_{k+1} is disjoint from that of $U_{k+1} - V_k$, there is a strictly *positive* lower bound to the distance between any point in the set $\varphi_{k+1}(V_{k+1}) \cap B_2$ and $B_2 - \varphi_{k+1}(V_k)$. As a consequence, the constructions from Appendix 1.2 can be used to obtain a smooth function on $B_2 \subset \mathbb{R}^{2n}$ that is equal to 1 on $\varphi_{k+1}(V_{k+1})$ and equal to zero on $B_2 - \varphi_{k+1}(V_k)$. Use θ to denote the composition of this function with the map φ_{k+1}. Then set $J^* = Je^{\theta Q}$. Note that using J^* to define $j^{(k+1)}$ on $M^{(k+1)} \cap U_{k+1}$ gives an almost complex structure that agrees with the definition given previously for $j^{(k+1)}$ on the set $M^{(k+1)} - (M^{(k+1)} \cap (V^k - V^{k+1})$; this is because $\theta\, Q$ is equal to Q on V_{k+1}.

Additional reading

- *Complex Geometry: An Introduction*, Daniel Huybrechts, Springer, 2004.
- *Lectures on Kähler Geometry*, Andrei Moroianu, Cambridge University Press, 2007.
- *Foundations of Differential Geometry, Volume II*, Shoshichi Kobayashi and Katumi Nomizu, Wiley Interscience, 1996.
- *Algebraic Curves and Riemann Surfaces*, Rick Miranda, American Mathematical Society, 1995.
- *Complex Manifolds*, James Morrow and Kunihiko Kodaira, American Mathematical Society, 2006.
- *Complex Manifolds without Potential Theory*, Shing-shen Chern, Springer, 1979.
- *Introduction to Symplectic Topology*, Dusa McDuff and Dietmar Salamon, Oxford University Press, 1999.

18 Holomorphic submanifolds, holomorphic sections and curvature

This chapter describes how complex manifolds and Kähler manifolds can appear as submanifolds in a larger complex or Kähler manifold.

18.1 Holomorphic submanifolds of a complex manifold

The examples here concern *holomorphic* submanifolds of a complex manifold. To set the stage, suppose that N is a complex manifold. Let j denote the almost complex structure on N that comes from its complex manifold structure. A submanifold $M \subset N$ is said to be holomorphic if j along M maps the subbundle $TN \subset TM|_N$ to itself. Thus, j maps any given tangent vector to M to some other tangent vector to M. Assume that M has this property. Then $j|_M$ defines an endomorphism of TM with square -1, so an almost complex structure, j_M, on TM. Moreover, the Nijenhuis tensor of this almost complex structure is zero by virtue of the fact that it is zero for j. To see this, remember that the Nijenhuis tensor vanishes if and only if the commutator of two sections of $T_{1,0}M$ is also a section of $T_{1,0}M$. Here, sections of $T_{1,0}M$ are viewed as $\mathbb{C}$-valued tangent vectors on M and $TM_{\mathbb{C}}$ is written as $T_{1,0}M \oplus T_{0,1}M$. This understood, remark that $T_{1,0}M \subset T_{1,0}N|_M$. Thus, the commutator of two sections of $T_{1,0}M$ is, a priori, in $T_{1,0}N|_M$; and thus in $T_{1,0}M$ since commutators of vectors tangent to M are tangent to M, whether complex or not. Given that the Nijenhuis tensor of j_M is zero, it follows from the Newlander–Nirenberg theorem (Theorem 17.5) that j_M comes from a complex manifold structure on M.

Now suppose that g is a Kähler metric on M. Then g's restriction to $TM \subset TN|_M$ defines a metric, g_M, on TM, and this is compatible with j_M since they are restrictions of g and j to a subspace in $TN|_M$. Let ω denote the associated almost Kähler form ω on N. Then the associated almost Kähler form, ω_M, for the pair g_M and j_M is the restriction of ω as a bilinear form on $TM \subset TN|_M$. This is just the pull-back via the inclusion map $M \to N$ of ω. It follows from this that if ω is closed, then ω_M is also closed. These last observations are summarized by the following:

Proposition 18.1 *Let* N *be a complex manifold and* M *a holomorphic submanifold of real dimension at least 2. Then* M *also has the structure of a complex manifold. Moreover, if* g *is a Kähler metric for* N, *then the restriction of* g *to the tangent space of* M *defines a Kähler metric for* M.

18.2 Holomorphic submanifolds of projective spaces

Chapter 17.3 explains how $\mathbb{CP}^n$ can be viewed as a complex manifold, and Chapter 17.7 describes its associated Kähler form. A plethora of holomorphic Kähler submanifolds can be found inside of $\mathbb{CP}^n$. What follows describes an archetype example: Let $z = (z_1, \ldots, z_{n+1})$ denote the coordinates for $\mathbb{C}^{n+1}$. Let h: $\mathbb{C}^{n+1} \to C$ denote a polynomial in the coordinates $(z_1, \ldots, z_{n+1})$ that is homogeneous in the sense that there exists a positive integer p such that $h(\lambda z) = \lambda^p\, h(z)$ for all $\lambda \in \mathbb{C}$. Here, λz is the point with coordinates $(\lambda z_1, \ldots, \lambda z_{n+1})$. Let $Z_h = \{z \in \mathbb{C}^{n+1} - \{0\} : f(z) = 0\}$. This subset is invariant under the action of $\mathbb{C}^*$ on $\mathbb{C}^{n+1}$ whose quotient gives $\mathbb{CP}^n$. Use $M_h \subset \mathbb{CP}^n$ to denote the image of Z_h under the projection.

The proposition that follows refers to the *holomorphic* differential of a $\mathbb{C}$-valued function. Here is the definition: Let $m \geq 1$, let $U \subset \mathbb{C}^m$ denote an open set, and let f: $U \to C$ denote a smooth function. The holomorphic differential of f is the projection of dh onto the span of the 1-forms $\{dw_i\}_{1\leq i\leq m}$. This is usually written as ∂f. To elaborate,

$$\partial f = \sum\nolimits_{1\leq i\leq m} \frac{\partial f}{\partial w_i} dw_i$$

where $dw_i = dx_{2i-1} + i dx_{2i}$ and $\frac{\partial}{\partial w_i} = \frac{1}{2}(\frac{\partial}{\partial x_{2i-1}} - i\frac{\partial}{\partial x_{2i}})$. Here, $w_i = x_{2i-1} + i x_{2i}$ is written in terms of its real and imaginary parts.

Note that a function f: $U \to \mathbb{C}$ is holomorphic as defined in Chapter 17.1 when $df = \partial f$. This is to say that $\frac{\partial f}{\partial \bar{w}_i} = \frac{1}{2}(\frac{\partial f}{\partial x_{2i-1}} - i\frac{\partial f}{\partial x_{2i}}) = 0$ for all i. If f is written as $f_1 + i f_2$ with f_1 and f_2 both ordinary $\mathbb{R}$-valued functions, then this says that f_1 and f_2 obey the Cauchy–Riemann equations for each of the coordinate pairs (x_{2i-1}, x_{2i}):

$$\frac{\partial f_1}{\partial x_{2i-1}} = \frac{\partial f_2}{\partial x_{2i}} \quad \textit{and} \quad \frac{\partial f_2}{\partial x_{2i-1}} = -\frac{\partial f_1}{\partial x_{2i}}.$$

Polynomials of the complex coordinates on $\mathbb{C}^{n+1}$ are examples of holomorphic functions on the whole of $\mathbb{C}^{n+1}$.

Proposition 18.2 *Suppose that* h *is a homogeneous polynomial of the coordinates on* $\mathbb{C}^{n+1}$ *whose holomorphic differential is nowhere zero along* $h^{-1}(0) \subset \mathbb{C}^{n+1}-\{0\}$. *Then the image,* $M_h \subset \mathbb{CP}^n$ *of* $h^{-1}(0)$ *is a compact, complex submanifold of* $\mathbb{CP}^n$ *of real dimension* $2n-2$. *Moreover, the complex manifold structure coming from* $\mathbb{CP}^n$ *is such that the induced metric from the Fubini–Study metric on* $\mathbb{CP}^n$ *makes* M_h *a Kähler manifold.*

This proposition about holomorphic submanifolds of $\mathbb{CP}^n$ is proved momentarily.

Examples for the case $n = 2$ are the polynomials $h = z_1^p + z_2^p + z_3^p$. The corresponding submanifold of $\mathbb{CP}^2$ is 2-dimensional, and so a surface. The genus of this surface is $\frac{1}{2} = (p^2 - 3p + 2)$. Examples in the case $n = 3$ are $h = z_1^p + z_2^p + z_3^p + z_4^p$. These are all simply connected manifolds of real dimension equal to 4.

The proof of the proposition requires the following preliminary lemma.

Lemma 18.3 *Let* $U \subset \mathbb{C}^m$ *denote an open set and let* $f\colon U \to C$ *denote a holomorphic function. Then* $f^{-1}(0)$ *is a smooth,* $(2m-2)$*-real dimensional submanifold near any point of* $f^{-1}(0)$ *where* $\partial f \neq 0$.

Proof of Lemma 18.3 View f as a map from $\mathbb{R}^{2n}$ to $\mathbb{R}^2$. Viewed in this light, the implicit function theorem from Chapter 1.2 tells us that $f^{-1}(0)$ is a submanifold of U near any point where the corresponding matrix of partial derivatives is surjective. This matrix has 2 columns and 2n rows, and its entries are the partial derivatives of the components of f. To see that this matrix is surjective under the assumptions of the lemma, suppose for the sake of argument that $\frac{\partial f}{\partial z_1}$ is nonzero at some point where f is zero. Then it is nonzero in a neighborhood of this point. Keeping this in mind, write f as $f_1 + i f_2$ with f_1 and f_2 real. Also, write $z_1 = x_1 + ix_2$ with x_1 and x_2 real. The 2×2 minor of the differential of f (as a map from $\mathbb{R}^{2n+1}$ to $\mathbb{R}^2$) corresponding to the (x_1, x_2) coordinates is

$$\begin{pmatrix} \partial_1 f_1 & \partial_1 f_2 \\ \partial_2 f_1 & \partial_2 f_2 \end{pmatrix}$$

where ∂_1 and ∂_2 here are the respective partial derivatives with respect to x_1 and x_2. This understood, it is enough to verify that the determinant of this matrix is nonzero. This determinant is $|\frac{\partial f}{\partial z_1}|^2$ when $\frac{\partial f}{\partial \bar{z}_1} = 0$.

Note that the proof of this lemma also establishes the following:

Lemma 18.4 *Let* $U \subset \mathbb{C}^m$ *denote an open set and let* $f\colon U \to \mathbb{C}$ *denote a holomorphic function. The point* $0 \in \mathbb{C}$ *is a regular value of* f *if and only if* $\partial f \neq 0$ *along* $f^{-1}(0)$.

Recall that a regular value of a map, $f: \mathbb{R}^p \to \mathbb{R}^q$ is a point $y \in \mathbb{R}^q$ with the following property: Let x denote any point with $f(p) = y$. Then the matrix with p columns and q rows whose entries are the partial derivatives of the coordinates of f is surjective.

Proof of Lemma 18.4 As seen in the proof of Lemma 18.3, the vanishing of $\bar{\partial} f$ at any given point $p \in U$ is equivalent to the assertion that the differential of f is zero at that point. Here, I view

$$f = \begin{pmatrix} f_1 \\ f_2 \end{pmatrix}$$

as a map from an open set in $\mathbb{C}^n = \mathbb{R}^{2n}$ to $\mathbb{C} = \mathbb{R}^2$ so as to view its differential as a matrix with 2n columns and 2 rows whose entries are partial derivatives of f.

By the way, Lemma 18.4 and Sard's theorem (Theorem 1.3) together supply the following additional conclusion:

Corollary 18.5 *Let* $U \subset \mathbb{C}^m$ *denote an open set and let* $f: U \to C$ *denote a holomorphic function. Then almost all values* $c \in C$ *are regular values of* f*. As a consequence,* $f^{-1}(c)$ *is a holomorphic submanifold of* U *for almost all values* $c \in \mathbb{C}$.

Recall that Sard's theorem says the following about a given map from $\mathbb{R}^p$ to $\mathbb{R}^q$: The set of nonregular values of the map has zero measure. Thus, the set of regular values in a given ball has full measure. This is what Corollary 18.5 means by the term *almost all*. To obtain the corollary, apply Lemma 18.4 to the function $f - c$ with c being a regular value of f.

18.3 Proof of Proposition 18.2, about holomorphic submanifolds of $\mathbb{CP}^n$

Lemma 18.3 guarantees that $f^{-1}(0)$ is a smooth submanifold of $\mathbb{C}^{n+1}-\{0\}$. Chapter 17.3 describes an open cover, $\{\mathcal{O}_k\}_{1\le k\le n+1}$ of coordinate charts, each diffeomorphic to $\mathbb{C}^n$ via a diffeomorphism that makes for holomorphic coordinate transition functions. By way of a reminder, the coordinate chart $\mathcal{O}_k$ consists of the set of complex lines through the origin in $\mathbb{C}^{n+1}$ on which the k'th coordinate $z_k \neq 0$. The coordinate chart map ϕ_k is defined so that the line spanned by a point $(z_1, \ldots, z_{n+1}) \in \mathbb{C}^{n+1}-\{0\}$ with $z_k \neq 0$ is sent to the point in $\mathbb{C}^n$ whose coordinates are $(z_1/z_k, \ldots, z_{k-1}/z_k, z_{k+1}/z_k, \ldots, z_n/z_k)$. A calculation, done in the coordinate charts $\mathcal{O}_k$ using the coordinates given by φ_k, shows that $M_h \cap \mathcal{O}_k$ is a smooth submanifold of O_k. For example, $\varphi_1(M_h \cap \mathcal{O}_1)$ is the submanifold of $\mathbb{C}^n$ given by

$$h(1, w_1, \ldots, w_n) = 0.$$

To see if this locus is a submanifold, it is enough to verify that at least one of the partial derivatives $\frac{\partial h}{\partial w_m}$ is nonzero at any given point along this locus. To see that they can't all vanish, remark that the inverse image of $\mathcal{O}_k$ in $\mathbb{C}^{n+1}$ has coordinates $(u, w_1, \ldots, w_n) \in \mathbb{C}^{n+1}$ which are related to the z-coordinates by the rule $z_1 = u$, $z_2 = uw_1, \ldots, z_{n+1} = uw_n$. In terms of the coordinates $(u, w_1, \ldots, w_n)$, the function h appears as

$$h(z_1, z_2, \ldots, z_{n+1}) = u^p h(1, w_1, \ldots, w_2).$$

If all of the partial derivatives of the function $w \to h(1, w_1, \ldots, w_n)$ vanish where this function is zero, then every partial derivative of the function $(u, w) \to u^p\, h(1, w_1, \ldots, w_n)$ is also zero, so every partial derivative of $z \to h(z_1, \ldots, z_{n+1})$ is zero too. But this is forbidden by the assumption that ∂h is nonzero on $\mathbb{C}^{n+1} - \{0\}$ where $h = 0$.

To see that M_h is a holomorphic submanifold, it is sufficient to verify that the almost complex structure on $\mathbb{C}^n$ preserves the tangent space to the zero locus of the function $w \to h(1, w_1, \ldots, w_n)$. To start this task, suppose for the sake of argument that $\frac{\partial h}{\partial w_1}$ is nonzero at a given point on this locus. Then it will be nonzero at all nearby points. This understood, it follows that the tangent space to the $h = 0$ locus in a neighborhood of the given point is spanned by the real and imaginary parts of the vectors

$$\left\{ \frac{\partial}{\partial w_j} - \left(\frac{\partial h}{\partial w_j}\right)\left(\frac{\partial h}{\partial w_1}\right)^{-1} \frac{\partial}{\partial w_1} \right\}_{2 \le j \le n}.$$

These vectors are mapped to themselves by j.

Note that the argument just given proves the following general fact:

Proposition 18.6 *Let* $U \subset \mathbb{C}^n$ *denote an open set and let* $f\colon U \to \mathbb{C}$ *denote a function that is holomorphic with respect to the coordinate functions of* $\mathbb{C}^n$*. The* $f^{-1}(0)$ *is a holomorphic submanifold of* $\mathbb{C}^n$ *on a neighborhood of any point where the holomorphic differential* $\sum_{1 \le j \le n} \frac{\partial f}{\partial w_j} dw_j$ *is nonzero.*

18.4 The curvature of a Kähler metric

Suppose that M is a complex manifold and g is a metric that makes the complex structure Kähler. As seen in Chapter 17.9, the Kähler form ω and J are both covariantly constant with respect to the Levi-Civita connection. To see what this implies, remember that if J is covariantly constant, then the Levi-Civita covariant derivative takes a section of $T_{1,0}M$ to a section of $T_{1,0}M \otimes T^*M_{\mathbb{C}}$.

This is to say that it defines a covariant derivative for $T_{1,0}M$. By way of reminder as to how this comes about, I can choose an orthonormal basis, $\{e^b\}_{1\le b\le 2n}$, of sections for T^*M on a neighborhood of a given set such that $Je^{2k-1} = e^{2k}$ and $Je^{2k} = -e^{2k-1}$. The action of ∇ is determined by its action on these basis 1-forms. This I write as $\nabla e^b = e^a \otimes W^{ab}$ where each W^{ab} is a 1-form, and where $W^{ab} = -W^{ba}$. This is to say that the collection $\{W^{ab}\}$ define a $2n \times 2n$ antisymmetric matrix of 1-form. (Recall that the vector space of $2n \times 2n$, antisymmetric matrices is the Lie algebra of SO(2n).) Meanwhile $\{\upsilon^k = e^{2k-1} + ie^{2k}\}_{1\le k\le n}$ is a basis for the dual space $T^{1,0}M \subset TM_{\mathbb{C}}$ to $T_{1,0}M$. Given that $J\nabla = \nabla J$, one finds that

$$\nabla\upsilon^k = \upsilon^m \otimes B^{mk}$$

where each B^{mk} is a section of $T^*M_{\mathbb{C}}$, and these are such that $\overline{B^{km}} = -B^{mk}$. Here,

$$B^{mk} = \frac{1}{2}\left(W^{(2m-1)(2k-1)} + W^{(2m)(2k)} + i\left(W^{(2m-1)(2k)} - W^{(2m)(2k-1)}\right)\right).$$

Note that ∇ and J commute if and only if $W^{(2m-1)(2k-1)} = W^{(2m)(2k)}$ and $W^{(2m-1)(2k)} = -W^{(2m)(2k-1)}$. The collection $\{B^{mk}\}$ defines an $n \times n$, anti-Hermitian matrix of 1-forms. (Recall that the vector space of $n\times n$, anti-Hermitian matrices is the Lie algebra of U(n).)

The induced covariant derivative ∇ on $T_{1,0}M$ has a curvature 2-form. When written near the given point used above in terms of the frame $\{\upsilon^k\}$, this curvature 2-form is an $n \times n$, anti-Hermitian matrix of 2-forms; its components being $\{(\mathcal{F}_\nabla)^{mk}\}$. It is defined so that

$$d_\nabla\nabla\upsilon^k = \upsilon^m \otimes (\mathcal{F}_\nabla)^{mk}.$$

The components $\{(\mathcal{F}_\nabla)^{mk}\}$ can be expressed in terms of the components, $\{R_{abcd}\}$, of the Riemann curvature tensor from Chapter 15.7 as follows: Write $d_\nabla\nabla e^b = e^a \otimes (F_\nabla)^{ab}$ where $(F_\nabla)^{ab} = \frac{1}{2} R_{abcd}e^c \wedge e^d$. Then

$$(\mathcal{F}_\nabla)^{mk} = \frac{1}{2}\left((F_\nabla)^{(2m-1)(2k-1)} + (F_\nabla)^{(2m)(2k)} + i(F_\nabla)^{(2m-1)(2k)} - (F_\nabla)^{(2m)(2k-1)}\right).$$

Note that the condition that ∇ and J commute implies that $(F_\nabla)^{(2m-1)(2k-1)} = (F_\nabla)^{(2m)(2k)}$ and also $(F_\nabla)^{(2m-1)(2k)} = -(F_\nabla)^{(2m)(2k-1)}$. The formula above guarantees that $\overline{(\mathcal{F}_\nabla)^{mk}} = -(\mathcal{F}_\nabla)^{km}$ and so the collection $\{(\mathcal{F}_\nabla)^{mk}\}$ does indeed supply the entries of an $n \times n$, anti-Hermitian matrix of 2-forms.

The proposition that follows describes two important features of $\mathcal{F}_\nabla$. Two remarks are needed to set the stage. Here is the first: The End($T_{1,0}$)-valued 2-form $\mathcal{F}_\nabla$ can be decomposed as $(\mathcal{F}_\nabla)_{2,0} + (\mathcal{F}_\nabla)_{1,1} + (\mathcal{F}_\nabla)_{0,2}$ where the subscripts correspond to the decomposition of $\mathrm{End}(T_{1,0}M) \otimes \wedge^2 T^*M_{\mathbb{C}}$ into the summands

$$\mathrm{End}(T_{1,0}M) \otimes (T^{2,0}M \oplus T^{1,1}M \oplus T^{0,2}M).$$

The proposition that follows asserts that $\mathcal{F}_\nabla$ has only a 1–1 component. The second remark concerns the trace of $\mathcal{F}_\nabla$, this now a closed 2-form on M. Recall from Chapter 14.5 that $\frac{i}{2\pi}\mathrm{tr}(\mathcal{F}_\nabla)$ represents the first Chern class of $T_{1,0}M$ in $H^2_{\mathrm{De\ Rham}}(M)$. The proposition writes this form in terms of the Ricci tensor from Chapter 15.7.

Proposition 18.7 *The* (0,2) *and* (2,0) *components of* $\mathcal{F}_\nabla$ *are zero. Meanwhile, the trace of* $\mathcal{F}_\nabla$ *is determined by the Ricci tensor of the metric* g *by*

$$\mathrm{tr}(\mathcal{F}_\nabla) = \Sigma_{1\le m\le n}(F_\nabla)^{mm} = -\frac{1}{4}\Sigma_{1\le m,k\le n}(\mathrm{Ric}_{(2m)(2k)}) + (\mathrm{Ric}_{(2m-1)(2k-1)})\upsilon^m \wedge \bar{\upsilon}k.$$

Proof of Proposition 18.7 The first point to make is that $(\mathcal{F}_\nabla)^\dagger_{0,2} = -(\mathcal{F}_\nabla)_{2,0}$ where the operation † here denotes the Hermitian adjoint. This is induced from the $\mathbb{R}$-linear automorphism of the bundle $\mathrm{End}(T_{1,0}M)$ that acts to send a given $\mathfrak{m} \in \mathrm{End}(T_{1,0}M)$ to the adjoint $\mathfrak{m}^\dagger$. The latter automorphism is then extended to give an automorphism of $\mathrm{End}(T_{1,0}M) \otimes \wedge^2 T^*M_{\mathbb{C}}$ by defining it on the reducible elements. In particular, it acts on $\mathfrak{m} \otimes \kappa$ to give $\mathfrak{m}^\dagger \otimes \mathbb{C}$.

To see that $(\mathcal{F}_\nabla)_{0,2} = 0$ near any given point, use a frame $\{\upsilon^k\}_{1\le k\le n}$ as introduced above. The formula $d_\nabla \nabla \upsilon^k = \upsilon^m \otimes \mathcal{F}_\nabla{}^{mk}$ means that

$$d_\nabla d_\nabla \upsilon^k = -(F_\nabla)^{mk} \wedge \upsilon^m.$$

This must be zero because ∇ is torsion free. The expression on the right is a $\mathbb{C}$-valued 3-form on M, and so has components of type (3, 0), (2, 1), (1, 2) and (0, 3). The latter is evidently zero. Meanwhile, the (1, 2) component is $-((\mathcal{F}_\nabla)^{mk})_{0,2} \wedge \upsilon^m$, and this is zero if and only if $((\mathcal{F}_\nabla)^{mk})_{0,2} = 0$ for all m and k.

To prove the claim about the trace of $\mathcal{F}_\nabla$, write the trace of $\mathcal{F}_\nabla$ in terms of F_∇ and so find it equal to

$$\mathrm{tr}(\mathcal{F}_\nabla) = \frac{i}{2}\Sigma_{1\le m\le n}((F_\nabla)^{(2m-1)(2m)} - (F_\nabla)^{(2m)(2m-1)} = i\Sigma_{1\le m\le n}(F_\nabla)^{(2m-1)(2m)}.$$

The right-most equality uses the fact mentioned earlier that $(F_\nabla)^{(2k-1)(2j)} = -(F_\nabla)^{(2k)(2j-1)}$. To continue, write $(F_\nabla)^{ab} = \frac{1}{2}R_{abcd}e^c \wedge e^d$ and then write e^{2m-1} as $\frac{1}{2}(\upsilon^m + \upsilon^{-m})$ and likewise write e^{2m} as $\frac{1}{2i}(\upsilon^m - \bar{\upsilon}^m)$. Use these identifications to write

$$\mathrm{tr}(\mathcal{F}_\nabla) = \frac{1}{2}(R_{(2m-1)(2m)(2k-1)(2j)})\upsilon^k \wedge \bar{\upsilon}^j.$$

This uses the identities mentioned earlier, $(F_\nabla)^{(2k-1)(2j-1)} = (F_\nabla)^{(2k)(2j)}$ and $(F_\nabla)^{(2k-1)(2j)} = -(F_\nabla)^{(2k)(2j-1)}$, in their guise as identities for R_{abcd}. This is to say that

$$R_{(2x)(2z)cd} = R_{(2x-1)(2z-1)cd} \quad \textit{and}$$
$$R_{(2x-1)(2z)cd} = -R_{(2x)(2z-1)cd} \quad \textit{for any } x, z \in \{1, \ldots, n\}\ .$$

A clever use of the identities $R_{abcd} + R_{acdb} + R_{adbc} = 0$ and $R_{abcd} = R_{cdab}$ from Chapter 15.7 with the identities just written will identify the expression above for $\mathrm{tr}(\mathcal{F}_\nabla)$ with the expression given in the proposition.

18.5 Curvature with no (0, 2) part

Suppose that M is a complex manifold and π: E $\to$ M is a complex vector bundle. Let ∇ denote a covariant derivative for sections of E. The associated curvature 2-form, F_∇, is a priori a section of $\mathrm{End}(E) \otimes \wedge^2 TM_{\mathbb{C}}$. As such, it has (2, 0), (1, 1) and (0, 2) components with respect to the decomposition of $\wedge^2 T^*M$ as $T^{2,0}M \oplus T^{1,1}M \oplus T^{0,2}M$. As is explained momentarily, a covariant derivative whose curvature has no (0, 2) part is especially significant. Note that if E has a Hermitian metric and ∇ is metric compatible, then $(F_\nabla)_{0,2} = (F_\nabla^\dagger)_{2,0}$ and so a metric compatible covariant derivative whose curvature has no (0, 2) part is purely of type (1, 1).

The significance of the vanishing of $(F_\nabla)_{0,2}$ stems from the following proposition:

Proposition 18.8 *Suppose that* M *is a complex manifold and* π: E $\to$ M *is a complex vector bundle of rank* q *with covariant derivative,* ∇, *whose curvature 2-form has no (0, 2) part. Then* E *has the structure of a complex manifold with the property that each fiber of* π *is a holomorphic submanifold, and as such a copy of* $\mathbb{C}^n$ *with its standard complex structure. In particular, addition of a given vector to other vectors in the same fiber, or multiplication of a vector by a complex number both define holomorphic maps from the fiber to itself.*

The assertion that a fiber of E is a complex manifold means the following: Let J_E: TE $\to$ TE denote the almost complex structure on E given by its complex manifold structure. Then J_E maps the kernel of π_* to itself, and so induces an almost complex structure on any given fiber. This makes the fiber into a complex manifold. The proposition asserts that this complex manifold structure is just that of $\mathbb{C}^n$.

A bundle E with a covariant derivative whose curvature has no (0, 2) part is said to be a *holomorphic* vector bundle over M. Holomorphic vector bundles E and E′ are said to be *holomorphically equivalent* if there is a bundle isomorphism φ: E $\to$ E′ whose differential intertwines the respective almost complex structures.

For example, if M is a Kähler manifold, then the covariant derivative on the bundle $T_{1,0}M$ has curvature of type (1, 1). As a consequence, the total space of $T_{1,0}M$ is a complex manifold of the sort that is described by the proposition.

Proof of Proposition 18.8 The first step is to define the almost complex structure J_E on TE. This is done by specifying $T^{1,0}E$. To start, fix attention on an open set $U \subset M$ where there is an isomorphism $\varphi: E|_U \to U \times \mathbb{C}^m$. Such an isomorphism identifies TE over $E|_U$ with the tangent space to $U \times \mathbb{C}^m$, and T^*E over $E|_U$ with $T^*(U \times \mathbb{C}^m)$. Use φ to write a section of $E|_U$ in terms of a map from U to $\mathbb{C}^m$. This is to say that if $\mathfrak{s}$ is a section, then $\varphi(\mathfrak{s})$ can be written as the section $x \to (x, \mathfrak{s}_U(x))$ where $\mathfrak{s}_U$ is a map from U to $\mathbb{C}^m$. By the same token, write $\varphi(\nabla\mathfrak{s})$ as a section $x \to (x, (\nabla\mathfrak{s})_U)$ of $(U \times \mathbb{C}^m) \otimes T^*M$ and write

$$(\nabla\mathfrak{s}_U) = d\mathfrak{s}_U + \mathfrak{a}\mathfrak{s}_U$$

where $\mathfrak{a}$ is an $m \times m$, complex valued matrix of 1-forms defined over U. Write the components of this matrix as $\{\mathfrak{a}^{ab}\}_{1 \le a,b \le m}$ so that each component is a 1-form. These component 1-forms enter the definition of $T^{1,0}E$ as follows: This vector subspace over U is the pull-back via the identification $\varphi: E|_U \to U \times \mathbb{C}^m$ of the vector space

$$T^{1,0}M|_U \oplus \text{Span}\{dz^a + \mathfrak{a}^{ab}z^b\}_{1 \le a \le m}.$$

Here, $\{z^a\}_{1 \le a \le m}$ are the standard complex coordinate functions for $\mathbb{C}^m$. Also, repeated indices are to be summed. Note that the definition given above is consistent with regards to the analogous definition made for $E|_{U'}$ when $U' \subset M$ intersects U. This is guaranteed by the manner by which the corresponding matrix of 1-forms $\mathfrak{a}'$ that defines ∇ in U' is related to the matrix of 1-forms $\mathfrak{a}$ that defines ∇ in U. See Chapter 11.2 if you are not remembering the relevant formula.

With $T^{1,0}E$ defined as above, the next task is to compute its Nijenhuis tensor. Since M is assumed to be complex, it is enough to verify its vanishing over U by seeing whether it maps the forms $\{\theta^a = dz^a + \mathfrak{a}^{ab}z^b\}_{1 \le a \le m}$ to zero. To see that such is the case, note that

$$\begin{aligned} d\theta^a &= d(dz^a + \mathfrak{a}^{ab}z^b) = d\mathfrak{a}^{ab}z^b - \mathfrak{a}^{ab} \wedge dz^b \\ &= (d\mathfrak{a}^{ab} + \mathfrak{a}^{ac} \wedge \mathfrak{a}^{cb})z^b - \mathfrak{a}^{ac} \wedge (dz^c + \mathfrak{a}^{cb}z^b). \end{aligned}$$

The expression on the far right here can be written so as to identify

$$d\theta^a = (F_\nabla)^{ab}z^b - \mathfrak{a}^{ac} \wedge \theta^c.$$

The (0, 2) part of this $((F_\nabla)^{ab})_{0,2}z^b$. This is to say that the Nijenhuis tensor vanishes if and only if F_∇ has zero (0, 2) part.

Granted that F_∇ has no (0, 2) part, the Newlander–Nirenberg theorem (Theorem 17.5) tells us that E has the structure of a complex manifold. To see that if the fibers of π are holomorphic submanifolds, it is enough to verify that J_E maps the the tangent space of any given fiber to itself. This follows from the fact that the vector fields $\{\frac{\partial}{\partial z_a}\}_{1\le a\le m}$ span a subbundle in $T_{1,0}E$ as each is annihilated by all 1-forms in $T^{0,1}E$.

18.6 Holomorphic sections

Let M and M′ denote even dimensional manifolds with respective almost complex structures j and j′. A map f: M → M′ is said to be *pseudoholomorphic* when its differential intertwines the almost complex structures. Thus $f_* j = j' f_*$. If J and J′ come from honest complex structures, then f is said to be *holomorphic*. This terminology is used for the following reason: Let U ⊂ M denote an open set and suppose that φ: U → $\mathbb{C}^n$ is an embedding onto an open set $\mathcal{O}$ whose differential sends j to the standard complex structure j_0 on $\mathbb{C}^n$. This is to say that $\varphi_* j = j_0 \varphi_*$. Suppose that U is such that $f(U) \subset U'$ where U′ is an open set M′ with an analogous embedding φ': U′ → $\mathbb{C}^{n'}$ onto an open set $\mathcal{O}'$. Let $f_U = \varphi' f \varphi$: $\mathcal{O} \to \mathcal{O}'$. Write the coordinates of f_U as $(f_1, \ldots, f_{n'})$. If the map f is holomorphic, then each of the coordinate function f_k is a holomorphic function of the coordinates on $\mathcal{O} \subset \mathbb{C}^n$. This is to say that $\frac{\partial f_k}{\partial \bar{z}_j}$ for all j ∈ {1,..., n}. Conversely, if the coordinates of f_U are holomorphic, and this is true for all charts U from a cover of M, then f is holomorphic. Note that if f embeds M in M′ and is holomorphic, then $f(M)$ is a holomorphic submanifold of M′.

An important example occurs in the following context: Let M denote a complex manifold and π: E → M denote a complex vector bundle with a covariant derivative ∇ whose curvature has vanishing (0, 2) part. As indicated above, E has a complex structure. When viewed as a complex manifold, the map π is *holomorphic* in the sense above. To see this, go to an open set U where there is an isomorphism φ: $E|_U \to U \times \mathbb{C}^m$. The map φ identifies $T^{1,0}E$ over $E|_U$ with

$$T^{1,0}M|_U \oplus \mathrm{Span}\{dz^a + a^{ab}z^b\}_{1\le a\le m}. \qquad (*)$$

The differential of π acts on this space by projecting to the $T^{1,0}U$ summand. This implies that π is holomorphic since it intertwines the complex structure endomorphism of TE with that of TM.

What follows describes another example. View a section, $\mathfrak{s}$, of E as a map from the complex manifold M to the complex manifold E. Viewed in this way, a section defines a holomorphic map between these two complex manifolds if and only if its covariant derivative, $\nabla\mathfrak{s}$, as a section of $E \otimes T^*M_{\mathbb{C}}$ has no part of type (0, 1). By way of reminder, ∇s is a section of $E \otimes T^*M_{\mathbb{C}}$ and so can be viewed as a section of the direct sum $(E \otimes T^{1,0}M) \oplus (E \otimes T^{0,1}M)$. The components of $\nabla\mathfrak{s}$ with respect to this splitting are written as $(\partial_\nabla\mathfrak{s}, \bar{\partial}_\nabla\mathfrak{s})$. The claim is that the section $\mathfrak{s}$ defines a holomorphic map when $\bar{\partial}_\nabla\mathfrak{s} = 0$. To see this, write $\phi(\mathfrak{s})$ as a map $x \to (x, \mathfrak{s}_U(x)) \in U \times \mathbb{C}^m$. The pull-back via this map of a form υ in the $T^{1,0}U$ summand in $(*)$ is υ, and so it lies in the $T^{1,0}M$ summand of $T^*M_{\mathbb{C}}$. Meanwhile, the pull-back of one of the forms $dz^a + a^{ab}z^b$ is

$$d\mathfrak{s}_U^a + a^{ab}\mathfrak{s}_U^b.$$

This is in $T^{1,0}M$ if and only if it has no $T^{0,1}M$ component. To see what this implies, remember that $\varphi(\nabla s)$ is the section of $(U \times \mathbb{C}^m) \otimes T^*M$ that sends x to $(x, d\mathfrak{s}_U + a\,\mathfrak{s}_U)$. What is written above is one of the $\mathbb{C}^m$ coordinates of the vector valued 1-form $d\mathfrak{s}_U + a\,\mathfrak{s}_U$. Thus, $\mathfrak{s}$ is holomorphic if and only if $d\mathfrak{s}_U + a\,\mathfrak{s}_U$ has no $T^{0,1}M$ part. This verifies the claim that $\mathfrak{s}$ defines a holomorphic map from M to E if and only if $\partial_\nabla\mathfrak{s} = 0$. Such a section of E is said to be a *holomorphic section*.

The notion of a holomorphic section is useful for the following reason:

Proposition 18.9 *Suppose that* $\mathfrak{s}$ *is a holomorphic section of* E *with the property that* $\partial_\nabla\mathfrak{s}$ *defines a surjective linear map from* $T_{1,0}M$ *to* E *at each point in* $\mathfrak{s}^{-1}(0)$. *Then* $\mathfrak{s}^{-1}(0) \subset M$ *is a holomorphic submanifold.*

Proof of Proposition 18.9 Fix a point $p \in \mathfrak{s}^{-1}(0)$ and fix an open neighborhood $U \subset M$ around p with an isomorphism $\varphi\colon E|_U \to U \times \mathbb{C}^m$. Write $\varphi(\mathfrak{s})$ as the map $x \to (x, \mathfrak{s}_U(x))$. Viewed as a map between real manifolds, this map is transversal to the submanifold $U \times \{0\}$ if and only if the map $\mathfrak{s}_U\colon U \to \mathbb{C}^m = \mathbb{R}^{2m}$ has surjective differential along $\mathfrak{s}_U^{-1}(0)$. The covariant derivative of $\mathfrak{s}$ is such that $\varphi(\nabla\mathfrak{s})$ is the section $x \to (x, d\mathfrak{s}_U + a\,\mathfrak{s}_U)$ of $(U \times \mathbb{C}^m) \otimes T^*M$. Where $\mathfrak{s}_U = 0$, this is just $x \to (x, d\mathfrak{s}_U)$. Thus, $\mathfrak{s}_U$ is transverse to 0 if and only if $d\mathfrak{s}_U$ is a surjective map from TM to $\mathbb{C}^m = \mathbb{R}^{2m}$ along $\mathfrak{s}_U^{-1}(0)$. If $\bar{\partial}_\nabla\mathfrak{s}_U = 0$, then a look at the real and imaginary parts of $\partial_\nabla\mathfrak{s}_U$ finds that surjectivity of $d\mathfrak{s}_U$ as a map from TM to $\mathbb{R}^{2m}$ occurs if and only if $\partial_\nabla\mathfrak{s}_U$ is surjective as a linear map from $T_{1,0}M$ to $\mathbb{C}^m$. Assuming that this is the case, then it follows that the tangent space to $\mathfrak{s}^{-1}(0)$ appears at a point $p \in U$ as the vector subspace of $T_{1,0}M$ that is sent to zero by $\partial_\nabla\mathfrak{s}_U$. This being the case, it follows that the almost complex structure of M preserves this subspace, and so $\mathfrak{s}^{-1}(0)$ is a holomorphic submanifold.

18.7 Example on $\mathbb{CP}^n$

View $\mathbb{CP}^n$ as S^{2n+1}/S^1 were $\lambda \in S^1$ acts on the unit sphere $S^{2n+1} \subset \mathbb{C}^{n+1}$ so as to send a given point $(z_1, \ldots, z_{n+1})$ to $(\lambda z_1, \ldots, \lambda z_{n+1})$. Here, S^1 is identified with the unit circle in $\mathbb{C}$. As noted in Chapters 10.5 and 10.6, the quotient map $\pi\colon S^{2n+1} \to S^{2n+1}/S^1 = \mathbb{CP}^n$ defines a principal S^1 bundle. A connection, A, on this principal S^1-bundle was described in Example 14.10. By way of reminder, the connection 1-form is the restriction to S^{2n+1} of the 1-form

$$A = \frac{1}{2}\sum\nolimits_{1\le a\le n+1}(\bar{z}_a dz_a - z_a d\bar{z}_a)$$

on $\mathbb{C}^{n+1}$. The pull-back of its curvature 2-form, π^*F_A, is thus the restriction to S^{2n+1} of the form

$$\pi^*F_A = \sum\nolimits_{1\le a\le n+1} d\bar{z}_a \wedge dz_a.$$

Keeping this in mind, recall from Example 17.11 that $\mathbb{CP}^n$ has a Kähler structure whose Kähler form, ω_{FS}, pulls back to S^{2n+1} via π as the restriction from $\mathbb{C}^{n+1}$ of the form

$$\frac{i}{2}|z|^{-2} dz_a \wedge d\bar{z}_a - \frac{i}{2}|z|^{-2}\bar{z}_a dz_a \wedge z_b d\bar{z}_b.$$

Here, repeated indices are implicitly summed over their values in the set $\{1,\ldots,n+1\}$. As explained next, the restriction of the latter to TS^{2n+1} is equal to $-\frac{i}{2}\pi^*F_A$. The way to see this is to note that the right-most term in the expression above for $\pi^*\omega_{FS}$ is zero when evaluated on any two tangent vectors to S^{2n+1}. Here is why: The 1-form $d|z|^2$ is zero on TS^{2n+1}, and the latter written out is $z_a d\,\bar{z}a + \bar{z}a\, dz_a$. Thus, $z_a d\,\bar{z}_a = -\,\bar{z}_a\, dz_a$ on TS^{2n+1}. As a consequence, the pull-back to TS^{2n+1} of the right-most term above is the wedge of a vector with itself.

Granted this identification of pull-backs, it follows that $F_A = -\frac{i}{2}\omega_{FS}$, and so F_A is of type (1, 1). Note in this regard that Chapter 17.7 exhibits ω_{FS} explicitly in complex coordinates which make manifest the fact that it is a 1–1 form.

By the way, it is not necessary to go through the exercise of writing out the Kähler form to verify that it is of type (1, 1). This is because the Kähler form on *any* Kähler manifold is automatically of type (1, 1). This can be seen as follows: Take an orthonormal frame for T^*M at any given point of the form $\{e^1,\ldots,e^{2n}\}$ such that $Je^{2k-1} = e^{2k}$ and $Je^{2k} = -e^{2k-1}$. Then $\omega = \sum_{1\le a\le n} e^{2a-1} \wedge e^{2a}$. Meanwhile $\{\upsilon^a = e^{2a-1} + ie^{2a}\}_{1\le a\le n}$ is a basis for $T^{1,0}M$, and this understood, it follows that $\omega = \frac{i}{2}\sum_{1\le a\le n}\upsilon^a \wedge \bar{\upsilon}\,a$.

It follows from the fact that F_A is of type (1, 1) that the induced covariant derivative ∇_A on any complex bundle associated to the principal bundle $\pi: S^{2n+1} \to \mathbb{CP}^n$ has curvature 2-form of type (1, 1). By way of an example, fix a nonnegative integer p, and consider the representation $\rho_p: S^1 \to Gl(1; \mathbb{C})$ that sends λ to λ^{-p}. The corresponding complex line bundle is then the equivalence class of points $((z_1, \ldots, z_{n+1}), \eta) \in S^{2n+1} \times \mathbb{C}$ under the equivalence relation that identifies the latter with $((\lambda z_1, \ldots, \lambda z_{n+1}), \lambda^p \eta)$. I use $E_p \to \mathbb{CP}^n$ to denote this complex line bundle.

A section of E_p is a map $h: S^{2n+1} \to \mathbb{C}$ that obeys

$$h(\lambda z_1, \ldots, \lambda z_n) = \lambda^p h(z_1, \ldots, z_n).$$

For example, the restriction to S^{2n+1} of a homogeneous polynomal of degree p on $\mathbb{C}^{n+1}$ defines a section of E_p. As explained next, such polynomials define holomorphic sections of E_p. To prove this assertion, I will take its directional derivative along the horizontal lifts of vectors from $T_{0,1}\mathbb{CP}^n$. To see what the latter look like, start with the vector field $\frac{\partial}{\partial \bar{z}_a}$ on $\mathbb{C}^{n+1}$. Its projection to TS^{2n+1} is

$$\bar{\sigma}_a = \frac{\partial}{\partial \bar{z}_a} - \frac{1}{2} z_a \left(z_b \frac{\partial}{\partial z_b} + \bar{Z}_b \frac{\partial}{\partial \bar{z}_b} \right).$$

The added term is to make $\bar{\sigma}_a \ |z|^2 = 0$. This vector field on S^{2n+1} is not horizontal. Its horizontal projection is

$$\bar{\sigma}_a - A(\bar{\sigma}_a) \left(z_b \frac{\partial}{\partial z_b} - \bar{z}_b \frac{\partial}{\partial \bar{z}_b} \right).$$

Now, $A(\bar{\sigma}_a) = -\frac{1}{2} Z_a$, so what is written above is

$$\frac{\partial}{\partial \bar{z}_a} - \frac{1}{2} z_a \left(z_b \frac{\partial}{\partial z_b} + \bar{z}_b \frac{\partial}{\partial \bar{z}_b} \right) + \frac{1}{2} z_a \left(z_b \frac{\partial}{\partial z_b} - \bar{z}_b \frac{\partial}{\partial \bar{z}_b} \right) = \frac{\partial}{\partial \bar{z}_a} - z_a \bar{z}_b \frac{\partial}{\partial \bar{z}_b}.$$

As I next demonstrate, the collection $\{\frac{\partial}{\partial \bar{z}_a} - z_a \bar{z}_b \frac{\partial}{\partial \bar{z}_b}\}_{1 \le a \le n+1}$ of $(n+1)$ vector fields on S^{2n+1} span the horizontal lift of $T_{0,1}\mathbb{CP}^{n+1}$. To see that this is the case, it is enough to prove that these are annihilated by the pull-back via the projection π of 1-forms from $T^{1,0}\mathbb{CP}^n$. Consider, for example, the part of S^{2n+1} over the open set $\mathcal{O}_{n+1} \subset \mathbb{CP}^n$ which is the image of the subset where $z_{n+1} \neq 0$. This set has complex coordinates $(w_1, \ldots, w_n)$ whose pull-back to $\mathbb{C}^{n+1} - \{0\}$ is $(z_1/z_{n+1}, \ldots, z_n/z_{n+1})$. The 1-forms $\{dw_1, \ldots, dw_n)$ span $T^{1,0}\mathbb{CP}^n$ over $\mathcal{O}_{n+1}$, and so their pull-backs are $\{z_{n+1}^{-1}(dz_a - z_a z_{n+1}^{-1} dz_{n+1})\}_{1 \le a \le n}$. Each of these annihilates all vectors in the set $\{\frac{\partial}{\partial \bar{z}_a} - z_a \bar{z}_b \frac{\partial}{\partial \bar{z}_b}\}_{1 \le a \le n+1}$.

Each of the vector fields in this set annihilates the polynomial h because h is the restriction to S^{2n+1} of a polynomial in the complex coordinates $\{z_a\}$ on $\mathbb{C}^{n+1}$. Thus, h is a holomorphic section of E_p.

Additional reading

- *Complex Geometry: An Introduction*, Daniel Huybrechts, Springer, 2004.
- *Lectures on Kähler Geometry*, Andrei Moroianu, Cambridge University Press, 2007.
- *Algebraic Curves and Riemann Surfaces*, Rick Miranda, American Mathematical Society, 1995.
- *Principles of Algebraic Geometry*, Phillip Griffiths and Joseph Harris, Wiley-Interscience, 1994.
- *Undergraduate Algebraic Geometry*, Daniel Perrin, Cambridge University Press, 1989.
- *Complex Algebraic Curves*, Frances Kirwan, Cambridge University Press, 1992.

19 The Hodge star

The holomorphic constraint for a section of a holomorphic bundle on a complex manifold is an example of a differential equation constraint on a section of a vector bundle. To describe perhaps the first such constraint, suppose now that M is a smooth manifold. The exterior derivative, d, is a linear map that takes a section of $\wedge^p T^*M$ to a section of $\wedge^{p+1}T^*M$. The p'th De Rham cohomology is, by definition, the vector space that is obtained by defining an equivalence relation on the kernel of d that posits $\omega \sim \omega'$ when ω' differs from ω by a form in the image of d. One can ask whether there exists a reasonably canonical choice of p-form to represent any given equivalence class. As explained momentarily, if M is compact, oriented and has a Riemannian metric, then there exists a canonical representation of each equivalence class.

19.1 Definition of the Hodge star

To set the stage, let n denote the dimension of M and let vol_M denote the metric's volume form as defined in Chapter 15.1. The first remark is that the metric with its associated volume form together supply a bundle isomorphism

$$*: \wedge^p T^*M \to \wedge^{n-p} T^*M$$

which is defined so that

$$\upsilon \wedge *\upsilon = |\upsilon|^2 vol_M.$$

Here, $|\cdot|$ denotes the norm on $\wedge^p T^*M$ that is defined by the given metric. More generally, if υ and υ' are any two p-forms, use $\langle \upsilon, \upsilon' \rangle$ to denote their inner product as defined by the metric. Then

$$\upsilon \wedge *\upsilon' = \langle \upsilon, \upsilon' \rangle \, vol_M.$$

Some linear algebra can be used to see that this does indeed characterize $*$. To give some idea as to what $*$ looks like, fix an oriented, orthonormal basis

$\{e^1, \ldots, e^n\}$ for T^*M at a given point. The volume form at this point is then $e^1 \wedge \cdots \wedge e^n$. This understood, then

$$*(e^1 \wedge \cdots \wedge e^p) = e^{p+1} \wedge \cdots \wedge e^n.$$

Other basis elements of $\wedge^p T^*M$ have the form $e^{i_1} \wedge \cdots \wedge e^{i_p}$ where $1 \leq i_1 < \cdots < i_p \leq n$. The image of this p-form under $*$ is

$$(\pm 1) e^{j_1} \wedge \cdots \wedge e^{j_{n-p}}$$

where $j_1 < \cdots < j_{n-p}$ is the ordered sequence of integers obtained from $\{1, \ldots, n\}$ by deleting $\{i_1, \ldots, i_p\}$. The sign, $+1$, or -1, is determined by the parity of the number of transpositions needed to reorder the ordered set $\{i_1, \ldots, i_p, j_1, \ldots, j_{n-p}\}$ as $\{1, \ldots, n\}$. The definition that two successive applications of the Hodge star defines an involution of $\wedge^p T^*M$, in particular $*^2 = (-1)^{p(n-p)}$.

For example, $*$ sends the volume form to the constant function 1 (this is a zero form on M). As a second example, note that in the case $\dim(M) = 2$, the Hodge $*$ maps T^*M to itself with square $*^2 = -1$ on 1-forms. This being the case, it defines an almost complex structure on the tangent space when the metric is used to identify TM with T^*M. This is the same almost complex structure that was introduced using the metric in Chapter 17.4.

19.2 Representatives of De Rham cohomology

Continue with the assumption that M is a compact, oriented manifold with a Riemannian metric. The Hodge theorem can be used to find a canonical representative of each De Rham cohomology class. What follows is the result:

Theorem 19.1 *Let* M *denote a compact, oriented, Riemannian manifold. Fix* $p \in \{0, \ldots, n\}$ *and let* $z \in H^p_{De\ Rham}(M)$ *denote a given class. Then there exists a unique p-form* ω *that obeys both* $d\omega = 0$ *and* $d{*}\omega = 0$, *and whose De Rham cohomology class is* z.

A form ω that obeys both $d\omega = 0$ and $d{*}\omega = 0$ is said to be *harmonic*. Let $\mathcal{H}^p$ denote the vector space of harmonic p-forms. This theorem asserts that $H^p_{De\ Rham}(M)$ is isomorphic to $\mathcal{H}^p$. Because $*$ maps $\mathcal{H}^p$ to $\mathcal{H}^{n-p}$, and because $\omega \wedge * \omega = |\omega|^2 \, \mathrm{vol}_M$, this theorem has the following corollary:

Corollary 19.2 *Let* M *denote a compact, oriented n-dimensional manifold. Then each* $p \in \{0, 1, \ldots, n\}$ *version of* $H^p_{De\ Rham}(M)$ *is isomorphic to* $H^{n-p}_{De\ Rham}(M)$. *In particular, there is an isomorphism with the following property:*

if $z \in H^p_{\text{De Rham}}(M)$ *and* $*z$ *is its partner in* $H^{n-p}_{\text{De Rham}}(M)$, *then* $\int_M \omega_z \wedge \omega_{*z} > 0$ *for any pair* ω_z *and* ω_{*z} *of representative forms.*

The corollary asserts the De Rham cohomology version of Poincaré duality.

19.3 A fairy tale

Before stating the Hodge theorem, consider for the moment the following fairy-tale scenario: Let V and V′ denote finite dimensional vector spaces with inner product, and let D: V → V′ denote a linear map. Since both V and V′ have inner products, there is a well-defined adjoint map $D^\dagger$: V′ → V. It is defined by the following rule:

$$\langle v, D^\dagger v'\rangle_V = \langle Dv, v'\rangle_{V'}$$

for all pairs $v \in V$ and $v' \in V'$. Here, $\langle\ ,\ \rangle_V$ and $\langle\ ,\ \rangle_{V'}$ denote the respective inner products on V and V′. What follows is a fundamental observation:

Let $\mathfrak{p}$: $V' \to V'/D(V)$ *denote the quotient projection. The restriction of* $\mathfrak{p}$ *to the kernel of* $D^\dagger$ *maps this kernel isomorphically onto* $V'/D(V)$.

To see that $\mathfrak{p}$ is 1–1 on the kernel of $D^\dagger$, suppose that $v' \in V'$ is such that both $D^\dagger v' = 0$ and $\mathfrak{p}(v') = 0$. If $\mathfrak{p}(v') = 0$, then $v' = Dv$ for some $v \in V$. As a consequence $D^\dagger Dv = 0$ and so $\langle v, D^\dagger Dv\rangle_V = 0$. But according to the definition of $D^\dagger$, this says the fact that $\langle Dv, Dv\rangle_{V'} = 0$, and so $v' = 0$. To see that $\mathfrak{p}$ maps the kernel of D′ surjectively onto V′/D(V), suppose that $\mathfrak{z}$ is an element in the latter space, and let $z' \in V'$ denote an element that projects to $\mathfrak{z}$. Let v′ denote the orthogonal projection of z′ onto the D(V). Then $\mathfrak{p}(v' - z') = \mathfrak{z}$. Meanwhile, $\langle Dv, v'\rangle_{V'} = 0$ for all $v \in V$, which says $D^\dagger v' = 0$.

To continue the fairy tale, suppose that $V = V'$ and that $D^2 = 0$. Note that this implies that $(D^\dagger)^2 = 0$ also. Let $\mathcal{H}$ denote the vector subspace $\{v \in V : Dv = 0$ *and* $D^\dagger v = 0\}$. Thus, $\mathcal{H} = \ker(D) \cap \ker(D^\dagger)$.

With the preceding as background, what follows is the finite dimensional analog of the Hodge theorem:

There is an orthogonal decomposition $V = \text{image}(D) \oplus \text{image}(D^\dagger) \oplus \mathcal{H}$.

Indeed, it follows from the definition of $D^\dagger$ that the three summands are pair-wise orthogonal. Two applications of the argument given two paragraphs back proves that they account for the whole of V. Indeed, a direct application asserts that there is the orthogonal decomposition $V = \text{im}(D) \oplus \ker(D^\dagger)$. Since $(D^\dagger)^2 = 0$, the image of $D^\dagger$ is a subspace of the kernel of $D^\dagger$. This understood, apply the argument of the preceding paragraph with V now replaced by the kernel of $D^\dagger$ and $D^\dagger$ then replaced by D's restriction to the kernel of $D^\dagger$. The argument in this context gives the orthogonal decomposition $\ker(D^\dagger) \oplus \mathcal{H}$.

19.4 The Hodge theorem

To see the relevance of this fairy tale, imagine replacing V with the vector space of $\Omega^*(M) = \bigoplus_{0\le p\le n} C^\infty(M; \wedge^p T^*M)$. Thus, the vector space of differential forms of all degrees. Define the inner product on this space as follows: If α and β have different degrees, then their inner product is zero. If they have the same degree, then their inner product is

$$\int_M \alpha \wedge *\beta.$$

This inner product on any summand $C^\infty(M; \wedge^p T^*M)$ in Ω^* is called the *L^2-inner product*. To complete the fairy-tale analogy, replace the fairy-tale's linear map D with the exterior derivative d.

With regards to this analogy, remark that the adjoint of the exterior derivative d as defined by this inner product is denoted by $d^\dagger$. It is usually called the *formal L^2 adjoint* of d. An application of Stoke's theorem finds that

$$d^\dagger = (-1)^{np} * d *$$

on the vector space of p-forms. Note that $d^\dagger$ maps p-forms to (p − 1)-forms and is such that $(d^\dagger)^2 = 0$.

Theorem 19.3 (The Hodge theorem) *Let* M *denote a compact, oriented Riemannian manifold. Then there is an orthogonal (with respect to the L^2-inner product) decomposition*

$$\Omega^* = \text{image}(d) \oplus \text{image}(d^\dagger) \oplus \mathcal{H}^*.$$

This means the following: Let $p \in \{0, 1, \ldots, n\}$ *and let* ω *denote any given* p*-form. Then there is a unique triple* (α, β, γ) *of* (p−1)*-form* $\alpha \in \text{im}(d^\dagger)$, *of* (p + 1)*-form* $\beta \in \text{im}(d)$, *and harmonic* p*-form* γ *such that* $\omega = d\alpha + d^\dagger\beta + \gamma$.

The proof that there is at most one such (α, β, γ) is obtained by copying the argument from the fairy tale. The proof of existence requires various tools from analysis that are beyond the scope of this book. Even so, I can pinpoint the reason for the complications: The space Ω^* with its L^2 metric is not a complete inner product space.

19.5 Self-duality

Suppose that M is an oriented, Riemannian manifold with even dimension 2n. Then $*$ acts on $\wedge^n T^*M$ as an involution with square $(-1)^n$. If n is even, this is 1, and so $\wedge^n T^*M$ decomposes as the orthogonal direct sum $\Lambda^{n+} \oplus \Lambda^{n-}$, where Λ^{n+} consists of the forms ω with $*\omega = \omega$, and Λ^{n-} consists of the forms ω with $*\omega = -\omega$. Each of these summands is a real vector bundle over M with fiber dimension $\frac{(2n)!}{2(n!)^2}$. Forms in Λ^{n+} are said to be *self-dual* and those in Λ^{n+} are said to be *anti-self-dual*.

Because $*$ maps the space $\mathcal{H}^n$ of harmonic n-forms to itself, this vector space also has a pointwise orthogonal direct sum splitting as $\mathcal{H}^n = \mathcal{H}^{n-} \oplus \mathcal{H}^{n+}$ where $\mathcal{H}^{n+}$ consists of forms ω with $d\omega = 0$ and $*\omega = \omega$. Meanwhile, $\mathcal{H}^{n-}$ consists of the closed, anti-self-dual n-forms. Neither the dimension of $\mathcal{H}^{n+}$ nor that of $\mathcal{H}^{n-}$ depends on the choice of metric. They depend only on the smooth structure of M. Here is why: The De Rham cohomology $H^p_{\text{De Rham}}(M)$ has the symmetric pairing that associates to classes z and z′ the number

$$Q(z, z') = \int_M \omega \wedge \omega'$$

as computed using any pair of representative closed n-forms ω and ω' for z and z′. As seen earlier in the chapter, the pairing is nondegenerate. Let n_+ denote the dimension of the maximal vector subspace in $H^n_{\text{De Rham}}(M)$ on which Q is positive definite. This is the dimension of $\mathcal{H}^{n+}$. The dimension of $\mathcal{H}^{n-}$ is that of the maximal subspace on which Q is negative definite. (As it turns out, the dimensions of $\mathcal{H}^{n+}$ and $\mathcal{H}^{n-}$ depend only on the underlying topological structure of M.) The difference $n_+ - n_-$ is called the *signature* of M.

Of special note with regards to self-duality is the case $n = 2$ and so $\dim(M) = 4$. In this case, the self-duality condition has been used extensively to find canonical covariant derivatives on bundles over M. To say more, suppose that $\pi: E \to M$ is a vector bundle and ∇ is a covariant derivative on M. The curvature 2-form of ∇ is a section, F_∇, of $\text{End}(E) \otimes \wedge^2 T^*M$. As such, it can be written as $(F_\nabla)^+ + (F_\nabla)^-$ where $(F_\nabla)^+$ is the part in $\text{End}(E) \otimes \Lambda^{2+}$ and $(F_\nabla)^-$ is the part in $\text{End}(E) \otimes \Lambda^{2-}$. The covariant derivative is said to be *anti-self-dual* when $(F_\nabla)^- = 0$; and it is said to be *anti-self-dual* when $(F_\nabla)^+ = 0$. As it turns out, these are very stringent conditions on a covariant derivative.

Simon Donaldson discovered how the spaces of self-dual connections can be used in certain cases to distinguish distinct smooth structures on a given topological 4-dimensional manifold. For more about this fantastic discovery, I recommend his book with Peter Kronheimer, *The Geometry of Four-Manifolds* (1997).

Additional reading

- *Riemannian Geometry and Geometric Analysis*, Jürgen Jost, Springer, 2008.
- *The Geometry of Differential Forms*, Shigeyuki Morita, American Mathematical Society, 2001.
- *The Laplacian on a Riemannian Manifold: An Introduction to Analysis on Manifolds*, Steven Rosenberg, Cambridge University Press, 1997.
- *Instantons and Four Manifolds*, Daniel Freed and Karen Uhlenbeck, Springer, 1990.
- *The Geometry of Four-Manifolds*, Simon Donaldson and Peter Kronheimer, Oxford University Press, 1997.

List of lemmas, propositions, corollaries and theorems

List of symbols

Index